New Techniques in Sheep Production

New Techniques in Sheep Production

Edited by

I. Fayez M. Marai
Department of Animal Production, Faculty of Agriculture,
Zagazig University, Egypt

J.B. Owen
Department of Agriculture, School of Agricultural and Forest Sciences,
University College of North Wales, Bangor, UK

Butterworths
London Boston Durban Singapore Sydney Toronto Wellington

First published 1987

© **Butterworth & Co. (Publishers) Ltd, 1987**

British Library Cataloguing in Publication Data

New techniques in sheep production.
  1. Sheep
  I. Marai, I. Fayez M.   II. Owen, John B.
  (John Bryn)
  636.3   SF375
  ISBN 0-408-10134-2

Library of Congress Cataloging-in-Publication Data

New techniques in sheep production.

  Bibliography: p.
  Includes index.
  1. Sheep. 2. Animal culture—technological
innovations. I. Marai, I. Fayez, M. (Ibrahim Fayez
Mahmoud) II. Owen, John B. (John Bryn), 1931–
SF375.N43   1987   636.3   86-33346
ISBN 0-408-10134-2

Phototypeset by Scribe Design, Gillingham, Kent
Printed and bound in Great Britain by
Butler and Tanner, Frome, Somerset

Preface

A wide range of new techniques and new scientific discoveries are of potential interest in the development of efficient sheep production systems. All too often scientific meetings at which relevant knowledge is discussed, and the scientific publications in which it is published, tend to be narrowly specialized and often do not consider the extension of the findings into the complex art of practical husbandry.

The aim of this publication and of the symposium upon which it is based was to cover a wide area of new knowledge by inviting leaders in these fields to give a brief account of the state of play in their specialism and in particular to consider the practical application into operating sheep systems at present and in the foreseeable future. This theme is a particular interest of the two institutions, the University of Zagazig in Egypt and the University College of North Wales, under whose wing the symposium was organized in Bangor in late August 1986. It was also intended that the discussion should emphasize both the problems of semiarid sheep production systems and those of the temperate systems.

Although it would be impossible to cover comprehensively all aspects of such a wide remit it is hoped that many of the most important new developments are considered. It is also inevitable that the developments discussed are very different and cover a wide range of approaches, from the basic to the applied. To help the reader the chapters are presented in what is hoped is a logical sequence, grouped into eight main sections, each dealing with a particular component of the main theme.

It is hoped that this book will be of value to the pioneer farmers, always thirsty for new possibilities, and their advisers and supporters in the associated marketing and supply organizations. We also hope that students at agriculture and veterinary universities and colleges will find this a useful and interesting account of what is going on in the sheep industry as we look to the end of the 20th century.

The response from the invited contributors has been excellent and the editors are most grateful to them for their cooperation. We hope that their efforts will be rewarded by the satisfaction of a wide platform for their ideas in the common goal of increasing efficiency of food production in difficult environments.

The photograph for the cover is by courtesy of Mr D.H. Roberts, of the University College of North Wales.

I.F. Marai
J.B. Owen

Contents

III Reproduction methods

IV Genetic improvement techniques

V Breed development

VI Feeding systems

VII   Management and health control

VIII   Evaluation of new techniques

Part I

Review of current sheep production systems and the scope for improvement

The world has a sheep population of some 1.2 billion (FAO, 1986) occupying largely the difficult environments for agriculture, particularly the mountainous and semiarid areas. Sheep are also particularly associated with traditional subsistence systems, especially in developing countries, that have remained basically unchanged for centuries. Sheep provide for their keepers meat, milk and wool as the main products, although much of the produce may be consumed by the family and not be easily accounted for in international statistics.

Although there is still a major role for the traditional systems as efficient converters of rangeland and byproducts, circumstances are changing in many respects, thus calling for a reassessment and modification of traditional sheep and sheep-keeping methods. At the extreme, where good land and various other inputs are possible, sheep genotypes are capable of much higher productivity, although not always at a higher rate of efficiency and profitability.

In between the two extremes lie many possibilities and it is the object of this section to look at the *status quo* in terms of sheep systems for producing meat, wool and milk and to consider what possibilities there are for improvements.

Chapter 1

Wool

M.L. Ryder

Summary

Considerable basic knowledge is available; its method of application depends on whether or not wool is the primary product. Emphasis should be placed on what the buyer pays for—fine fibres in clothing wools, freedom from kemp and pigment in carpet wools. Price per fleece is more important to the farmer than price per kg, and so in all systems a heavy fleece weight is important.

Most fleece characteristics are highly heritable so improvements in weight and quality can be made through selective breeding. Better nutrition to increase wool production is only economical through pasture improvement, but improvements aimed at increasing meat production also give more wool.

Although the Merino breed is unique in combining fine fibres with a heavy fleece weight, and its wool is in greatest demand because it has the widest range of uses, it is not the best breed to use in a first cross with 'hair' sheep. In order to eliminate kemp the first cross should be with a longwool (such as the Wensleydale) and the Merino should be used in a second cross.

The slowness of selective breeding to improve carpet wools might be overcome by crossing with hairy mutants of the New Zealand Romney breed in which a carpet-type fleece is controlled by a single gene.

Introduction

The intention is to outline the main questions relating to world wool production and to indicate where some of the answers might be found. Only those closest to particular problems can apply the basic knowledge available in the most appropriate way. As indicated by Doney (1983) it is not possible for any single recommendation to be made.

There is such a strong tendency to look on wool solely in terms of a crop, which is usually a byproduct, that other aspects of the skin and fleece tend to be ignored. The skin is the largest organ of the body and is where the animal meets its environment. Wool, therefore, forms an important component of the anatomy and physiology of the sheep. The fleece gives mechanical as well as insulative protection. A 5 cm long fleece will provide heat insulation equivalent to a

3

temperature gradient of 40 °C, and this applies in a hot as well as a cold climate. There is also the question of the partition of, and competition for, nutrients between wool, meat and milk production. This aspect is not usually considered in the many investigations of body growth for meat production. Finally, wool is a unique textile fibre, and this brings us back to the fleece as a crop, since a major theme of this paper will be how desired wool properties can be produced during fleece growth. Howe and Turner (1984) have underlined the economic importance of wool in developing countries by pointing out that they produce 30% of world wool production from over half of the world sheep population. The proportion of world sheep meat produced in these countries is 40% and of milk, 50%.

Considerable basic knowledge of wool biology is available for application (Ryder and Stephenson, 1968; Ryder, 1975; Black and Reis, 1979; Doney, 1983).

Doney (1983) pointed out that the way in which this knowledge would best be applied depends on the farming system. Only in low-cost extensive systems, with low capital and labour inputs, is it economical to produce wool as a primary product. A distinction must also be made between the more valuable, fine (Merino) clothing wool, and coarser, usually less valuable, carpet wool to supply the local carpet industry. This situation contrasts with eastern Europe and the USSR, for example, where clothing wool is so important, and there is a need to avoid imports, that it is produced as a major product in relatively intensive systems. In the USSR the payment received by farmers for wool is up to 18 times more than that in Britain (Sheifer, 1981).

In northern Europe, notably Britain, wool is very much a byproduct of intensive lamb production for meat, yet much is of intermediate 'crossbred' type, having unique speciality uses. High meat prices and subsidies, however, mean that wool accounts for less than 10% of the income from sheep flocks. In southern Europe and parts of the Middle East, milk is the primary product of the ewe, and this pushes wool (usually of carpet type) into third place, so that it often produces no more than 5% of the income.

What does the wool buyer pay for?

The reasons why farmers have been slow to apply available knowledge to wool improvements are: first, that in few breeds is it economical to concentrate on wool; and second, the vagueness of textile manufacturers about which attributes of wool are most desirable, which includes the difficulty of defining these characteristics in biological terms.

The vagueness is not surprising when one considers the wide variety of wool and the diversity of its uses. The reaction of wool biologists has been to carry out objective investigations on the factors that determine price. In a survey of Merino wool carried out in Australia during the 1950s it was found that price per kg was mainly determined by quality number, and that the next most important characteristics were length and colour. This suggested that the greatest return would be gained from animals producing a heavy fleece of wool with a high quality number, moderate length and white colour.

Quality number (for example, 64s or 70s) is a buyer's estimate of fineness, defined as the number of hanks, each of 457 m long, that might be spun from 0.45 kg of the wool. The number of crimps per unit length, which increases with decreasing fibre diameter, is used as a guide to fineness, and in the Australian survey it was found that crimp number could be substituted for quality number as a factor determining price.

In contrast, in a similar survey carried out in the United States in 1966, crimp number had no effect on price. Buyers there paid 5½ cents per kg for each increase in quality number, and 2½ cents more for each extra 2½ cm of staple length. Today with Merino wool in Australia, crimp number is no longer used for appraisal, since it is realized that fibre diameter is the single most important character in processing. The main aims for the Merino breeder are therefore: a heavy clean fleece weight, a specified mean fibre diameter, and whiteness.

A survey of the 1973 British Price Schedule (Ryder, 1975) (which gives guaranteed prices based on the price that each grade gained at auction in the previous year) showed that in the British pricing system increased wool weight brings a greater return than increased quality number. In 1973 for instance the 'Pick' grade of fine wools with 58s quality fetched 61.9p per kg, while 'Super', the next lower grade, with 56s quality, was priced at 59.1p per kg. With an average fleece weight of 2¼ kg, the price differential per fleece was 6.5p. But only ¼ kg more of 'Super' grade wool would have given twice this amount (13.4p). The British 'grade' refers to the degree of excellence rather than the quality number and incorporates penalties for faults.

Although comparisons are always made on price per kg, the price per fleece is actually more important. Thus the relatively high (1985) price of 145.8p per kg for fine Shetland wool is transformed into a low return per sheep by the low fleece weight of only 1 kg. Conversely, a heavy fleece (5.5 kg) of Lincoln lustre longwool converts the lower price per kg of 117.2p kg into a gain of 644.6p per fleece. Most price differentials in British wool are such that the more readily achieved object of ¼–½ kg more wool is likely to bring a greater return than the more difficult change to a different grade. Only the Merino among world breeds has a good price per kg *and* a high fleece weight, so Merino fleeces tend to be twice as valuable as the average of non-Merino fleece.

Anomalies exist in Britain when certain grades receive a high price owing to increased demand, regardless of quality. For instance, the coarsest Scottish Blackface wool has long received the highest price of that type because of the demand for it in southern Europe as a mattress filling. This grade in 1985 fetched 121.7p per kg compared with 102.6p per kg for the best Blackface carpet grade.

Increasing fleece weight and improving wool quality

There are three broad types of wool: Merino, crossbred (which is not necessarily from crossbred sheep), and carpet. Although Merino and crossbred types are clothing wools, Merino wool is unique, the approach to crossbred and carpet wools being similar.

Heredity contributes far more to fleece type than does the environment, which covers such factors as day-length, as well as nutrition. Genetic differences occur between breeds, between flocks, between individuals, between different parts of the body and even within the wool staple. Hereditary differences can be manipulated by selective breeding, which is a slow process, but it is fortunate that most wool characteristics are highly heritable. The Merino, however, is virtually the only breed in which it is economically viable to apply selective breeding solely to wool characteristics. In most other breeds meat or milk provide such a high proportion of the income from the sheep that wool takes second place and so the rate of progress is reduced. But this should not deter farmers from attempting to increase fleece weight and to eliminate wool faults. With clothing wool an alternative to selective breeding is to crossbreed with Merino sheep, and this has been done in eastern Europe to upgrade local populations towards Merino type.

Environmental variation in the fleece can be exploited through sheep husbandry, and improved husbandry can lead to an immediate response in the fleece. Although it is not economical to control nutrition solely for wool, where sheep housing is common there is the possibility of controlling other seasonal factors such as day-length.

Clean fleece weight has the high heritability of 0.45, and it is illuminating to look at the components of fleece weight with the following simple formula:

$$W = L \times A \times N \times D \times S$$

where L = the mean fibre length,
$\quad\quad A$ = the mean cross-sectional area (measured as fibre diameter),
$\quad\quad N$ = the mean number of fibres per unit area (fibre density in the skin),
$\quad\quad D$ = the density (specific gravity) of wool substance, and
$\quad\quad S$ = the area of the skin bearing wool.

Theoretically, selecting for an increase in fleece weight could increase fibre length and diameter (and this happened in some Australian experiments) as well as increasing the density of fibres in the skin and the area of the body growing wool. The specific gravity of wool keratin has the constant value of 1.31.

An increase of the fibre diameter in the Merino is undesirable, but in other Australian experiments increased fleece weight was brought about entirely by an increase in fibre density, which was associated with a decrease in mean fibre diameter of $1\,\mu m$. In the Merino, selective breeding for increased fleece weight should be accompanied by selection for fibre fineness.

Extra food stimulates wool production and the increased fleece weight is brought about entirely by an increase in the length and diameter of the fibres. Better nutrition can also increase the size and therefore the skin area (S) of the sheep, but after the birth of the lamb this greater surface area has no effect on the wool production since the result is merely a lower wool follicle density. The maximum number of follicles that will develop in a fetus is determined genetically, but poor nutrition before birth can reduce the number of secondary follicles that mature and grow fibres. The nutritional handicap of being a twin causes a reduction in the number of follicles great enough to reduce adult fleece weight by at least 3%.

Better nutrition is only likely to be economical for wool production if carried out through pasture improvement. Doney (1983) showed that a system designed to increase stocking rate for greater meat production also increased wool output. In two experiments designed to reduce the nutritional limitations to lamb production in hill areas, increases in total lamb weight of 51% and 94% were accompanied by increases in total wool output of 31% and 84% respectively. The increased wool production was brought about by a combination of increased stocking rate and greater fleece weight.

Whereas a management system designed solely to increase wool production might increase lamb growth, it would have little effect on reproductive performance, on which total lamb production depends. In contrast, improvements aimed at maximizing lamb production could have the additional benefit of increased wool production.

The successive first-crosses of the British sheep-stratification system (see below, p. 8), allow the possibility of heterosis in such characteristics as growth rate, which is not normally available to hill farmers who usually keep purebred flocks. In order to remedy this, cyclical crosses, first with one breed and then with another, have been proposed. When this was tried between the Scottish Blackface and the Swaledale breeds, an increase in fleece weight of 7% was obtained in addition. This cross between breeds having a similar fleece type would have little effect on fleece grade, but if dissimilar breeds such as the Blackface and Cheviot were to be used, the fleece would be likely to be heavily penalized as not fitting in with recognized grades.

Fleece characteristics

Increased fleece weight in the Merino must not be brought about through increase in fibre diameter, which is the single most important characteristic. Fleece weight can be increased by increased wool length, provided that, in the Merino for instance, the increased staple length does not change the class of wool. When Merino wool becomes too long for its type, the animals have to be shorn more frequently. Like fibre diameter, wool length is affected by diet, but staple length has the relatively high heritability of 0.3–0.6. In breeds other than the Merino, there is considerable scope for reducing the variability of individual fibre length within the staple, and of staple length between sheep. Lack of uniformity is a common complaint among wool users and since this usually means variation in length and diameter, it can be tackled by selective breeding. With regard to wool quality, fibre diameter is more important than quality number and crimp number, both of which have greater heritabilities than fibre diameter. Because of the effect of nutrition, fibre diameter values can have a heritability from 0.5 to as low as 0.12. Even so it is best to approach diameter directly instead of through the quality and crimp. Selection on fibre diameter will be assisted by a computerized method described by Hutchings and Ryder (1985) which for the first time speeds the precision method of measuring fibre diameter in small fleece samples.

Wool faults

Important genetic faults are colour and kemp. Natural black pigment is rare in the Merino, but can be a serious fault in crossbred and carpet wools. Pigment and kemp are entirely genetic in origin and so can be tackled by choosing rams lacking in kemp and pigment, and at the same time culling badly affected ewes. Hairiness, too, is basically genetic in origin, but there is an environmental factor in that the medulla of hairs is only fully formed during the summer; hairs become narrow and tend to lose their medulla during the winter, when they appear like wool fibres.

Since hairy fleeces are commonly used in carpets, the impression is gained that the hairiness (coarse fibres with a wide medulla) is a desirable feature in carpet wools. In fact hairy fleeces are used because of their cheapness, and many such fleeces have too many brittle kemps or coloured fibres for this purpose. The main requirement for carpets is relatively coarse non-medullated wool and, in modern factory carpet manufacture, this is achieved by blending wools of several types (Ince and Ryder, 1984).

Faults such as vegetable matter and external parasites can be tackled through management, but other faults due to the environment are less fully understood and more difficult to control. Tenderness of the fibre can be caused by an attack from microorganisms and can occur in shorn fleeces if they are stored in damp conditions. Break, which occurs only at one point in the staple, results from seasonal narrowing, worsened by poor nutrition, but stress from cold or disease is another cause of break and actual wool loss.

Buyers penalize all yellow stains on wool since they cannot distinguish those that will scour out from canary stain, which cannot be removed. Canary stain is common in hot countries and might be due to alkaline suint (sweat), but its seasonal occurrence allows shearing to be timed before the worst period. Weathering from too much sun or rain can cause a harsh handle to the staple tip and alters the affinity to dyes.

How meat production affects wool grade

In the past, when British wool grading began, most sheep were kept in purebred flocks. Now, however, as a result of the integrated stratification system to produce fat lamb, most British lowland sheep are crossbred animals. Hill ewes produce four or five crops of lambs, after which they are drafted to better land where they can produce one or two more crops of lambs. Draft ewes from the hills are crossed with a larger, longwoolled ram. The crossbred ewes produced inherit the thrift and milking ability of their hillbreed mother and the increased growth rate and probably fecundity from their longwool father. The ewes from this cross are sold to a lowland farmer, who crosses them in turn with a Down ram to produce early maturing fat lamb.

One consequence of this stratification system is that longwool and Down breeds are now kept in small numbers mainly to produce rams for crossbreeding. A consequence of this decrease in the amount of medium and finer British wool produced has been an increase in the proportion of carpet wool in the national clip.

Whereas the longwool–cross wool comes from the mothers of fat lambs, the Down–cross wool actually comes from the fat lambs. Therefore, unless the farmer shears his lambs before sale, he gets little return from the wool of the lambs sent for slaughter, and the finest part of the British clip reaches the trade as skin wool, which is sold outwith the main wool marketing system.

Advantages of crossing with a longwool, in particular the Wensleydale breed, are the increase in fleece weight and the elimination of kemp. This has applications in countries wishing to grow a wool fleece on kempy 'hair' sheep. The usual advice is to cross with a Merino, but Merino crosses retain kemp. The correct sequence is to use the Wensleydale first to eliminate kemp and give a fleece, and then to cross with the Merino to make the wool finer. Crossing with a longwool could be advantageous in the improvement of carpet fleeces, but the crossbred is likely to have wool in the clothing category, which might make it too expensive for use in carpets.

Possibilities for change in Britain

Despite the declining fineness of British wool, it still has a ready market for speciality uses, and so many people think that no change is desirable. Others believe that changes could be made towards carpet wool, and yet others (including myself) consider that the British clip could be made finer by the introduction of Merino influence.

With the stratification system of crossbreeding already described, carcass improvements are injected into the system at either the terminal, Down–cross stage or at the preceding longwool–cross stage. But wool improvements would have to be injected earlier, that is, at the longwool–cross and hill stages. This means that the final and crucial Down–cross stage could remain unaltered and that one could aim at the growth of a finer fleece by the mother of fat lambs.

Merino influence

Merino influence would have to be injected at the hill stage, and Cheviot–Merino cross fleeces for example are much finer. The second introduction would take place at the longwool–cross stage. Although there have been experimental Merino sheep in Britain for some time, large-scale commercial trials have yet to be carried out. Survival of the Merino in Britain has been adequately demonstrated and any shortcomings of the carcass can be overcome (Boaz, Ryder and Tempest, 1977). A further objection has been a failure of most Merino strains to reach the required 200% lambing, but this can now be overcome by the use of the Australian Booroola Merino. The advantages of the Merino are its combination of fine wool (high price per kg) with a heavy fleece weight so that fleece value for the British farmer could be doubled. Any Merino wool grown in Britain would reduce the millions of pounds spent annually on importing this type. Merino wool has the widest range of uses and is always in steady demand.

Carpet wool

With carpet wool, one either has the possibility of crossing with an improved type, or of selective breeding within existing breeds. The possibility of crossbreeding was investigated at the Animal Breeding Research Organization, Edinburgh, after the import of rams from the Drysdale and Tukidale hairy strains of New Zealand Romney. Hairiness in these breeds had been reported to be controlled by a single dominant gene so that crosses with the ordinary Romney always gave a hairy fleece. These strains have the additional advantage of a much greater fleece weight and a complete lack of pigmented fibres. Crosses in Britain indicated that these hairy strains had an unacceptably high incidence of kemp, and with breeds other than the Romney, the hairiness appeared to be intermediate between that of the parents, rather than dominant (Ryder, 1983). A 20-year selective breeding experiment in the Scottish Blackface, initiated by the late A.F. Purser at ABRO, demonstrated the way in which fleece characteristics can be manipulated. Animals were selected on a medullation index at the age of 8 weeks and put into a *hairy* or *fine* line. The hairy line eventually acquired very coarse hair fibres (not kemps) in each primary follicle, and the secondary (wool) fibres also acquired a medulla, as well as becoming less numerous (a true correlated response). In the fine line the lateral primaries began to grow wool fibres in place of hairs, but the centrals surprisingly retained kemps. The secondary follicles became finer, and increased in number. At the end of the experiment the hairy, fine and control lines were bred with each other in a search for heterosis or single-gene effects like that reported for hairy New Zealand Romney. There was no evidence of a single-gene effect, but some evidence of heterosis in medullation such that crosses of hairy with fine animals are likely to have more hairy fleeces than the intermediate value. The hairy and fine lines were also crossed with the Wensleydale and it was found that, whereas the Wensleydale eliminated kemps from the fine line, it had much less effect on the medullation of the coarse hairs of the hairy line (Ryder, 1985).

Basic information of this nature makes it possible to predict the likely outcome of crosses between sheep types similar to those investigated, in different situations.

References

BLACK, J.L. and REIS, P.J. (1979). *Physiological and Environmental Limitations to Wool Growth.* Armidale, Australia: University of New England Publishing Unit

BOAZ, T.G., RYDER, M.L. and TEMPEST, W.M. (1977). A role for the Merino in British sheep production. *Journal of the Royal Society of England,* **138**, 98–115

DONEY, J.M. (1983). Factors affecting the production and quality of wool. In *Sheep Production,* edited by P.W. Haresign. London: Butterworths

HOWE, R.R. and TURNER, H.N. (1984). Sheep breed resources in developing countries. In *Proceedings of the 2nd World Congress on Sheep and Beef Cattle Breeding.* Bloemfontein: South African Stud Book and Livestock Improvement Association

HUTCHINGS, N.J. and RYDER, M.L. (1985). The automation of the projection-microscope method of fibre diameter measurement. *Journal of the Textile Institute,* **76**, 295–299

INCE, J. and RYDER, M.L. (1984). The evaluation of carpets made from experimental wools. *Journal of the Textile Institute,* **75**, 47–59

RYDER, M.L. (1975). Producing desired wool properties during fleece growth. In *Contributions of Science to the Development of the Textile Industry,* edited by M. Cordelier and P.W. Harrison, pp. 18–31. Institute of Textile France, Textile Institute Great Britain

RYDER, M.L. (1983). Wools for carpets. *ABRO Report,* 23–27

RYDER, M.L. (1985). Cross-breeding studies with selected fleece lines of Scottish Blackface sheep. *Journal of the Textile Institute,* **76**, 362–376

RYDER, M.L. and STEPHENSON, S.K. (1968). *Wool Growth.* London: Academic Press

SHEIFER, O.Y. (1981). *The Production of High Quality Fleeces.* Moscow: Rossel'khozizdat

Chapter 2

Meat

G.E. Pollott and J.B. Kilkenny

Summary

The place of sheepmeat production in sheep farming is extremely variable depending not only on the country involved but also on the region. The systems of sheepmeat production operating in any such environment depend on both the physical and the economic climate. The various physical and economic conditions under which sheep are kept in Britain are described as an example and conclusions drawn about the scope for improvement.

The United Kingdom produced 300 000 tonnes of sheepmeat in 1985, 16% as mutton and the rest as lamb; 63% of the lamb was produced from lowland flocks with the remainder split equally between the hill and upland sectors. The major genetic make-up of slaughtered lambs consisted of 41% from hill breeds and 34% from Down breeds. The sale of lambs dominates output from British flocks accounting for 80–90% of gross output, net of ewe subsidies, in most production systems.

A wide range of production systems exist in Britain, each containing considerable variation in performance. In most systems of production, ewe stocking rate and the number of lambs reared per ewe mated are the two key factors which contribute most to profitability. These have increased in recorded flocks over the last 15 years. The variety of breeds available in Britain provides scope for breed substitution which can be very rapid on the male side and less so for ewes. However, changes in the ewe breeds are evident.

A range of recent improvements in sheep production are identified. These include a better understanding of ruminant nutrition, improved grazing management, more objective breeding methods, better marketing and effective parasite control. The scope for future improvements is discussed in relation to extension work, improved lamb-rearing techniques, lower-cost grazing systems, live animal carcass assessment and between-flock genetic comparisons. Many of these developments have an application in other sheep production industries.

Introduction

The production of sheepmeat is carried out in a wide range of environments using many different systems of production throughout the world. The importance of

TABLE 2.1 Relative production of sheep products by region (adapted from Croston and Pollott, 1985) (sheepmeat = 100)

	Europe	Africa	North and Central America	South America	Asia	Oceania	USSR	United Kingdom	France	Greece	Italy
Wool	23	27	31	115	25	97	55	21	13	11	22
Milk	306	93	–	9	195	–	12	1	629	775	920
Skins	15	35	14	29	19	24	33	13	8	19	22

meat production compared with the other major sheep products—wool, milk and skins is a further important factor. The importance of the four major sheep products in different parts of the world and in the major sheep-keeping countries of Europe is shown in *Table 2.1* using the ratio of the weight of each product relative to sheepmeat production. Even using such a crude method as this it is obvious that sheepmeat production is of greater importance in North and Central America, for example than in South America. Milk production is high in Europe compared to Oceania where wool and meat assume equal importance. *Table 2.1* also shows the extreme position of the United Kingdom in world sheep production with no milk production, wool and skins virtually a byproduct and the dominance of sheepmeat as the main sheep product. One consequence of this situation is the variety of sheepmeat production systems that have developed in Britain and the concentration of research and development effort into their improvement. In practice only economically viable systems are taken up and used by the industry.

Sheepmeat production in Britain

Sheep flocks in Britain produced 300 000 tonnes of meat in 1985. *Table 2.2* shows the quantities produced, numbers of animals and average carcass weight of this meat for lamb and mutton. Lamb is the major part of sheepmeat production and originates from all sectors of the British sheep industry. Estimates based on 1983 data show 63% of the lamb carcass weight originating in lowland flocks, with the remainder equally split between upland and hill flocks (*Table 2.3*). A certain proportion of the hill and upland quantities include the sale of store lambs which are finished on upland or lowland farms.

Although there are over 50 pure breeds in Britain only the hill breeds are numerically important and almost 50% of all ewes are crossbreds. Because of the stratified structure of the industry in Britain certain breed types contribute disproportionately to the genetic make-up of the lambs slaughtered (*Table 2.4*). The hill breeds contribute 57% of the genes of slaughter lamb mothers and 24% of their fathers, giving them on average the greatest influence on the carcasses of all lambs slaughtered. The predominance of the Suffolk as the main terminal sire

TABLE 2.2 Sheepmeat production in the United Kingdom, 1985

	Lambs	Mutton
Number of animals ('000 head)	14 030	1576
Quantity ('000 tonnes)	250.5	47.4
Average carcass weight (kg)	18.0	30.1

TABLE 2.3 Relative output from hill, upland and lowland sectors (from Croston and Pollott, 1985)

	Breeding ewes 1983 ('000 head)		Finished lambs sold/ewe	Average carcass weight (kg)	Total carcass weight ('000 tonnes)	
Hill	6509	(43%)	0.45	15	43.9	(18%)
Upland	2539	(17%)	1.0	18	45.7	(19%)
Lowland	6091	(40%)	1.3	19	150.4	(63%)
Total	15 139				250	

TABLE 2.4 Estimated genetic contribution (%) of various purebred groups to the weight of lamb carcass meat produced, to the relative numbers of lambs slaughtered and parent stock (source: MLC sheep breed survey 1972)

Breed group	Lamb carcass meat (%)	Lambs slaughtered (%)	Dams of lambs slaughtered (%)	Sires of lambs slaughtered (%)
Hill	36	41	57	24
Upland	8	7	11	4
Lowland*	6	6	8	3
Longwool	13	12	15	9
Down	37	34	9	60

*Including Devon and Romney breeds

TABLE 2.5 Structure of gross output of main classes of flocks in Great Britain 1985

	Per cent of gross output derived from				Excluding subsidies		
	Sales of lambs	Sales of other stock	Sales of wool	Ewe subsidies	Sale of lambs	Sale of wool	Ratio of lamb sales to wool sales
Intensive production—out-of-season and early lamb production	82	8	4	6	87	4	20:1
Lowland grass-based flocks	80	10	4	6	85	5	17:1
Upland flocks	74	9	4	13	85	4	21:1
Hill/mountain flocks	58	12	4	26	78	6	13:1

TABLE 2.6 Guide prices 1986

	Average* price (p per kg carcass weight)	Return for 18 kg carcass (£)
January	239	43.02
February	254	45.72
March	260	46.80
April	259	46.62
May	250	45.05
June	227	40.86
July	204	36.72
August	201	36.20
September	201	36.20
October	202	36.36
November	208	37.44
December	223	39.91

*Guide prices are fixed on a weekly basis; the monthly average based on these is used for illustrative purposes only

breed makes it the most important contributor to the genes of the sires of slaughter lambs.

The economics of sheep production in Britain amply demonstrate why the production of meat has overshadowed wool production. A breakdown of gross output from four production systems (*Table 2.5*) shows wool accounting for 4% in all systems with lamb sales ranging from 58% to 82%. Even when direct subsidies are excluded wool only accounts for 6% of gross output in hill flocks.

Since 1980 when the EEC sheepmeat regime was introduced, British sheep producers have benefited from the predetermined scale of guide prices. This has enabled producers to plan their production in advance, with some certainty of the returns likely, well before the beginning of the sheep year. *Table 2.6* shows the value of an 18 kg carcass in different months of 1986. The year can be divided into four main periods based on guide prices:

March–April	High prices
May–July	Falling prices
August–October	Low prices
November–February	Rising prices

These form one of the basic criteria for classifying production systems in Britain.

The other effect of the EEC sheepmeat regime has been increased sheepmeat production in Britain through an increase in the breeding flock. This increased production has resulted in a rising level of both exports and consumption of home produced lamb (*Figure 2.1*). Both markets require lean lamb in the weight range 15–20 kg. It is worth noting that fat is not an issue in either production or consumption in the majority of world markets. However, in the major western markets and importers of sheepmeat it is an important factor.

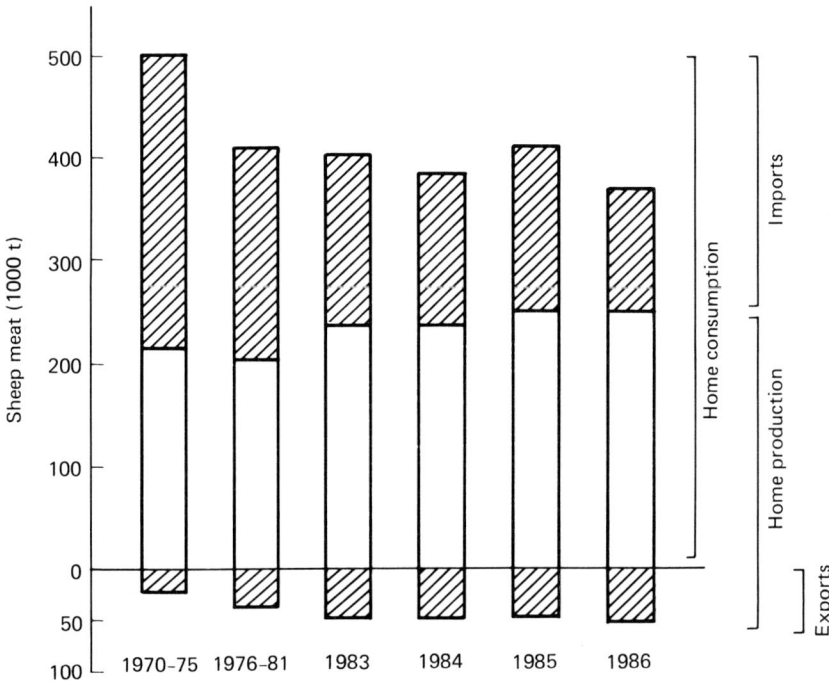

Figure 2.1 Quantities of sheepmeat imported, exported, produced and consumed in the United Kingdom

Current British systems of lamb production

British lamb production is characterized not only by a wide variety of systems but also by considerable variation within systems. The average and top third results for a range of systems is shown in *Table 2.7* and demonstrates considerable within-system variation in all situations.

There are three main criteria for describing British production systems: altitude, time of lambing, and time of lamb sale. A further criterion based on the type of grazing offered to the ewes, is important in some situations. The characteristics of the main lowland and upland systems are shown in *Table 2.8*.

The choice of system will depend on individual farm circumstances, but amongst the most important factors to be considered are:

(1) seasonal availability of labour;
(2) availability of suitable winter housing;
(3) area and type of grassland;
(4) potential for growing catch crops;
(5) availability and nature of crop byproducts suitable for sheep feed.

TABLE 2.7 Within-system variation in 1985

| | Lowland | | | | Upland | | Hill | |
| | Early lambing | | Spring lambing | | | | | |
	Average	Top third	Average	Top third	Average	Top third	Average	Top third
Lambs born alive/ 100 ewes	150	152	156	165	138	152	110	123
Rearing %	141	144	148	147	131	145	105	120
Stocking rate (ewes/ha)	15.5	21.5	12.0	14.0	10.1	12.6	—	—
Gross returns (£/ewe)	70.07	71.17	61.56	66.45	57.48	63.81	40.46	50.74
Variable costs (£/ewe)	32.28	26.93	20.96	19.39	15.51	15.75	9.29	8.67
Gross margin/ewe (£)	29.30	36.84	31.06	38.93	34.53	40.75	29.56	40.16
Gross margin/ha (£)	454	792	373	561	349	513	—	—
N use (kg/ha)	170	165	160	184	90	121	—	—

TABLE 2.8 Characteristics of various systems of sheepmeat production

Lambing time	Early	Early	Spring	Spring	Spring	Spring
Sale time	Spring	Spring	Early	Summer	Autumn	Winter
Method of lamb finishing	Concentrates	Grass	Grass and creep	Grass	Any	Any
Lamb growth	Fast	Fast	Fast	Moderate	Moderate	Low
Stocking rate	Very high	Moderate	Moderate	High	High	High
Lamb sale price/kg	Very high	High	High	Low	Moderate	High
Costs	Very high	High	High	Low	Low	Low
Ram size	Small	Small	Medium	Medium	Medium	Large
Concentrates use/ lamb	Very high	High	Moderate	None	None	None

Intensive early lamb production

The high guide price from February through to May offers the opportunity to produce indoor intensively reared lambs profitably. This is a high cost specialist system which depends on having ewes that will breed reliably to lamb in December/January and produce a large crop of lambs. Ewes must be fed generously to ensure an unrestricted milk supply and lambs introduced to creep feed at an early stage. The objective is to have lambs eating a complete diet concentrate so that they can be weaned by 6 weeks and finished off the concentrate for sale between 10 and 14 weeks of age. Dry ewes can be heavily stocked on the available grassland during the time they are not housed and the system may have some attraction for arable farms which have suitable buildings and skilled labour available in the winter months. In common with all intensive systems of production it requires skilled and careful management as failure to achieve the production targets means that high costs are incurred without realizing the high returns per lamb that are essential for success.

Early lamb from grass

This system also aims to produce lambs during the period of high prices in the spring but relies on grass as the main lamb feed. It can only be operated where grass growth starts early. Lamb returns, stocking rates and concentrate uses are usually lower than that for the intensive early lamb system.

Intensive spring lamb production

The guide price stays at a relatively high level until late June; this means that provided lambs can be kept growing rapidly, they can be finished from a late February/early March lambing before the price drops. This provides the basis for a system which utilizes generously fertilized grass and unrestricted creep feeding of lambs. Sale weights of lambs are lower than would be expected from slower grown lambs but the higher price per kg compensates for the lower weight, and the end product is a very attractive carcass carrying a light cover of fat. To ensure success a large lamb crop and compact lambing are essential and the management of the lambs at grass must be extremely well organized. Particular care must be given to the drawing of lambs for slaughter at frequent intervals. This is a system which might be applied on arable farms with a limited amount of grassland which they are prepared to exploit fully. It has the attraction that lambs are away before the drop in grass growth in midsummer and it avoids any delay in establishing autumn sown crops.

Summer lamb production

Lambs can be sold when the guide price is at its lowest in the late summer. Because returns are at their lowest, costs must also be kept low and a useful margin achieved using grass as the only source of lamb feed. This system is favoured in the wetter livestock areas of the west where grass growth is less impaired by summer drought and the cereal harvest is of less importance.

Autumn/winter lamb production

Lambing in late March/early April provides the best opportunity for normal crossbred commercial ewes such as the Mule and Halfbreds to produce the maximum possible number of lambs. Provided that the flock can be maintained economically on homegrown forages and arable byproducts this can be the basis for producing heavier lambs for sale from November onwards when the guide price begins to rise. Such a system will allow the potential of catch crops such as stubble turnips after winter barley to be exploited, but it does require heavy stocking rates through the spring and summer on grassland if an acceptable return per unit of land is to be achieved. Grass and forage crops dominate the feed for sheep (*Table 2.9*). With the exception of the early lamb flocks, often housed and always heavily concentrate-fed, grassland provides overwhelmingly the greatest part of lowland sheep feed.

TABLE 2.9 Contribution of grass to sheep feed requirements (source: MLC)

Type of production	Contribution to total feed requirements (%)*				Concentrates kg per ewe
	Grass	Other forage	Other bulk feeds	Concentrates	
Early lamb	58	5	2	35	95
Spring lamb	86	2	2	10	50
Autumn/winter lamb	84	5	2	9	55
Young sheep	82	7	2	9	60

*Based on metabolizable energy, ME, for ewe/lamb unit, the grass contribution is the residual after deducting the ME from other sources from the total requirements

It follows from this that the stocking rate of the sheep is crucial and the farmer's management in this matter is all-important. MLC's annual *Sheep Production Yearbook* provides information on the results from its Flockplan recording scheme which highlights the large variation in performance that is achieved between farms and the close relationship between physical performance and financial returns.

The latest results from finished spring lambing flocks in 1985 are described below (p.19). These show a consistent pattern with previous years and the main points are equally relevant for different types of lowland flocks and upland and hill flocks.

Detailed results for lowland spring lambing flocks, 1985

The performance of 398 lowland spring lambing flocks was recorded in 1985. A summary of results from those flocks is shown in *Table 2.10*. Top third flocks were arrived at by ranking all 389 flocks by their gross margin per hectare and selecting the third at the top of the ranking. The average of the results of these flocks was taken as the top third average.

The top third can be used to demonstrate the attributes of the more profitable flocks in any particular category. Table 2.11 shows the factors which contributed to top third superiority; the biggest factor was stocking rate. In 1985 this contributed 40% to the difference between average and top third flocks, and demonstrated once again that the most important component of profitability in lowland flocks is the ability to get as much out of grass as possible. The second most important

TABLE 2.10 Results for 398 lowland spring lambing flocks selling most of their lambs off grass in summer and autumn 1985

Financial results (£ per ewe)	Average	Top third
Output		
Lamb sales*	54.37	58.89
Wool sales	2.98	3.09
Ewe premium and LDA subsidy	4.21	4.47
Gross returns	61.56	66.45
Less Flock replacements	9.54	8.13
Output	52.02	58.32
Variable costs		
Ewe concentrates**	7.26	6.86
Lamb concentrates***	1.34	1.50
Purchased forage	0.91	0.59
Fertilizer	5.83	5.35
Other forage costs	0.76	0.57
Rented grass keep	0.40	0.46
Total feed and forage	16.40	15.33
Veterinary costs and medicine	3.28	2.99
Miscellaneous and transport	1.28	1.07
Total variable costs	20.96	19.39
Gross margin (output − variable costs)	31.06	38.93
Gross margin per grass ha (£)	401	592
Gross margin per ha (£)	373	561
Physical results		
Average flocks size (ewes to ram)	562	515
Ewe to ram ratio	39	39
Ewe lambs in breeding flock (%)	8	8
Ewes (per 100 ewes to ram)		
Empty	6	5
Dead	5	5
Lambed	92	93
Lambs (per 100 ewes to ram)		
Born dead	10	9
Born alive	156	165
Dead after birth	8	8
Reared	148	157
Retained for breeding	7	8
Sold finished	82	96
Sold or retained for feeding	56	49
Grazing stocking rate (ewes/ha)		
Summer grazing	12.8	15.1
Overall grass	12.9	15.2
Overall grass and forage	12.0	14.4
N fertilizer per hectare (kg)	160	184
N fertilizer per ewe (kg)	12	12
*Average return per lamb (£)	36.74	37.51
Estimated return per kg lamb carcass (£)	2.04	2.08
**Ewe concentrate cost per tonne (£)	135	137
Ewe concentrates per ewe (kg)	53	50
***Lamb concentrate cost per tonne (£)	168	167
Lamb concentrates per ewe (kg)	8	9

TABLE 2.11 Contribution to top third superiority in gross margin per hectare for lowland spring lambing flocks

	Per cent contribution
Stocking rate	40
Number of lambs reared	25
Lamb sale price/head	9
Flock replacement cost	11
Feed and forage cost	8
Other factors	7

component of top third superiority was the number of lambs reared per ewe which contributed to 25% of the difference between average and top third flocks. Lamb sale price per head and flock replacement cost had similar values of 19% and 11% respectively. The difference in feed and forage cost was lower, at 8%, reflecting the fact that, despite the higher performance of the top third flocks, they spent less than average on feed.

Breed variation

The large number of breeds found in Britain provide a wide range of performance levels to fit into the different production systems, described above. Some examples of breed performance for mature weight, rearing percentage and carcass weight are shown in *Table 2.12*. Variation of this type allows scope for breed substitution to take place in order to achieve particular production objectives.

TABLE 2.12 Between-breed variation in British sheep

Breed	Mature weight (kg)	Rearing percentage	Fat class 3 carcass weight of lamb from Suffolk ram (kg)
Purebreds in pedigree flocks			
Welsh Mountain	35	113	16.9
Clun Forest	62	146	19.7
Devon Closewool	69	140	18.8
Dorset Horn	72	139	20.8
Border Leicester	83	162	22.2
Oxford Down	89	135	22.9
Wensleydale	103	168	24.6
Crossbreds in commercial flocks			
Welsh Halfbred	58	136	19.6
Greyface	70	150	20.8
Mule	73	151	20.5
Scots Halfbred	77	149	21.5

Variation in feeds

Despite the prominence of grass as a feed in British sheep flocks other crops and a variety of methods of conserving grass are used. Root crops and brassicas are grown as autumn and winter feed for both ewes and lambs whilst byproducts from arable crops are used; these include straw, sugar beet waste and vegetable waste.

Grass is conserved as hay or silage whilst dried grass has become too costly to produce in recent years.

Recent improvements

Progress within any particular sector of a livestock industry often follows a pattern. Problems of disease and animal health are tackled first. This leads to changes in animal nutrition and a range of other management factors. Finally, attention to marketing and quality control are required to capitalize on the extra production achieved. These stages are not discrete and overlap to varying degrees in different countries.

TABLE 2.13 Lambs reared per ewe and stocking rates for lowland and upland flocks, 1970–85

	1970	1979	1985
Lowland flocks			
Lambs reared per ewe	1.32	1.37	1.48
Stocking rate (ewes/ha)	8.9	11.1	12.0
Upland flocks			
Lambs reared per ewe	1.22	1.29	1.31
Stocking rate (ewes/ha)	8.7	8.9	10.1

Improvements in British sheepmeat production over recent years can be crudely demonstrated from MLC's commercial recording scheme (*Table 2.13*). The two most important factors affecting profitability have increased in the last 15 years by up to 35%. It is impossible to attribute these increases to particular improvements but a range of innovations have been introduced in that period, which include:

(1) an improved understanding of ruminant nutrition;
(2) better grazing management;
(3) improved breeding methods;
(4) better marketing;
(5) performance monitoring and advice;
(6) effective parasite control.

Nutrition

Elucidation of the metabolizable energy (ME) system and the difference between rumen degradability of different feeds has led to the improved understanding of ruminant nutrition in recent years. Rationing systems based on these concepts have been developed which allow a more accurate matching of the nutrients offered with those required for maintenance and production. Two further techniques, condition scoring and pregnancy scanning, have enabled feeding to be more accurately matched to the requirements of the ewe throughout the year. The recent increase in the silage feeding of ewes has led to reduced feed costs and better ewe nutrition during pregnancy.

Grazing management

The recent increases in nitrogen fertilizer usage has been a contributing factor in increasing output from grass. This has been supplemented by the improved integration of conservation and grazing possible with silage making, as opposed to the more traditional weather-dependent hay-making. Systems of parasite control using clean grazing have been introduced and can lead to better growth rates from lambs.

Breeding

Recent studies have shown the relationship between breed and carcass quality. This has allowed a more accurate matching of breed type to market requirements. A further significant change in British sheep breeding has been the substitution of the Mule for the larger, less prolific Suffolk cross type of ewe (*Table 2.14*). The use of selection indexes for early lamb growth and ewe performance in pedigree flocks has helped concentrate breeding into areas of commercial performance. Artificial insemination techniques have been improved over recent years, but the use of frozen semen is still limited by the poor conception rates achieved. Very recent innovations using fertility enhancing preparations have been shown to be effective but have had a limited impact on commercial production at present.

TABLE 2.14 Lowland ewe breed changes between 1979 and 1983 (% of category)

	1979	1983	Per cent change
Scottish Halfbred	8	6	−2
Mule	24	42	+18
Welsh Halfbred	5	5	0
Welsh Mule	<1	1	+1
Masham	2	2	−1
Greyface	7	6	−1
Romney	4	3	−1
Suffolk × Scottish Halfbred	7	2	−5
Various Suffolk crosses	16	9	−7

Marketing

The stability and quality requirements produced by the EEC sheepmeat regime has been a big factor in the improved marketing of lambs. It has provided the incentive to match production to market requirements which in turn has stimulated improved lamb selection on farms. The stimulation of increased demand has been attempted by preparing lamb in ways which are more attractive to the consumer.

Performance monitoring

Onfarm management decisions depend on quality information if they are to be effective. The increase in performance monitoring schemes for both breeders and lamb producers has provided a great deal of useful information. This type of information has an application for both individual farmers and others in the industry which can aid efficient sheep production.

Parasite control

The advent of effective anthelmintics and an improved understanding of the life cycle of certain internal parasites have led to improved lamb output.

Scope for further improvements

The variation within systems already described in this chapter points to the gap that exists between what is possible and what is achieved on many farms. The wider dissemination of the improved techniques described above should lead to greater efficiency in sheepmeat production. This challenge should be coupled with the need to broaden the market for lamb both in Britain and abroad.

There is scope in certain areas for technical advance as well. Techniques to control ovulation are important provided that lamb-rearing techniques are developed to cope with the extra lambs produced. This is a key issue since the number of lambs reared per ewe has a direct effect on profitability in any sheepmeat production system.

In situations where output per hectare is important the scope for future research lies in developing lower cost systems based on clover or improved varieties of grass with greater yield.

The scope for improved breeding by using frozen semen has been mentioned above and will depend on improving conception rates above the current poor levels. The objective assessment of carcass quality on the live animal must have a high priority in sheepmeat producing systems both for breeding–replacement–selection and for selecting stock for slaughter. In addition the ability to match breeding stock to production requirements will remain limited until methods of comparing stock from different flocks are readily available.

Although this chapter has concentrated entirely on sheep production in Britain many of the techniques described have a place in any industry.

Reference

CROSTON, D.C. and POLLOTT, G.E. (1985). *Planned Sheep Production*. London: Collins

Chapter 3

Milk

T.T.Treacher

Summary

Total production of sheep's milk was 8.3 million tonnes in 1984, about 1.7% of total world milk production from all species. Production is greatest in countries bordering the Mediterranean and Black Seas and in the Near and Middle East. Although intensive production occurs in Israel, France and Spain, most milk is produced from extensive systems in Mediterranean or semidesert environments. Sheep's milk is manufactured into yoghurt, cheeses and, occasionally, butter, and has in the past commanded a price about three times that of cows' milk because of consumer preference and its high solids content. There are now indications that sheep's milk products are not so competitive with those from cows' milk, or a mixture of cows' and sheep's milk. If the market for ewes' milk can be expanded, the potential for increasing production is very great. In the majority of breeds very little recording and scientific selection for yield has taken place, although in some breeds there has been positive selection for ease of milking. Production from extensive systems is very dependent on seasonal and year-to-year variations in herbage production from natural pasture, and large responses have been demonstrated to simple improvements in pasture and small inputs of fertilizers. The future of dairy sheep production depends as much on factors such as status of shepherds, development of rural facilities, and availability of other employment, as on economic returns and technical developments within the systems themselves.

Introduction

The statistics from the United Nations Food and Agriculture Organization (FAO, 1985) for the total production of ewes' milk, and as a percentage of the total production of milk from all species (cow, buffalo, goat and sheep), given in *Table 3.1* must be treated with considerable caution. There are many instances of data for individual countries fluctuating widely from year to year or being adjusted retrospectively (for example, Morocco between 1972 and 1984). Many of the estimates are likely to be very inaccurate, as in many countries sheep's milk is produced in small quantities for use within the family or village and does not enter the commercial food distribution chains. India, with a sheep population of 41

25

TABLE 3.1 Total production of milk from cows, buffalo, sheep and goats (1000 × tonnes) and production from sheep (1000 × tonnes) for the countries with the highest production of sheep's milk (from FAO, 1984)

	Total production from cows, buffalo, sheep and goats	Production from sheep	(%)*
Africa	14 218	735	5
of which			
Algeria	890	180	20
Egypt	1780	18	1
Somalia	530	100	19
Sudan	1561	136	9
North and Central America	80 738	0	0
South America	24 030	37	<1
of which			
Bolivia	124	29	23
Asia	74 983	3812	5
of which			
Afghanistan	900	245	27
China	4498	525	12
Iran	2617	705	27
Iraq	543	165	30
Syria	1133	525	46
Turkey	5805	1300	22
Europe	190 800	3649	2
of which			
Bulgaria	2590	345	13
France	34 835	1060	3
Greece	1716	596	35
Italy	11 543	647	6
Romania	4650	350	8
Spain	7125	255	4
USSR	97 620	90	<1
World	496 241	8323	1.7

*As percentage of total milk production from cows, buffalo, sheep and goats

million, records no milk production from sheep, yet Acharya (1982) states that five breeds, with a combined population of 2.4 million head, are milked in north-west India.

The statistics indicate that 1.7% of total milk production in the world is from ewes. In northern and central America and Oceania there is no milk production from sheep, and production in South America and the USSR is low. In Europe only 1% of milk production is from sheep, but in a few Mediterranean and east European countries production is high. France is the second largest producer in the world, and in Greece and Bulgaria sheep's milk is an important proportion of total milk production. In Africa and Asia about 5% of production comes from sheep but in Afghanistan, Iran, Iraq, Syria, Turkey, Algeria and Somalia sheep's milk is more than 20% of total production.

These statistics show that milk production from ewes is important mainly in countries bordering the Mediterranean and Black Seas, and in the Near and Middle East, with China, Sudan and Somalia the only major producers outside these areas. There are, however, a few other countries where, although total sheep's milk production is low, production per head of population is high (*Table 3.2*). It must be borne in mind that in Africa and Asia the average annual consumption of milk from all species is only 28 kg.

TABLE 3.2 Milk production (1000 tonnes) from sheep and goats in the countries with the highest production of sheep's milk per head of human population (from FAO, 1984)

	Production from		Sheep and goat milk as per cent of total milk production* (%)	Sheep milk production kg/head of population
	Sheep	Goats		
Greece	596	420	59	60
Syria	525	79	53	52
Cyprus	26	40	55	39
Bulgaria	345	72	16	39
Mauritania	66	74	61	36
Mongolia	55	39	30	30
Turkey	1300	625	33	26
France	1060	475	5	19
Aghanistan	245	55	33	17
Iran	705	223	35	16
Romania	350	—	8	15
Iraq	165	69	43	11
Yemen Arab Republic	38	125	73	6
Sudan	136	415	35	6
Israel	22	25	6	5
Mali	32	37	36	4

*Total production of milk from cows, buffaloes, sheep and goats

Existing systems of sheep's milk production

In a considerable number of the countries, where milk production from sheep is important, milk production is the main output from sheep. It is obtained from systems, often of considerable antiquity, adapted to obtaining some human food from harsh Mediterranean or semidesert environments with great seasonal variations in rainfall and herbage production, in which cattle only survive with very low levels of productivity. A low level of milk production from milked ewes can be maintained in environments which will not sustain the suckling ewe and her lamb adequately.

In most sheep's milk production systems the ewes also rear their lambs, but there are great differences in the length of time that lambs are suckled, and Flamant and Casu (1978) suggested length of suckling as a basis for differentiating the main systems (*Figure 3.1*).

System 1

This system is used for production of heavy slaughter lambs in which the ewes rear their lamb or lambs for 3 or 4 months and are then dried off when the lambs are weaned.

System 2

In central and eastern Europe it is common for ewes that have suckled lambs for 3–4 months to be milked for about a month after the lambs have been weaned.

Germany — milking throughout lactation

Israel and Cyprus — partial milking in first 2 months

Nomadic flocks — partial milking from 1 month after lambing

Traditional Mediterranean system — lambs weaned at 1 month

Central and Eastern Europe — suckling followed by 1 month of milking

Northern Europe — suckling only

Lambing   Months of lactation

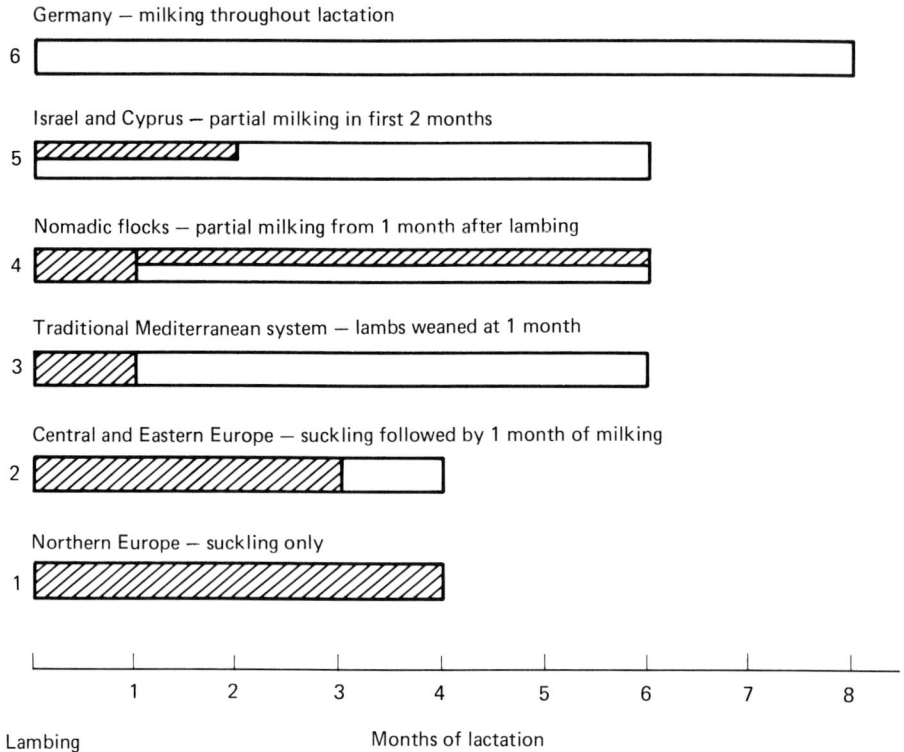

Figure 3.1 Different milking production systems for sheep

System 3

Weaning of lambs takes place at about 4 weeks of age and the ewes are then milked for about 5 months, initially twice daily but then once daily as yield falls at the end of lactation. This is the main system in the areas with Mediterranean climate, where it is well adapted to a climate with winter rainfall. Lambing takes place in early winter (November–December) when there is generally adequate herbage growth after the autumn rain. The lambs are removed before herbage growth is restricted by low temperature in midwinter and generally sold at a high price for immediate slaughter to provide Christmas lamb. In France and, to a lesser extent, in Spain, the lambs are grown to higher slaughter weights on concentrates but elsewhere the very high price commanded by the young milk lamb prevents the development of concentrate feeding of these lambs.

System 4

In nomadic flocks in the desert and steppe areas of Asia, the Near East and North Africa the young lambs are kept close to the tents while the ewes are away grazing during the daytime. Then about 4 weeks after lambing the ewes are milked once a day before the lambs are allowed to suckle.

System 5

In intensive systems with high levels of nutrition and breeds with high-yield potential, milking once a day, after the lambs have been shut away from the ewes for some hours, is started soon after lambing and continued until the lambs are weaned at about 6–8 weeks. In Israel and Cyprus milking starts 2 or 3 days after lambing, and in Spain at about 2 weeks after lambing in intensive units.

System 6

Very occasionally in high-yielding ewes, notably the East Friesland in northern Germany, the lambs are removed within a few hours of birth and the ewes milked for up to 10 months. This system has generally been adopted in the small sheep's milk industry that has started in recent years in Britain and Holland.

Milk products

Sheep's milk has a dry matter content of about 18.5% with 6.5–7.5% fat, 5.8% protein and 4.4% lactose. It is rarely used in the liquid form but is generally transformed into yoghurt, soft cheeses and lactalbumin (precipitated from whey), which can be stored for a few days, or hard cheeses and butter oil, which can be stored for many months. The lactose is generally either fermented or removed so that the product does not cause illness in the large proportion of the people in Africa and Asia, who lack the enzyme lactase to digest the lactose in whole milk.

The high solids content results in a much higher yield of cheese per litre of milk from sheeps' milk in comparison with cows' or goats' milk. This, together with a strong preference by consumers in Mediterranean and Middle Eastern countries for sheep's milk products, has resulted in ewes' milk commanding a price approximately three times that of cows' milk. In the 1980s this differential has been reduced in some European countries. For example, in Sardinia (Idda, Gutierrez and Usai, 1984), the major region for sheep's milk production in Italy, the ratio between the price of sheep's and cows' milk narrowed to 2:1 between 1981 and 1984. In France and Italy, in response to a combination of increased milk production and difficulty in selling more of their traditional cheese, the factories are diversifying to produce other cheeses.

Competition from dairy products made from cows' milk, or a mixture of cows' and sheep's milk by similar methods and in the same forms as traditional sheep's milk cheeses, is becoming a major problem as consumers, mainly in urban communities, become more conscious of price and are less inclined to buy high-priced, traditional products for everyday consumption. There is a great danger that these consumers will eventually entirely lose the taste for traditional sheep's milk products or, even worse, actually come to prefer the cows' milk products, which tend to be less strongly flavoured. At present legal controls of production only exist for some sheep's milk cheeses in France, Italy and Rumania (IDF, 1983). In other countries, legislation to protect the purity, quality and manufacture of their traditional cheeses and dairy products would be a major step in defining the true market for sheep's milk products and possibly in preserving it.

In southern Europe, where sheep's milk cheeses are almost entirely manufactured in factories, with very little processing now being carried out by the shepherds, the seasonality of sheep's milk production increases the costs of production. Although the labour is employed on a seasonal basis, the capital costs have to be spread over a season of no more than 8 months.

Scope for technical advance

The potential for an increase in milk production from sheep is very great. At present most sheep's milk comes from traditional, extensive systems which have changed very little in decades, if not centuries. Natural pasture, often on land owned by the village, provides most of the nutrition. The flocks are always shepherded and milked by hand in temporary folds or, in desert areas, with the ewes tethered together by their necks in long lines. So capital investment, other than in the sheep themselves and transport (donkey or van) to carry the milk to the village or milk collection point, is negligible. Yet at the other end of the range there are examples of very intensive systems. In the Roquefort region of France, production has become steadily more intensive in the last 30 years with increases in flock size and individual yield and in investment in housing and machine milking. Although a considerable amount of concentrates is fed, improved grassland still contributes a large part of the nutrient requirements as both grazing and conserved forage. In Israel, and a few large units in Greece, Italy and Spain and the Near East, pasture has become a very minor part of the diet. Production is based on concentrate feeding in houses or yards and milking is carried out in large sophisticated parlours, including rotaries with automatic cluster removal.

Breeds

Most ewes' milk is produced from a wide variety of local breeds. Actual yields, excluding milk obtained by the lambs in the suckling period, vary enormously with the level of nutrition and the length of the period of milking, which ranges from 1–10 months (Boyazoglu, 1963). Yields range from about 10 kg in Zackel and Merino breeds milked for about a month at the end of lactation in eastern Europe, through a large number of breeds with yields in the range 80–150 kg from 5 months of milking, to about 500 kg in East Friesland in Germany and the selected Awassi breed in Israel.

In most breeds there has been little selection for milk production. The extensive nature of most production systems makes recording of yields difficult and most shepherds have little contact with advisory services. Also until recently, improvement of the traditional sheep production systems was given a very low priority. There are exceptions, notably the improvements made in the Awassi in Israel (Finci, 1957), and the Lacaune in the Roquefort region of France (Rouquette and Vareilles, 1981). Although in both cases recording, selection and introduction of breeding programmes were accompanied by improvements in nutrition and management and the introduction of machine milking, the results give an indication of the potential for increasing yield that must exist in other breeds.

Attempts have been made to improve milk production by crossing local breeds with the East Friesland imported from Germany, but losses from respiratory and other diseases were great, even when nutrition had been improved. In Israel, however, the Assaf breed was created by crossing these with the Awassi breed. In some regions of Greece crosses of local breeds with East Friesland are used because of the immediate improvement in yield that is achieved (Katsaonnis and Zygoyiannis, 1986). In general, however, the East Friesland has had little effect on the indigenous breeds, in spite of repeated importations, and attention is increasingly turning to evaluating the full potential of the indigenous breeds that are adapted to the environment and resistant to the endemic diseases.

Nutrition

With the majority of sheep's milk being produced from flocks grazing natural pasture or range there are considerable fluctuations in production from season to season and within seasons. Although many studies have demonstrated that great improvements in herbage yields and animal performance can be made by pasture improvement and small inputs of fertilizers, very little pasture improvement has been carried out, other than occasional, and often illegal, burning to control invasion of shrubs. This results from the clear separation between flock owners and shepherds, and the farmers and agriculturalists. Individual ownership of grazing land is rare, and most uncultivated land belongs to the local administration or village for grazing in common. If regulation of numbers of stock or periods when grazing is permitted are not maintained it is common for these areas to become overstocked, leading to degradation of the pasture and eventually erosion. This is a major problem in many areas including central Turkey but particularly in the Near East and North Africa (Draz, 1983), where the ancient Hema system of restricted grazing rights in certain areas is no longer enforced.

Where grazing land is in individual ownership the potential for improvement is great. Responses to phosphate fertilization are large and to nitrogenous fertilizer in winter economic (see, for example, Bullitta, 1980). On marginal hill land in Sardinia pasture improvement on about 10% of the area enabled the flock to remain on the farm throughout the winter, when formerly they had to move to lowland grazing (Bullitta and Casu, 1980).

Irrigated areas offer an enormous potential for increased animal production. Casu (personal communication) suggests that on irrigated land, with two-thirds of the area in long-term grassland and a third in short-term forage crops, yields from the Sardinian breed of 200 kg per ewe and 5000 kg/ha are easily achieved with an input of concentrates of about 50 kg per ewe. The real potential is even higher as higher yielding genotypes, such as East Friesland crosses, could be used. These systems have the great advantage of extending milk production through the period in summer and into autumn when there is little production from the traditional systems. The major problem, however, to the development of these systems is the traditional separation of the skills of the shepherd and the agriculturalist, and it will need much technical advice and education to train a new generation of farmers with skills for managing pastures and growing forage crops as well as managing and milking the flock.

Machine-milking

The major labour requirement in dairy sheep systems, after herding the flock on unfenced land, is milking. Most flocks are hand-milked which is physically demanding and often carried out exposed to the weather. There are, however, large differences between breeds in ease of hand-milking. The number of ewes milked per man-hour varies from about 20 in, for example, the Lacaune, to 55 in the Manech and the other breeds in the French Pyrenees, to 80 in the Sardinian and Corsican breeds. It appears that very positive selection for ease of milking has been made by the shepherds in Sardinia and Corsica, resulting in udders with large cisterns. A study of the ejection characteristics of eight major breeds from six countries (Labussiere, 1983) shows that the Sardinian breed is also easy to machine-milk.

Machine milking started in France in the 1930s. The recommended characteristics for ewes' milking machines (Le Du, 1981; Roberts and Treacher, 1985) mainly results from work on the Lacaune and Prealpes du Sud breeds in France and it is not known if some modification is needed for efficient milking of other breeds, which are often far easier to milk by hand than these French breeds.

In spite of the availability of effective milking machines for many years, machine milking has not been widely adopted. It is most common in systems in France and Spain where milking traditionally took place in a sheep house. Machine-milking is almost unknown in the extensive systems, which often utilize widely dispersed grazing areas in different seasons, even if they are not full transhumant systems with separate summer and winter grazing areas. Vallerand (1983) has shown that attempts to machine-milk in these extensive systems necessitate too many other changes in the system, even if electricity and clean water are available.

The future

The future size of the sheep's milk industry will depend on its ability to maintain or increase the size of its markets in competition with similar products made from cows' milk or a mixture of cows' and sheep's milk. Expansion of the market, which, at least on any major scale, can only be at the expense of cow products, would seem unlikely. In northern Europe, however, there are some indications that there is a new luxury market. The very small sheep's milk industry in Britain with approximately 6000 ewes is at present successfully developing into that market but there is also some competition from imported cheeses and yoghurts. There is some potential for reduction of production costs in dairy factories in the main production areas by reducing the seasonality of milk production. This could be achieved by an increase in intensive systems using concentrate feeding or of systems on irrigated pastures.

These developments necessitate contact with advisory and technical services that are truly orientated to the shepherds' problems. Above all, however, the continuation of dairy sheep systems depends on the status of flock owners and shepherds, the development of rural facilities and the availability of alternative employment in industry or tourism.

References

ACHARYA, R.M. (1982). Sheep and goat breeds of India. *FAO Animal Production and Health*, Paper No. 30

BOYAZOGLU, J.G. (1963). Aspects quantitatif de la production latière des brebis. *Annales de Zootechnie*, **12**, 237–296

BULLITTA, P. (1980). Pascoli della Sardegna. Situazione e prospettive. *L'Italia Agricola*, **117**, 109–118

BULLITTA, P. and CASU, S. (1980). Risultati di una prova di intensificazione foraggera su terreni marginale della collina sarda. *Rivista di Agronomia*, **16**, 126–128

DRAZ, O. (1983). Rangeland conservation and development. *World Animal Review*, **47**, 2–14

FAO (1985). *Production Yearbook*

FINCI, M. (1957). *The Improvement of the Awassi Breed of Sheep in Israel*. Jerusalem: The Weizmann Science Press of Israel

FLAMANT, J-C. and CASU, S. (1978). Breed differences in milk production potential and genetic improvement on milk production. In *Milk Production in the Ewe*, edited by J.G. Boyazoglu and T.T. Treacher, pp. 1–20. EAAP Publication 23

IDDA, L., GUTIERREZ, M. and USAI, R. (1984). La cooperazione nel settore lattiero-caseario. *Camera di Commercio, Industria, Artigianato e Agricoltura*, Quaderno No. 13, p. 84. Sassari: Gallizzi

IDF (1983). Production and utilization of goat's and ewe's milk. In *International Dairy Federation Bulletin*, Document 158, 1983

KATSAONNIS, N. and ZYGOYIANNIS, D. (1986). The East Friesland sheep in Greece. *Research and Development in Agriculture*, **3**, 19–30

LABUSSIERE, J. (1983). Etude des aptitudes latières et de la facilité de traité de quelques races de brebis du 'Bassin Mediterranéen'. Resultats preliminaires obtenus au 16 Mai 1983. *Tercero Symposium Internacional de Ondeño Mecanico de Pequenos Rumiantes*, Vallodolid, pp. 730–792

LE DU, J. (1981). La machine a traité les brebis: son incidence sur la traité. *Sixième Journées de la Recherche Ovine et Caprine*, pp. 115–128

ROBERTS, J.R. and TREACHER, T.T. (1985). Sheep milking systems. *Proceedings of the Sheep Veterinary Society*, 43–46

ROUQUETTE, J.L. and VAREILLES, M. (1981). Evolution des structures de production et des résultats économiques dans le Rayon de Roquefort. *Sixième Journées de la Recherche Ovine et Caprine*, pp. 487–515

VALLERAND, F. (1983). Les problemes de mécanisation de la traité dans les systèmes laitiers extensifs. *Tercero Symposium Internacional de Ondeño Mecanico de Pequenos Rumiantes*, Vallodolid, pp. 316–327

Part II

Methods of increasing fecundity

Increasing fecundity—the number of young produced annually per female—is the major driving force in sheep productivity. It can be influenced by several means, the main ones being covered by the contributors to the present section.

A major factor is the ovulation rate of the ewe and the associated embryo mortality. Major new discoveries have recently been made both on the genetic front and on the physiological front, particularly the discovery of the role of immunization.

Another means of increasing fecundity is to achieve more frequent breeding. Many of the components of frequent breeding systems have long been familiar, and the major challenge is how to develop systems that are not only technically feasible but practicable and profitable for the producer.

Another factor is the stage at which puberty is reached in the young sheep. Because of the seasonal nature of sheep breeding, failure to achieve puberty at a certain age may delay the start of useful breeding activity by 12 months. Achieving early breeding not only offers scope for improving productivity but also enhances the opportunity for genetic improvement by reducing generation intervals.

Chapter 4

Genetic variation in ovulation rate in sheep

J.P. Hanrahan

Summary

Ovulation rate sets the upper limit to litter size, and accounts for an overwhelming proportion of the genetic variation in litter size both among and within breeds of sheep. Because most of the variation in litter size attributable to embryo mortality is non-genetic this masks genetic variation in litter size due to ovulation rate, with the result that heritability of litter size is lower than the heritability of ovulation rate. Where within-breed selection is the only available option for genetic improvement of prolificacy it is shown that the annual rate of genetic improvement in litter size can be doubled when ovulation rate is used as the selection criterion. The demonstration that a single gene with a large effect on ovulation rate (about 1.5 ova) is segregating in Booroola Merino sheep makes it possible to substantially alter the reproductive rate of a target population by introducing this gene. Backcrossing with selection for the gene would allow its incorporation while recovering the genetic background of the target population. Recent evidence suggests that genes with effects of similar magnitude are segregating in Icelandic sheep, in the Cambridge breed and in Javanese sheep. Another approach to breed transformation is to use genetic material from prolific breeds like the Finn and Romanov, for which the evidence suggests that high ovulation rate is not due to a single major gene. The utility of these breeds may be enhanced by selection for further increases in ovulation rate. Results are presented for a Finn line in which selection has increased ovulation rate by 1.5 ova. Use of this line rather than unselected Finn sheep allows a reduction of about 50% in the proportion of Finn genes required for a given increase in the ovulation rate of a target population.

Introduction

Ovulation rate sets the upper limit to the number of offspring produced per pregnancy and the latter is a major determinant of the biological efficiency of meat production, especially for ruminants (Dickerson, 1970). There is considerable variation among breeds of sheep with respect to prolificacy and most of this variation is attributable to corresponding differences in ovulation rate (Hanrahan, 1982; Hanrahan and Quirke, 1986). The development of procedures for the determination of ovulation rate in sheep by endoscopy (Roberts, 1968; Thimonier and Mauleon, 1969) has facilitated extensive research on sources of variation in ovulation rate in sheep, and especially in relation to the relative importance of genetic and environmental effects.

The literature on genetic variation in ovulation rate in sheep has been examined in a number of recent reviews (Hanrahan, 1982; Piper and Bindon, 1984; Hanrahan and Quirke, 1986). The purpose of this chapter is to summarize published information on variation in ovulation rate in sheep with reference to breed differences and within-breed variation, including evidence in relation to single gene effects. The role of ovulation rate in the genetic improvement of prolificacy is also considered including the importance of variation in embryo survival.

Ovulation rate

Breed differences

Comparative studies have demonstrated the existence of a very large reservoir of genetic variation among breeds as demonstrated by the results summarized in *Table 4.1*. Crossbreeding studies involving the high ovulation rate breeds (identified in *Table 4.1*) with low ovulation rate breeds have generally indicated that there is little heterosis for this trait (Land, Russell and Donald, 1974; Ricordeau *et al.*, 1976;

TABLE 4.1 Examples of variation in ovulation rate among sheep breeds

Source	Number of pure breeds	Range in mean ovulation rate	Breed with maximum
Bradford, Quirke and Hart (1971)	3	1.43–3.31	Finn
Land *et al.* (1973)	3	1.13–2.57	Romanov
Hanrahan and Quirke (1975)	3	1.53–4.03	Finn
Lahlou-Kassi and Marie (1985)	2	1.09–2.85	D'man

Lahlou-Kassi and Marie, 1985). However, evaluation of crosses between the Finn and Galway breeds in Ireland gave conflicting results in relation to heterosis with significant positive estimates for two-tooth ewes (Hanrahan, 1974a), but significant negative estimates were obtained from data on mature ewes based on a comparison of F1 and mid-parent values. However, another data set involving mature F1 and F2+ ewes gave a significant positive estimate for heterosis (Hanrahan and Quirke, 1982).

Intrabreed variation

Estimates of the heritability of ovulation rate in sheep are summarized in *Table 4.2* and show that there is a considerable amount of additive genetic variation, especially within the Finn breed.

Comparison of estimates for breeds with low and high mean ovulation rates are complicated by the discrete nature of the trait. Thus, it can be shown that in populations like the Finn the estimate from the observed data is almost identical to that which would be obtained if ovulation rate could be measured on a continuous

TABLE 4.2 Heritability estimates for ovulation rate

Breed	Estimate	Source
Romanov	0.27±0.10	Ricordeau et al. (1986)
Finn	0.50±0.09	Hanrahan and Quirke (1985)
Galway	0.32±0.16	Hanrahan and Quirke (1985)
Merino	0.07±0.03	Piper et al. (1984)

underlying scale, whereas estimates for a breed like the Galway (most ewes having either one or two corpora lutea) are considerably lower on the observed scale than on the assumed underlying continuous scale.

The evidence for a large amount of additive genetic variation for ovulation in Finn ewes has been confirmed by divergent selection on ovulation rate in this breed. Selection was initiated in 1976 based on the sum of the ovulation rates at two consecutive cycles at 18 months of age (Hanrahan and Quirke, 1982). There has been a significant response to selection, and the performance of the ewes from the selection lines and a contemporary random-bred control line (1985 breeding season) is summarized in *Table 4.3*.

TABLE 4.3 Effect of selection on ovulation rate in Finn sheep (data for 1985)

| Ewe age | Line | | | Divergence |
	High	Low	Control	(H–L)
0.75	3.29	1.78	2.07	1.41±0.23
1.5	4.53	2.25	3.00	2.18±0.25
4.5	5.27	3.35	3.80	1.92±0.41

These results show that a divergence of about two ova has been generated between the high and low lines. Most of the response has been in the high line which is probably a reflection of the greater selection differential in this line due to the skewed nature of the distribution of ovulation rate. Another feature of the results shown in *Table 4.3* is that the divergence observed in adult ewes is also reflected in the ovulation rate of ewe lambs when expressed relative to the mean of contemporary control line animals. This shows that the age at selection could be reduced to 8–9 months with a consequent reduction in the generation interval.

Single gene effects

Merino sheep usually exhibit a low litter size with a correspondingly modest ovulation rate. The exceptionally high prolificacy of the Booroola Merino, 2.29 versus 1.22 for controls (Piper and Bindon, 1982), has been demonstrated to be attributable to a single gene (called the F-gene) acting via effects on ovulation rate (Piper and Bindon, 1982; Davis et al., 1982). The evidence for the single gene basis for the high prolificacy of the Booroola has been reviewed and summarized by Piper, Bindon and Davis (1985). The effect of the gene on ovulation rate is summarized in *Table 4.4* based on New Zealand data on crosses derived from Booroola Merino by local Merino.

The estimate of the ovulation rate difference between the two homozygotes (FF−++ = 2.89±0.08) is slightly smaller than the estimate of 3.30±0.29 obtained

TABLE 4.4 Mean ovulation rate for ewes with zero, one or two copies of the Booroola gene (F)

Genotype	Mean ovulation rate			Unweighted
	A	B	C	average
FF	4.51	4.39	4.08	4.33
F+	2.78	3.05	2.56	2.80
++	1.49	1.51	1.32	1.44
FF−++	3.02	2.88	2.76	2.89±0.08

A = Davis *et al.* (1982); B = Davis *et al.* (1984);
C = Owens, Johnstone and Davis (1985)

by Piper, Bindon and Davis (1985) using data from a Booroola Merino flock in Australia. Pooling these estimates yields a mean value of 1.46 for the average effect of one copy of the Booroola gene. This estimate is at the upper end of the range for a set of 21 estimates, based on the difference between F+ and ++ genotypes, plotted by Piper, Bindon and Davis (1985). These authors concluded that there was no association between the mean ovulation rate of the ++ animals and the effect of one copy of the Booroola gene on this trait. In a study of Booroola × Romney ewes involving F+ and ++ genotypes the effect of the F-gene was to increase ovulation rate from 1.65±0.13 to 3.58±0.13 (Quirke *et al.*, unpublished observations).

It is apparent from the results summarized in *Table 4.4* that the effect of the F-gene on ovulation rate is essentially additive as shown by Piper, Bindon and Davis (1985). The effect of the gene on litter size is likely to depend on the genetic background and non-additive effects on litter size would be expected given the non-linear association between ovulation rate and litter size (Hanrahan, 1982).

The discovery of the Booroola gene led to examination of other sheep populations with the aim of determining whether genes with similar effects were segregating. Jonmundsson and Adalsteinsson (1985) reported evidence for the segregation of a gene which increased litter size by 0.64 in presumed heterozygous condition. These authors suggest that the gene may be associated with reduced fertility when ewes are homozygous. If the relationship between ovulation rate and litter size described by Hanrahan (1982) applies to Icelandic ewes then the effect of the putative gene on ovulation rate may be estimated to be +1.2 ova using the litter size data given by Jonmundsson and Adalsteinsson (1985). The effect is therefore similar to that already described for the Booroola F-gene.

In a study of ovulation rate and litter size in Indonesian sheep, Bradford *et al.* (1986) found evidence for a gene with an equally large effect on ovulation rate (+1.3 ova). The between-year repeatability of ovulation rate was 0.64 while the repeatability of ovulation rate between consecutive cycles was 0.80.

The Cambridge breed is a highly prolific composite formed from foundation ewes selected on prolificacy records and representing genetic material from 11 British breeds (Owen, 1976; 1982). The foundation ewes were crossed with Finnish Landrace rams, and were subsequently backcrossed to half-Finn rams born to foundation ewes (Owen *et al.*, 1986). The population of sheep formed from this foundation material was selected for high litter size with some emphasis also on preweaning growth of lambs. The present Cambridge population has a litter size of about 2.8 for mature ewes (Owen *et al.*, 1986) and up to eight lambs per litter have been recorded (B. Maund, personal communication). The ovulation rate of Cambridge ewes was investigated recently (Hanrahan and Owen, 1985) and revealed a variation (within age groups) of two to 13 ova. The repeatability of ovulation rate, between years, was 0.86 which, together with the extreme variability (CV = 0.55), strongly suggests that a gene with a large effect on ovulation is segregating in this breed. This hypothesis was examined by cluster analysis of ovulation data on 70 ewes each with two records. Assuming that three genotypes were present in the data the analysis yielded three clusters with mean ovulation rates of 2.6, 4.6 and 8.0. This preliminary evidence suggests that a gene with an effect greater than two ova is segregating in the Cambridge breed.

Another breed with remarkable prolificacy is the D'man in Morocco which has been reported as having a range in ovulation rate from one to eight with a repeatability, over nine cycles, of 0.59 (Lahlou-Kassi and Marie, 1985). From data reported by these authors the coefficient of variation for ovulation rate is about 0.45. These results taken together suggest that a gene with a large effect on ovulation rate is also present in the D'man breed.

Distribution of ovulation rate

The distribution of ovulation rate together with the incidence of embryo mortality will determine the distribution of litter size at birth. The influence of these two factors has been examined by Hanrahan (1986) while the impact of various litter size distributions on lamb output has been examined by Bradford (1985) who stressed the need for greater uniformity of litter size under extensive husbandry conditions. There is some evidence for breed differences in the distribution of ovulation which, when combined with high rates of embryo survival at a mean ovulation rate of 2.06, would yield mean litter sizes of 1.9, but with a difference of 0.13 in the proportion of ewes producing twins.

Embryo survival

The probability of embryo survival declines in an essentially linear manner as the number of eggs shed increases and data from embryo transfer studies show that this relationship is essentially non-genetic in nature (Hanrahan, 1980; Hanrahan, 1982; Hanrahan and Quirke, 1985). While most earlier studies suggested little variation among breeds with respect to embryo survival, given equal numbers of embryos entering the uterus, there is good evidence that some breeds show an enhanced ability to support embryos to term (Ricordeau et al., 1976; Ricordeau, Razungles and Lajous, 1982; Meyer et al., 1983; Hanrahan and Owen, 1985; Bradford et al., 1986). Estimates of the probability of embryo survival, as a function of ovulation rate, for various breeds are given in Table 4.5. These estimates, while not strictly

TABLE 4.5 Estimates of the probability of embryo survival in various sheep breeds

Breed or type of ewe	Number of corpora lutea			Source of data
	2	3	4	
Various	0.82	0.73	0.65	Hanrahan (1980)
Various	0.85	0.71	—	Kelly and Johnstone (1983)
Romney	0.74	—	—	Meyer et al. (1983)
Border Leicester	0.76	—	—	Meyer et al. (1983)
Romanov	0.94	0.87	0.82	Ricordeau, Razungles and Lajous (1982)
Javanese	0.93	0.88	0.90	Bradford et al. (1986)
Cambridge	1.00	0.84	0.81	Hanrahan and Owen (1985)
Finn	0.82	0.74	0.71	Hanrahan (unpublished observations)

comparable among sources of data, were derived from information on pregnant ewes using the equation of Hanrahan (1980). It is evident that many of the world's most prolific breeds also have high embryo survival (*see Table 4.5*) although the Finn is clearly an exception. Available evidence on the Booroola Merino also suggests that embryo survival is similar to that in controls (Bindon *et al.*, 1978, and unpublished data). Regardless of the breed it is evident from *Table 4.5* that embryo survival declines as the number of ova increases.

There has only been limited investigation of variation among animals within a breed with respect to their ability to support a given number of embryos. Hanranhan (1982) reported an estimate of 0.07 for the repeatability of this trait in multiple ovulating ewes while Ricordeau *et al.* (1986) obtained a repeatability estimate of 0.04, in Romanov ewes, for embryo survival unadjusted for ovulation rate. A slightly negative estimate for the heritability of embryo survival was quoted by Hanrahan (1982) for Booroola cross ewes and a similar result was obtained by Ricordeau *et al.* (1986) for Romanov ewes. However, the latter group obtained a non-significant positive estimate in ewe lambs (0.09 ± 0.08). It seems reasonable to conclude from these data together with information from lines selected for litter size (which show that responses can be adequately explained in terms of ovulation rate) that the capacity to support a given number of embryos is not an important source of variation among ewes with respect to litter size.

Selection on ovulation rate

The use of ovulation rate as a selection criterion when increased litter size is the objective has been proposed by Hanrahan (1974b; 1980). It has been shown that a single ovulation rate measurement is more efficient than a single litter size record and that the advantage in favour of ovulation rate increases with mean ovulation rate (Hanrahan, 1980). In addition, ovulation rate may be measured a number of times prior to mating whereas only one litter size can be obtained. The magnitude of the advantage to be gained from using ovulation rate as the selection criterion will depend on the alternative breeding programme designs and on the mean ovulation rate in the population under selection. As an example of the likely benefits two schemes have been compared and the results are summarized in *Table 4.6*.

The population for selection was assumed to have a mean ovulation rate of 1.4 at 18 months of age and 1.6 at 30 months increasing to 1.8 at 42 months. The coefficient of variation for litter size was taken to equal 0.33. In the case of selection

TABLE 4.6 Comparison of two approaches to selection for increased litter size

Item	Selection criterion	
	Ovulation rate	Litter size
Number of measurements per ewe	4	2
Age of ewes at selection (years)	2.5	3.0
Proportion selected		
ewes	0.35	0.40
rams*	0.10	0.10
Generation interval (years)	2.5	3.0
Annual genetic gain in litter size (% of the mean)	2.7	1.3

*Selected on dam's record

based on ovulation rate each ewe is measured twice in each of 2 years—that is, at around 18 months and again around 30 months of age. Following measurements at 30 months of age selected ewes and their female progeny, born at 24 months, are retained. Ram lambs are selected on the basis of their dam's ovulation rate records (best 10%). In the case of selection based on litter size the assumption was made that performance over two breeding seasons would be used as the selection criterion—the same time over which ovulation rate data was collected. Ewe lambs born to 2-year-old ewes were not retained until after these ewes had lambed again at 3 years of age—this gave a slightly longer generation interval (3.0 versus 2.5).

The results in *Table 4.6* show that the expected annual genetic gain is more than doubled when ovulation rate is used as the selection criterion. If the mean ovulation rate in the population was increased to around 1.9, the advantage in favour of ovulation rate would be increased by around 16%.

Discussion

It is clear from the results of studies on ovulation rate over the past 15 years that there is a wealth of genetic variation available both among breeds and within breeds. This variation can be exploited to obtain increased prolificacy in target populations. The demonstration of the single gene basis for the high ovulation rate of Booroola Merinos has opened up the additional possibility that prolificacy of a target population can be raised by gene migration, while at the same time the genetic background of the target population is retained. It has been shown by Elsen *et al.* (1985) that it takes about 12 years to backcross the Booroola F-gene into another breed to the point where a reasonable number of FF homozygotes with 15/16 recipient breed ancestry can be obtained. It is likely that such a population would only be of interest for crossbred ewe production since a population homozygous for the F-gene is likely to have too high an ovulation rate to be of interest in all but the most intensive systems.

Genetic material from prolific breeds, such as the Finn and Romanov, in which the high prolificacy is not apparently due to a single gene, has been used in many countries to raise the prolificacy of local breeds or as the basis of new composite breeds. This approach has the advantage that since the high prolificacy is multifactorial, any desired increase in prolificacy may be obtained by using the appropriate proportion of ancestry from the prolific breed. A disadvantage is that the resulting crossbred stock will usually be differentiable from the local breed and may be unacceptable or may be poorly adapted to local conditions as a consequence of the proportion of exotic ancestry. One way to tackle these difficulties is to select the prolific breed for an even higher ovulation rate so as to reduce the proportion of exotic genes required for a given increase in prolificacy. The utilization of the Finn line selected for increased ovulation rate can be used as an example with the results shown in *Table 4.7* (assuming an additive genetic model).

The increased ovulation rate as a result of selection has been sufficient to allow a halving of the proportion of Finn genes necessary for a given increase in litter size. This advantage could be further exploited by intense selection of the low fecundity breed to produce an elite group for crossing with the exotic rams. This example shows how current genetic resources can be exploited for genetic upgrading of the performance of a population without the need for extensive recording.

TABLE 4.7 Mean ovulation rate and resulting litter size expected from crossing with two Finn lines

| Ewe type | Source of Finn genes | | | |
| | High line | | Control line | |
	Ovulation rate	Litter size	Ovulation rate	Litter size
Finn	5.2	—	3.7	—
½ Finn	3.45	2.44	2.70	2.09
¼ Finn	2.58	2.03	2.20	1.82
⅛ Finn	2.14	1.78	1.95	1.67
Local	1.7	1.50	1.7	1.50

[+]Expected litter size

Where increasing prolificacy by using genetic material from a more prolific source is not an option, then within-breed selection represents the only avenue to genetic improvement. Using ovulation rate rather than litter size as the selection criterion can at least double the rate of genetic improvement. Additional gains are possible if the selected individuals are induced to leave more progeny by using either PMSG or immunization against gonadal steroids to enhance ovulation rate. Since ovulation rate measurements can be obtained on ewe lambs (provided they attain puberty) the age at selection could be reduced from that shown in *Table 4.6* with the possibility of some shortening of the generation interval provided ewe lambs can be successfully bred to lamb around 1 year of age.

Selection on ovulation rate rather than litter size ignores any genetic variation in embryo survival, and Bradford (1985) has expressed reservations on the effectiveness of this approach. It has been shown, by daughter–dam regression, that selecting ewes on ovulation rate significantly increases the litter size of their daughters (Hanrahan, 1984). It is also clear that selection for increased litter size in sheep has resulted in increased ovulation rate in all cases studied (summarized by Hanrahan and Quirke, 1985). In addition, estimates of the repeatability and heritability of embryo survival indicate that repeatability is less than 0.1 and the heritability estimate was negative in two out of three cases (Hanrahan, 1982; Ricordeau *et al.*, 1986). An examination of the likely contribution of ovulation rate and embryo survival to genetic variation in litter size makes clear that the contribution of embryo survival will increase with mean ovulation rate, and that at mean ovulation rates of 1.7 or less embryo survival is unlikely to account for more than about 5% of the genetic variation in litter size. Thus in most situations wherein increased litter size is a likely selection objective, ovulation rate will be the principal source of genetic improvement in litter size and selection on this criterion will be considerably more efficient than relying on litter size. Such an approach may only be practicable in centralized nucleus-type improvement programmes.

References

BINDON, B.M., PIPER, L.R., CHEERS, M. and CURTIS, Y. (1978). Uterine capacity of low and high fecundity Merinos. *Proceedings of the Australian Society for Reproductive Biology*, **10**, 83

BRADFORD, G.E. (1985). Selection for litter size. In *Genetics of Reproduction in Sheep*, edited by R.B. Land and D.W. Robinson, pp. 3–18. London: Butterworths

BRADFORD, G.E., QUIRKE, J.F. and HART, R. (1971). Natural and induced ovulation rate of Finnish Landrace and other breeds of sheep. *Animal Production*, **13**, 627–635

BRADFORD, G.E., QUIRKE, J.F., SITORUS, P., INOUNU, I., TIESNAMURTI, B., BELL, F.L., FLETCHER, I.C. and TORELL, D.T. (1986). Reproduction in Javanese sheep: evidence for a gene with large effect on ovulation rate and litter size. *Journal of Animal Science*, **63**, 418–431

DAVIS, G.H., ARMSTRONG, G.R. and ALLISON, A.J. (1984). Ovulation rates and litter sizes of Booroola ewes classified homozygous, heterozygous and non-carrier for the Booroola gene and the transfer of the Booroola gene from the Merino into another breed. *Proceedings of the Second World Congress on Sheep and Beef Cattle Breeding*, pp. 721–724

DAVIS, G.H., MONTGOMERY, G.W., ALLISON, A.J., KELLY, R.W. and BRAY, A.R. (1982). Segregation of a major gene influencing fecundity in progeny of Booroola sheep. *New Zealand Journal of Agricultural Research*, **25**, 525–529

DICKERSON, G. (1970). Efficiency of animal production: moulding the biological components. *Journal of Animal Science*, **30**, 849–859

ELSEN, J.M., VU TIEN, J., BOUIX, J. and RICORDEAU, G. (1985). Linear programming model for incorporating the Booroola gene into another breed. In *Genetics of Reproduction in Sheep*, edited by R.B. Land and D.W. Robinson, pp. 175–181. London: Butterworths

HANRAHAN, J.P. (1974a). Crossbreeding studies involving Finnish Landrace and Galway sheep. In *Proceedings of the Working Symposium on Breed Evaluation and Crossing Experiments, Zeist*, pp. 431–444

HANRAHAN, J.P. (1974b). Ovulation rate as the selection criterion for fecundity in sheep. *First World Congress on Genetics Applied to Livestock Production*, **3**, pp. 1033–1038. Madrid: Editorial Garsi

HANRAHAN, J.P. (1980). Ovulation rate as the selection criterion for litter size in sheep. *Proceedings of the Australian Society of Animal Production*, **13**, 405–408

HANRAHAN, J.P. (1982). Selection for increased ovulation rate, litter size and embryo survival. *Second World Congress on Genetics Applied to Livestock Production*, **5**, pp. 294–309. Madrid: Editorial Garsi

HANRAHAN, J.P. (1984). Results on selection for increased litter size and ovulation rate in sheep. *Proceedings of the Second World Congress on Sheep and Beef Cattle Breedings*, pp. 483–493

HANRAHAN, J.P. (1986). Reproductive efficiency in sheep. In *Exploiting New Technologies in Animal Breeding: Genetic Developments*, edited by C. Smith, J.W.B. King and J. McKay, pp. 59–70. Oxford: Oxford University Press

HANRAHAN, J.P. and OWEN, J.B. (1985). Variation and repeatability of ovulation rat in Cambridge ewes. *Animal Production*, **40**, 529

HANRAHAN, J.P. and QUIRKE, J.F. (1975). Repeatability of the duration of oestrus and breed differences in the relationship between duration of oestrus and ovulation rate of sheep. *Journal of Reproduction and Fertility*, **45**, 29–36

HANRAHAN, J.P. and QUIRKE, J.F. (1982). Selection on ovulation rate in sheep aided by the use of superovulation and egg transfer. *Proceedings of the World Congress on Sheep and Beef Cattle Breeding*, edited by R.A. Barton and W.C. Smith, **2**, pp. 329–385. Palmerston North, New Zealand: The Dunmore Press Ltd

HANRAHAN, J.P. and QUIRKE, J.F. (1985). Contribution of variation in ovulation rate and embryo survival to within breed variation in litter size. In *Genetics of Reproduction in Sheep*, edited by R.B. Land and D.W. Robinson, pp. 193–201. London: Butterworths

HANRAHAN, J.P. and QUIRKE, J.F. (1986). Breeding season and multiple births in small ruminants. *Third World Congress on Genetics Applied to Livestock Production*, **11**, pp. 30–45. Lincoln: University of Nebraska

JÓNMUNDSSON, J.B. and ADALSTEINSSON, S. (1985). Single genes for fecundity in Icelandic sheep. In *Genetics of Reproduction in Sheep*, edited by R.B. Land and D.W. Robinson, pp. 159–168. London: Butterworths

KELLY, R.W. and JOHNSTONE, P.D. (1983). Influence of site of ovulation on the reproductive performance of ewes with one or two ovulations. *New Zealand Journal of Agricultural Research*, **26**, 433–435

LAHLOU-KASSI, A. and MARIE, M. (1985). Sexual and ovarian function of the D'man ewe. In *Genetics of Reproduction in Sheep*, edited by R.B. Land and D.W. Robinson, pp. 245–260. London: Butterworths

LAND, R.B., PELLETIER, J., THIMONIER, J. and MAULEON, P. (1973). A quantitative study of genetic differences in the incidence of oestrus, ovulation and plasma luteinizing hormone concentration in the sheep. *Journal of Endocrinology*, **58**, 308–317

LAND, R.B., RUSSELL, W.S. and DONALD, H.P. (1974). The litter size and fertility of Finnish Landrace and Tasmanian Merino sheep and their reciprocal crosses. *Animal Production*, **18**, 265–271

MEYER, H.H., CLARKE, J.N., HARVEY, T.G. and MALTHUS, I.C. (1983). Genetic variation in uterine efficiency and differential responses to increased ovulation rate in sheep. *Proceedings of the New Zealand Society of Animal Production*, **43**, 201–204

OWEN, J.B. (1976). *Sheep Production*, pp. 366–376. London: Baillière Tindall

OWEN, J.B. (1982). Selection for prolificacy in the Cambridge sheep. *Proceedings of the European Association for Animal Production*, Leningrad: Genetics and Sheep, pp. 1–7

OWEN, J.B., CREES, S.R.E., WILLIAMS, J.C. and DAVIES, D.A.R. (1986). Prolificacy and 50-day lamb weight of ewes in the Cambridge sheep breed. *Animal Production*, **42**, 355–363

OWENS, J.L., JOHNSTONE, P.D. and DAVIS, G.H. (1985). An independent statistical analysis of ovulation rate data used to segregate Booroola Merino genotypes. *New Zealand Journal of Agricultural Research*, **28**, 361–363

PIPER, L.R. and BINDON, B.M. (1982). The Booroola Merino and the performance of medium non-peppin crosses at Armidale. In *The Booroola Merino*, Proceedings of a workshop held at Armidale NSW 24–25 August 1980, edited by L.R. Piper, B.M. Bindon and R.D. Nethery, pp. 9–19. Melbourne: CSIRO Division of Animal Production

PIPER, L.R. and BINDON, B.M. (1984). Genetic and non-genetic manipulation of reproduction rate in sheep. *Proceedings of the Second World Congress on Sheep and Beef Cattle Breeding*, pp. 502–512

PIPER, L.R., BINDON, B.M., ATKINS, K.D. and ROGAN, I.M. (1984). Ovulation rate as a selection criterion for improving litter size in Merino sheep. In *Reproduction in Sheep*, edited by D.R. Lindsay and D.T. Pearce, pp. 237–239. Cambridge: Cambridge University Press

PIPER, L.R., BINDON, B.M. and DAVIS, G.H. (1985). The single gene inheritance of the prolificacy of the Booroola Merino. In *Genetics of Reproduction in Sheep*, edited by R.B. Land and D.W. Robinson, pp. 115–125. London: Butterworths

RICORDEAU, G., RAZUNGLES, J. and LAJOUS, D. (1982). Heritability of ovulation rate and level of embryonic losses in Romanov breed. *Second World Congress on Genetics Applied to Livestock Production*, **7**, pp. 591–597. Madrid: Editorial Garsi

RICORDEAU, G., POIVEY, J.P., LAJOUS, D. and EYCHENNE, F. (1986). Genetic aspects of ovulation rate and embryo mortality in Romanov ewes. *3rd World Congress on Genetics Applied to Livestock Production*, **11**, pp. 90–95. Lincoln: University of Nebraska

RICORDEAU, G., TCHAMITCHIAN, L., EYCHENNE, F. and RAZUNGLES, J. (1976). Performance de reproduction des brebis Berrichonnes du Cher, Romanov et croisées. I. Activité sexuelle en debut de saison et à contre saison. *Annales de Génétique et Selection Animale*, **8**, 9–24

ROBERTS, E.M. (1968). Endoscopy of the reproductive tract of the ewe. *Proceedings of the Australian Society of Animal Production*, **7**, 192–194

THIMONIER, J. and MAULEON, P. (1969). Variations saisonnières du comportement d'oestrus et des activités ovarienne et hypophysaire chez les ovines. *Annales de Biologie Animale, Biochimie, Biophysique*, **9**, 233–250

Chapter 5

Increasing prolificacy of ewes by means of gonadotrophin therapy and treatment with Fecundin

R.J. Scaramuzzi, B.K. Campbell, Y. Cognié and J.A. Downing

Summary

Sheep immunized with Fecundin and mated in the autumn have increased ovulation and lambing rates. The use of Fecundin immunization in association with other techniques for the control of reproduction was evaluated in a series of experiments.

Ovulation rates can also be increased by exogenous gonadotrophins, and this effect was found to be additive with the effect of Fecundin on ovulation rate. Combined treatment of Fecundin and superovulatory doses of exogenous gonadotrophin did not, however, increase the number of 'healthy' embryos recovered from Merino ewes when compared to the use of gonadotrophins alone.

Fecundin was shown to improve the lambing performance of spring-mated Border Leicester × Merino ewes, induced to ovulate using a combined progestagen–PMSG treatment. The time interval between booster immunization and ovulation should be about 3 weeks for optimum results.

Three methods of oestrous synchronization were compared in Fecundin-treated ewes and found to have no influence on the ovulation rate response to fecundin.

The use of Fecundin in conjunction with methods of oestrous synchronization and progestagen–PMSG treatment would appear compatible. However the use of Fecundin to enhance the superovulatory response to exogenous gonadotrophins requires additional evaluation in view of apparent breed effects.

Introduction

Sheep managed under extensive systems of husbandry are often afflicted by problems of reduced prolificacy and poor fertility. These problems to some extent can be managed by the selective use of available nutrition (Egan, 1984) and by the introduction of genes from prolific genotypes such as the Romanov (Cornu and Cognié, 1985), or the Booroola Merino (Bindon, 1984). Nevertheless, there has existed for some considerable time the need for inexpensive, convenient and effective therapeutic methods of improving both prolificacy and fertility of extensively managed flocks.

Fecundin® (Glaxo Animal Health Pty Ltd) is a product developed specifically to fill such a need at least as far as prolificacy is concerned (Geldard, Scaramuzzi

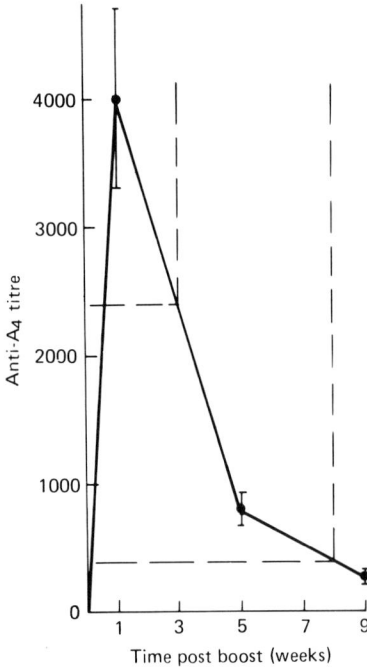

Figure 5.1 The androstenedione-antibody response (reciprocal mean ± SD) of Fecundin-treated ewes. The ewes were boosted by an injection of 2 ml of Fecundin at time zero. The broken vertical lines represent the recommended mating period of 3–8 weeks post-boosting over which time the antibody titre declines from about 1/2500 to about 1/500 as represented by the broken horizontal lines. (Data adapted from Cox, Wilson and Wong, 1984)

and Wilkins, 1984). Increasing prolificacy by increasing ovulation rate invariably also improves fertility and indeed can be regarded as the key to improving fertility (Robinson and Scaramuzzi, 1986). Initial trials evaluated the use of steroid immunity to improve the prolificacy of Merino ewes mated in the autumn (Cox *et al.*, 1982). Primary immunization with Fecundin does not produce a detectable androstenedione-antibody response in any ewes. However if a booster immunization is given between 3 and 5 weeks post-primary there is a rapid increase in the levels of circulating androstenedione-antibodies reaching a maximum level from 7 to 14 days post-boost. The antibody level then declines exponentially reaching levels of about a tenth of the maximum by 9 weeks post-booster (*Figure 5.1*). With further booster immunizations a similar pattern of response is produced with, however, a tendency towards increased maximum antibody responses. Early results suggested that beneficial increases in lambing percentages could be expected when ewes were mated as the antibody response to immunization was declining. However, it has now become apparent that the poor nutrition and minimal husbandry of typical extensive management systems also limits and in some circumstances totally blocks increased lambing percentages in response to Fecundin (Scaramuzzi *et al.*, 1983).

Fecundin, although initially intended for the extensively managed Australian Merino also has potential applications in intensive husbandry (Robinson and Scaramuzzi, 1986). The use of Fecundin in conjunction with programmes of superovulation, of out-of-season breeding, and of synchronization of oestrus and artificial insemination need to be evaluated; this forms the subject matter of this chapter.

Superovulatory responses in Fecundin-treated ewes

The use of gonadotrophins such as pregnant mare serum gonadotrophin (PMSG) and porcine follicle stimulating hormone (FSHp) to induce superovulation in embryo transfer programmes is a widespread practice. The use of Fecundin to further improve superovulatory responses to PMSG and FSHp has been evaluated in a series of three experiments.

For the first two experiments Merino ewes were injected with PMSG, at the same time as intravaginal progestagen-releasing sponges containing 60 mg of medroxy-progesterone acetate (Repromap®, Upjohn Pty Ltd) were removed after a 12-day period of treatment. Ovulation rates were determined by counting the number of corpora lutea present at an endoscopic examination of the ovaries 7–10 days after oestrus. Half of the ewes had been immunized with 2 ml of Fecundin 2 and 6 weeks before sponge removal.

In both experiments there was an increase in ovulation rate due to Fecundin treatment alone, 0.76 of an ovulation in 1984 and 0.75 in 1985. For both years irrespective of treatment there were parallel increases in ovulation rate with increasing doses of PMSG and from these data we can conclude that the effects of PMSG and Fecundin treatment are additive over the dose range tested in these experiments (*Table 5.1*). Similar results have been reported for the oestrone-immune ewe (Smith, Tervit and Cox, 1983). The enhanced response to low doses, 300 IU, of PMSG administered to the oestrone-immune ewe (Scaramuzzi and Hoskinson, 1984) was not observed in Fecundin-treated ewes given similar low doses, 250 IU, of PMSG (*see Table 5.1*). The data contain one anomalous result, the response to 750 IU of PMSG in experiment 1 does not conform to the general conclusion and for this no explanation can be offered. The effects of these treatments on embryo quality and embryo recovery rates were not determined. In a third experiment the effect of exogenous gonadotrophin treatment on embryo recovery rate and embryo quality was examined.

A group of 26 ewes, treated with Repromap sponges for 12 days, were injected with 16 mg (Armour equivalents) of FSHp. Half of the ewes had been immunized with Fecundin, 7 and 3 weeks before sponge removal. The FSHp was given as four injections spread over 2 days and in decreasing doses of 6, 5, 3 and 2 mg. The first injection was 24 h prior to sponge removal and then at 12 hourly intervals, with the

TABLE 5.1 The effect of Fecundin treatment on the ovulation rate of Merino ewes treated with pregnant mare serum gonadotrophin (PMSG) administered at the time of removal of progestagen sponges. Each experiment of 200 ewes consisted of eight groups of 25 ewes. Experiments 1, February 1984; and 2, February 1985

Treatment			Ovulation rate		
	0	250 IU PMSG	500 IU PMSG	750 IU PMSG	1500 IU PMSG
Control: experiment 1	1.29	1.39	1.96	3.42	—
Immunized: experiment 1	2.05	2.20	2.56	2.96	—
Control: experiment 2	1.35	—	1.48	2.04	6.08
Immunized: experiment 2	2.10	—	2.56	3.76	8.00

TABLE 5.2 The effect of porcine FSH administered around the time of progestagen sponge removal on embryo recovery from Fecundin-treated ewes; values are mean±SD. (Experiment 3)

	Control	Immunized
Number of ewes	13	13
Ewes in oestrus (%)	8 (61.5)	13 (100)
Interval—sponge removal to oestrus (h)	42.5±5.4	47.1±4.0
Ovulation rate (range)	6.88	9.08*
	(1–15)	(2–17)
Number of embryos recovered (%)	5.5±1.6	4.9±0.9
	(79.9)	(53.3)*
Number of eggs fertilized (%)	4.0±1.8	2.9±0.8
	(72.7)	(60.2)
Number of viable embryos (%)	3.4±1.5	2.2±0.7*
	(84.5)	(73.0)

*$P<0.05$, Fisher's Exact Test

final injection at 12 h after sponge removal. Entire rams, approximately one ram per ten ewes, were introduced at sponge removal and subsequent mating times recorded at 4 hourly intervals. Embryo recovery under pentothal-induced anaesthesia (0.6 g/50 kg liveweight) was attempted 5 days after the detection of oestrus. The recovered embryos were assessed for normality following microscopic examination. The ovulation rate was recorded at the time of embryo recovery.

The results of experiment 3 (*Table 5.2*) support the findings of experiments 1 and 2 that the effects of exogenous gonadotrophin and Fecundin immunization on ovulation rate are additive. Fecundin treatment did not suppress the expression of oestrus nor did it delay its onset. The recovery rate of day 5 embryos was reduced by Fecundin treatment, and the proportion of one cell unfertilized ova recovered (39.8%) was higher than in non-immunized ewes (27.3%). Overall there was a significant reduction in the number of viable embryos recovered in Fecundin-treated ewes superovulated with FSHp.

Ewes immunized with Fecundin but not treated with exogenous gonadotrophins also have reduced embryo recovery rates on day 2 after mating (Boland *et al.*, 1986). This effect on embryo recovery rate appears to be associated with the magnitude of the androstenedione-antibody response to Fecundin, since it diminishes as the antibody titre declines (Boland *et al.*, 1986); however, the problem may well be exacerbated by exogenous gonadotrophin. Similar experimentation conducted in France, using crossbred Lacaune ewes, yielded an increased number of viable embryos from Fecundin-treated ewes given an identical regime of FSHp treatment (Y. Cognié, unpublished data). This difference, while most difficult to explain, suggests some type of breed effect. Clearly the use of Fecundin in programmes of superovulation requires further investigation.

The use of Fecundin in out-of-season breeding

Ewes can be successfully bred out-of-season if induced to ovulate using combined progestagen–PMSG therapy (Robinson, Smith and Scaramuzzi, 1987). The commonest means of progestagen treatment is the intravaginal sponge or pessary. A series of experiments was conducted to explore the use of Fécundin in a traditional out-of-season breeding programme using flugestone acetate-impregnated sponges (Chrono-gest®, Intervet International) and PMSG (Robinson and Scaramuzzi, 1986).

Two experiments were carried out in the spring, one in November (experiment 4) and the other in December (experiment 5). Groups of anoestrous Border Leicester × Merino ewes were allocated to experimental groups in two experiments of factorial design. All ewes were treated for 12 days with 30 mg Chrono-gest sponges and given PMSG at the time of sponge removal. The doses of PMSG used were 250, 500, 750 and 1000 IU in experiment 4, and 500 and 750 IU in experiment 5. Half of the ewes at each dose of PMSG were immunized with Fecundin 15 and 43 days prior to sponge removal in experiment 4 and 23 and 51 days in experiment 5.

Following removal of sponges and PMSG injection the ewes were mated using an intensive mating system, which facilitated the precise detection of oestrus and provided optimal conditions for fertilization (Robinson, Smith and Scaramuzzi, 1987).

Pregnancy was diagnosed at endoscopy conducted 19–20 days post mating (Robinson, Smith and Scaramuzzi, 1987). This method, based on the knowledge that treated ewes return to anoestrus following regression of the corpora lutea in non-pregnant ewes, is extremely reliable in diagnosing pregnancy of spring-mated crossbred ewes (Robinson, Smith and Scaramuzzi, 1987) and compares most favourably with the traditional method of measuring blood progesterone levels 18 days after mating.

Close supervision of the flock at parturition allowed lambing ewes to be identified and their litter size to be recorded.

The results of these two experiments are presented in *Figure 5.2* and *Tables 5.3* and *5.4*. When the interval between booster immunization and sponge removal was 15 days (experiment 4), the incidence of oestrus was reduced (*see Table 5.3*) and its onset delayed (*see Figure 5.2*). These undesirable effects were not observed if the above interval was extended to 23 days (experiment 5; *see Table 5.4* and *Figure 5.2*).

The effect of Fecundin on fertility and lambing performance was also strongly influenced by the interval between booster immunization and sponge removal.

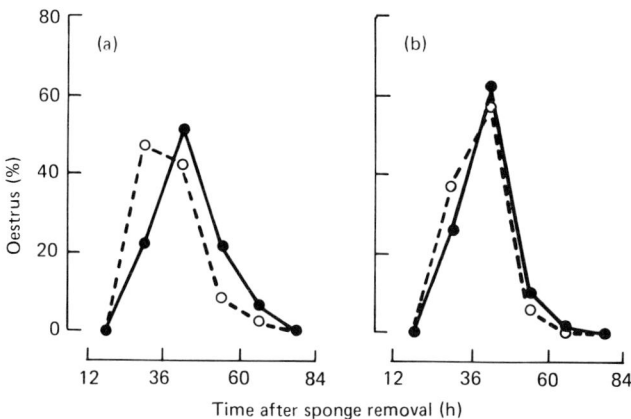

Figure 5.2 The percentage distribution of the onset of oestrus in Fecundin-treated ●————●, and untreated ○---○ anoestrus ewes treated with progestagen–PMSG. The data are from experiment 4 (a) and experiment 5 (b)

TABLE 5.3 The effect of Fecundin treatment 15 days prior to progestagen sponge removal on the reproductive characteristics of BLM ewes treated with progestagen–PMSG during anoestrus. (Experiment 4)

Main effect	Number of ewes	Ewes mated (%)	Ovulation rate	Fertility (%) (EL/EJ)*	Prolificacy (LB/EL)**	Lambing performance (LB/EJ)† (fertility × prolificacy)
Immunization status						
Immunized	168	93.5	2.92	26.2	1.80	0.47
Non-immunized	168	98.2	2.23	33.9	1.42	0.48
P		=0.05	<0.001	NS	<0.01	NS
Dose of PMSG (IU)						
250	84	89.3	1.70	22.6	1.26	0.28
500	84	96.4	2.12	25.0	1.57	0.39
750	84	100.0	2.45	36.9	1.68	0.62
1000	84	97.6	4.05	35.7	1.70	0.61
P		<0.02	<0.001	<0.05	<0.01	<0.01

There was no interaction between immunization treatment and dose of PMSG.
*   ewes lambing / ewes joined   **   lambs born / ewes lambing   †   lambs born / ewes joined

TABLE 5.4 The effect of Fecundin treatment 23 days prior to progestagen sponge removal on the reproductive characteristics of BLM ewes treated with progestagen–PMSG during anoestrus. (Experiment 5)

Main effect	Number of ewes ewes	Ewes mated	Ovulation rate	Fertility (%) (EL/EJ)*	Prolificacy (LB/EL)†	Lambing performance (fertility × prolificacy
Immunization status						
Immunized	193	97.4	2.82	64.2	1.81	1.16
Non-immunized	199	98.5	1.84	65.5	1.34	0.88
P		NS	<0.001	NS	<0.02	<0.05
Dose of PMSG (IU)						
500	196	97.4	1.97	61.9	1.71	1.06
750	196	98.5	2.73	67.5	1.44	0.98
P		NS	<0.001	NS	NS	NS

There was no interaction between immunization treatment and dose of PMSG.*
*   ewes lambing / ewes joined   †   lambs born / ewes lambing

With the shorter interval, while litter size was increased the overall lambing performance was not, because of a reduction in fertility (*Table 5.3*). The longer interval on the other hand resulted in an increased litter size and lambing performance, there being no effect on fertility (*see Table 5.4*). The interval between booster immunization and sponge removal did not influence the size of the increase in ovulation rate or litter size in response to immunization.

Injection of PMSG over the dose range 250–750 IU produced linear increases in the incidence of oestrus, ovulation rate, fertility, prolificacy and overall lambing performance (*see Table 5.3*). The ovulation rate leading to optimum reproductive performance appears to be about 2.5 or at least to be between 2 and 3, and doses of PMSG above 1000 IU will cause ewes to exceed this figure and so appear to be counterproductive (*see Table 5.3*). Fecundin may well have a PMSG-sparing effect as the optimum dose of PMSG which gives an ovulation rate of 2–3 will be reduced in Fecundin-treated ewes since the effects of PMSG and Fecundin were additive and there were no significant interactions.

Fecundin, if used alone during the anoestrus of highly seasonal breeds, will be ineffective since it can only act on ewes that are ovulating naturally or that have been induced to ovulate. When combined with methods for inducing ovulation in anoestrous animals, such as the ram effect (Martin, Scaramuzzi and Lindsay, 1981), or progestagen–PMSG treatment (*see Table 5.4*), Fecundin is highly effective. The interval between a booster immunization with Fecundin and ovulation ideally should be between 21 and 35 days; shorter intervals lead to decreased fertility which may offset gains in prolificacy, while with longer intervals the effect of Fecundin may be diminished as the antibody levels decline (*see Figure 5.1*).

Oestrus synchronization and artificial insemination

The use of artificial insemination during the normal breeding season is greatly facilitated by oestrus synchronization. The period of labour-intensive activities associated with artificial insemination can as a result be reduced to a few days and the cost effectiveness of artificial insemination can be enhanced further by increasing prolificacy. The use of Fecundin in conjunction with techniques for the synchronization of oestrus was investigated in an experiment (experiment 6) involving 200 Merino ewes and conducted in the autumn.

Three different methods of oestrus synchronization were compared in Fecundin-treated and untreated ewes. Fecundin was injected 36 and 55 days before the end of the synchronizing treatment. The methods compared were: Repromap intravaginal sponges inserted for a 12-day period, an injection of 125 µg of the synthetic prostaglandin cloprostenol (Estrumate®, ICI Pty Ltd) and the use of two subcutaneous progesterone releasing silastic implants (Silestrus®, Abbott Laboratories). The sponges and implants were removed and prostaglandin injected on the same day, and at the same time the flock was joined with a group of 20 vasectomized Merino rams fitted with Sire Sine® harnesses and crayons. The flock was examined for the presence of oestrous ewes every 6 h between 24 and 96 h after ram introduction.

The ovulation rate was recorded at endoscopy carried out 7 days later during the luteal phase of the following oestrous cycle.

The results presented in *Table 5.5* and *Figure 5.3* show that Fecundin treatment advanced the onset of oestrus in ewes treated with Repromap intravaginal progestagen sponges but had no effect on the onset of oestrus in ewes synchronized using prostaglandin or progesterone implants. The method of synchronization did not influence the ovulation rate. The incidence of oestrus was reduced in Fecundin-treated ewes given prostaglandin. The results of experiment 6 are curious in so much as they show an effect of Fecundin treatment advancing the onset of oestrus in ewes synchronized using Repromap intravaginal sponges. We would have expected any effects of Fecundin treatment to be in the opposite direction (Robinson and Scaramuzzi, 1986). The ewes used in experiment 6 had been previously synchronized so that a single prostaglandin treatment could be given during the mid-luteal stage of the oestrous cycle; as a result the sponges and implants used on the other ewes were inserted late in the luteal phase of the

TABLE 5.5 The effect of Fecundin on the synchronization of oestrus induced by three different methods. (Experiment 6)

Method of oestrus synchronization	Treatment of ewes	Mean time interval to onset of oestrus (h)	Mean ovulation rate	Oestrus (%)
Repromap sponges	Immunized	43.8	1.82	30 (100)
	Non-immunized	53.6	1.16	30 (97)
Silestrus implants	Immunized	31.4	1.85	33 (97)
	Non-immunized	32.9	1.23	31 (100)
Prostaglandin	Immunized	42.2	1.86	26 (81)
	Non-immunized	42.9	1.25	32 (100)·
Total	Immunized	—	1.84	88 (93)
	Non-immunized	—	1.22	93 (99)

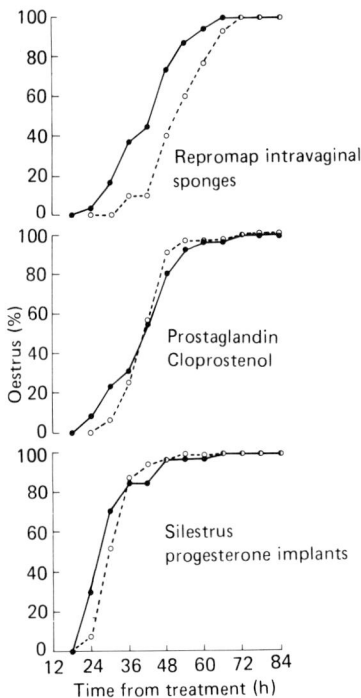

Figure 5.3 The cumulative percentage onset of oestrus in Fecundin-treated (●———●) and untreated (○---○) ewes in which oestrus was synchronized by three different methods: (1) the insertion of progestagen-releasing pessaries (Repromap, Upjohn Pty Ltd) for 12 days; (2) the injection of 125 μg cloprostenol, a synthetic analogue of prostaglandin $F_{2\alpha}$ (Estrumate, ICI Pty Ltd) during the mid-luteal phase of the oestrus cycle; (3) the subcutaneous insertion of two progesterone-releasing silastic implants (Silestrus, Abbott Laboratories) for 12 days

preceding cycle and at which time these ewes were also treated with prostaglandin, and may account for the unusual effect with Repromap sponges. These results also suggest that the delayed oestrus and increased anoestrus often seen in Fecundin-treated ewes (Cox *et al.*, 1982) might be due to increased endogenous progesterone production from the corpus luteum, since only prostaglandin-treated immunized ewes showed any evidence of anoestrus, and only ewes given prostaglandin would have had increased corpora lutea present on their ovaries at synchronization.

The occurrence of oestrus following oestrus synchronization was earliest when Silestrus implants were used and the latest when Repromap sponges were used. This result is perhaps a reflection of the rate of disappearance of the circulating levels of progesterone or medroxyprogesterone acetate. However, it may also reflect the differential effects of medroxyprogesterone acetate and progesterone on follicular development at least in the case of the Repromap sponges.

Programmes of artificial insemination may be based on oestrus detection or fixed time artificial insemination. In either case success will depend on the precision of oestrus synchronization. An important feature of these results is that Fecundin use has not seriously reduced the precision of oestrus synchronization. Although we as yet have no data on the effectiveness of artificial insemination in conjunction with Fecundin the above results provide some basis for optimism.

Systems of sheep husbandry and production exist in great variety in the many countries in which sheep are farmed. The evaluation and adaptation of new technologies such as Fecundin is of necessity tedious and slow. These results suggest that Fecundin originally intended for the extensively managed Australian Merino also has applicability in a variety of intensive management systems. The full extent of these applications, however, remain to be investigated.

References

BINDON, B.M. (1984). The reproductive biology of the Booroola Merino. *Australian Journal of Biological Sciences*, **37**, 163–189

BOLAND, M.P., NANCARROW, C.D., MURRAY, J.D., SCARAMUZZI, R.J., SUTTON, R., HOSKINSON, R.M. and HAZELTON, I.G. (1986). Fertilization and early embryonic development in androstenedione-immune Merino ewes. *Journal of Reproduction and Fertility*, **78**, 423–431

CORNU, C. and COGNIÉ, Y. (1985). The utilization of Romanov sheep in a system of integrated husbandry. In *Genetics of Reproduction in Sheep*, edited by R.B. Land and D.W. Robinson, pp. 383 390. London: Butterworths

COX, R.I., WILSON, P.A. and WONG, M.S.F. (1984). Change of ovulation rate with time in ewes immunized against androstenedione. *Proceedings of the Australian Society for Reproductive Biology*, **16**, 64

COX, R.I., WILSON, P.A., SCARAMUZZI, R.J., HOSKINSON, R.M., GEORGE, J.M. and BINDON, B.M. (1982). The active immunization of sheep against oestrone or androgens to increase twinning. *Animal Production in Australia*, **14**, 511–514

EGAN, A.R. (1984). Nutrition for reproduction. In *Reproduction in Sheep*, edited by D.R. Lindsay and D.T. Pearce, pp. 262–268. Canberra: Australian Academy of Sciences

GELDARD, H., SCARAMUZZI, R.J. and WILKINS, J.F. (1984). Immunization against polyandroalbumin leads to increases in lambing and tailing percentages. *New Zealand Veterinary Journal*, **32**, 2–5

MARTIN, G.B., SCARAMUZZI, R.J. and LINDSAY, D.R. (1981). Induction of ovulation in seasonally anovular ewes by the introduction of rams: effect of progesterone and active immunization against androstenedione. *Australian Journal of Biological Sciences*, **34**, 369–375

ROBINSON, T.J. and SCARAMUZZI, R.J. (1986). Immunization against androstenedione and out-of-season breeding. *Animal Production in Australia*, **16**, 323–326

ROBINSON, T.J., SMITH, C. and SCARAMUZZI, R.J. (1987). The time of mating and of LH release and subsequent fertility of anoestrous Border Leicester × Merino ewes treated with progestagen and Pregnant Mare Serum Gonadotrophin. *Animal Reproduction Science*, **13**

SCARAMUZZI, R.J. and HOSKINSON, R.M. (1984). Active immunization against steroid hormones for increasing fecundity. In *Immunological Aspects of Reproduction in Mammals*, edited by D.B. Crighton, pp. 445–474. London: Butterworths

SCARAMUZZI, R.J., GELDARD, H., BEELS, C.M., HOSKINSON, R.M. and COX, R.I. (1983). Increased lambing percentages through immunization against steroid hormones. *Wool Technology and Sheep Breeding*, **31**, 87–97

SMITH, J.F., TERVIT, H.R. and COX, R.I. (1983). Effect of immunization and dose of PMSG on ovulation rate and fertilisation in ewes. *Annual Report for 1981/82 of Agricultural Research Division, New Zealand Ministry of Agriculture and Fisheries*, p. 39

Chapter 6

Frequent lambing systems

D.E. Hogue

Summary

Five frequent lambing schedules are discussed and compared to a normal once-a-year lambing system. These include a twice a year, three lambings in 2 years, four lambings in 3 years and the CAMAL and STAR systems. The CAMAL (*Cornell Alternate Month Accelerated Lambing*) system allows ewes the maximum opportunity to lamb at intervals of less than 12 months but is more difficult to manage than the STAR system. The STAR system is based on the fact that a calendar year (365 days) contains exactly five, 73 day periods each of which is exactly half of a sheep pregnancy (146 days). By expressing the calendar in circular fashion with each 73 day period marked off and connecting alternate points a perfect star is formed, hence the name STAR. Ewes are managed to breed at every third point thus allowing 73 days for lambing, lactation and weaning and 146 days for pregnancy. Optimally, ewes can lamb five times in 3 years. The system fits into sheep biology and the calendar year, and because lambing and breeding dates are exactly coincident management of the sheep is easier than most systems.

After working on frequent lambing systems for approximately 15 years at Cornell, the author considers the STAR system the frequent lambing sheep production system of choice.

Introduction

The number of lambs that are reared to market age or weight per ewe per year is usually considered the most important effector of profit in a well-managed sheep enterprise. In some cases, the optimal number of lambs may depend on limitations of feed or other resources that preclude high levels of reproduction. An example would be in arid areas of the world where the nutrient supply will only support wool growth such as parts of Australia. In the United States, lambs are the primary saleable product and often exceed income from wool by 5–15-fold.

The number of lambs born per year can be increased by two methods. These are (1) increasing the number of lambs born per lambing, and (2) increasing the number of lambings per ewe per year or decreasing the lambing interval. The increase in number of lambs per lambing is achieved by selecting for twinning or using fecund breeds of sheep such as the Finnsheep or, more recently, the Booroola Merino. Increasing the number of lambings per ewe per year may be done by selecting aseasonal breeds of sheep and managing them in specific frequent or accelerated lambing production systems. More frequent lambing can also be achieved by using exogenous hormone therapy or by using photoperiod control. In this chapter, only frequent lambing production systems are discussed.

In addition to the increased reproductive efficiency that may be achieved with a frequent lambing system, there are some other advantages:

(1) a more uniform supply of lamb throughout the year which is an advantage to processors and retail markets;
(2) increased efficiency of use of facilities such as lambing barns, feeding pens, etc.;
(3) a more uniform yearly labour requirement;
(4) spreading the risk of bad storms or unfavourable markets throughout the year; and
(5) providing producers with a uniform cash flow throughout the year.

Once frequent lambing systems are developed, the major difficulty encountered may be the natural lack of cooperation by the sheep. Sheep are normally seasonally polyoestrus and breed during the shortening days of late summer, autumn and early winter in the temperate climates. In the tropical regions, they are less seasonal. Some breeds are less seasonal in the temperate regions.

Example of frequent lambing systems

The natural methods of producing sheep usually include once-a-year lambing, with ewes lambing in the spring of the year with a 12 month lambing interval between lambings. This is considered a normal system and interval and is used for comparison, and is shown in *Table 6.1*. This table includes six lambing frequencies each in its own distinct system. Each is identified by number of lambing periods per year and by a common name. In each of these systems, the sheep may be managed in groups and each lambing period is distinct, usually limited to a 4 week or a 1 month period for the purposes of this chapter. These are in contrast to a continuous lambing programme where rams are simply allowed to be with the ewes throughout the year with ewes lambing whenever they please. The programme of continuous lambing is characteristic of many wild or primitive sheep flocks as well as of sheep production in many developing countries. This schedule may be advantageous, or at least not disadvantageous, under these conditions.

The various systems outlined in *Table 6.1* all repeat each 12 months or multiple of 12 months. This is generally considered advantageous in practical farming conditions, and is easier to understand and follow than systems based on schedules that do not fit the calendar year. Each of the six systems or schedules will be described and discussed briefly.

Normal 12 month lambing interval

This is normal sheep production where ewes usually lamb once a year. If a ewe does not conceive and is kept in the flock she automatically moves to a 24 month possible lambing interval, or failing that a 36 month interval.

TABLE 6.1 Comparison of several frequent lambing systems and a normal once-a-year system of production

Lambing periods per year	Common name	Lambing interval*			Lambings/ewe/year		
		Bred at first opportunity	Bred at second opportunity	Bred at third opportunity	Bred at first opportunity	Bred at second opportunity	Bred at third opportunity
1	Normal	12	24	36	1.0	0.5	0.33
2	Twice a year	6	12	18	2.0	1.0	0.67
3	Three in 2	8	12	16	1.5	1.0	0.75
4	Four in 3	9	12	15	1.33	1.0	0.80
5	STAR**	7.2	9.6	12.0	1.67	1.25	1.0
6	CAMAL†	6	8	10	2.0	1.50	1.20

*Assumes ewes are grouped and moved to the next group if they do not conceive at one breeding

**See Figure 6.1

†Cornell Alternate Month Accelerated Lambing system

Twice a year

If all ewes in the flock would breed the month after lambing, they could effectively be managed on a 6 month basis and this may be the preferred frequent lambing system. The flock could be divided to provide additional lambing periods throughout the year and a very uniform and consistent level of production achieved. Very few ewes, even those selected from the more non-seasonal breeds, will achieve this in temperate zones and the system is generally considered impractical. The CAMAL system discussed later has some of the advantages of the twice-a-year system.

Three lambings in 2 years

This system was proposed about 25 years ago and is the system usually thought of when frequent or accelerated systems are mentioned. In this programme, a flock is usually divided into two subflocks with one or the other lambing every 4th month. Each ewe is given the opportunity to lamb every 8 months, or failing that, at 12 months. Optimally, each ewe can lamb three times in 2 years.

The breeding and lambing schedule can be depicted over 24 months as follows:

	Year 1													Year 2											
Month	J	F	M	A	M	J	J	A	S	O	N	D		J	F	M	A	M	J	J	A	S	O	N	D
Flock A	L		B					L		B								L		B					
Flock B				L		B								L		B					L		B	

L = Lamb   B = Breed

January, May and September are lambing months although the schedule could be shifted to other periods provided the proper intervals are maintained. Lambs are usually weaned the second month post-partum and the ewes are, therefore, bred the month post-weaning. Since all ewes have an opportunity to lamb for a 12 month interval if they do not conceive for the 8 month interval, any ewes that do breed the first time represent an increase in efficiency over the 12 month interval.

Four lambings in 3 years

This system is designed around a 9 month lambing interval with four lambing periods per year. It was used for several years by USDA at Beltsville, Maryland where the Morlam sheep was developed. Each ewe is put with the ram the 4th month after lambing and every 3 months thereafter until bred. As used with the Morlams, January, April, July and October were the lambing months and May, August, November and February the breeding months. This system could be combined with twice-a-year lambing so that ewes would have an opportunity to lamb after 6, 9 or 12 month intervals.

CAMAL (Cornell Alternate Month Accelerated Lambing system)

This system is intentionally considered ahead of the STAR system. Basically, the sheep from the three lambings in 2 years system are divided into four rather than two flocks with one flock lambing every other month.

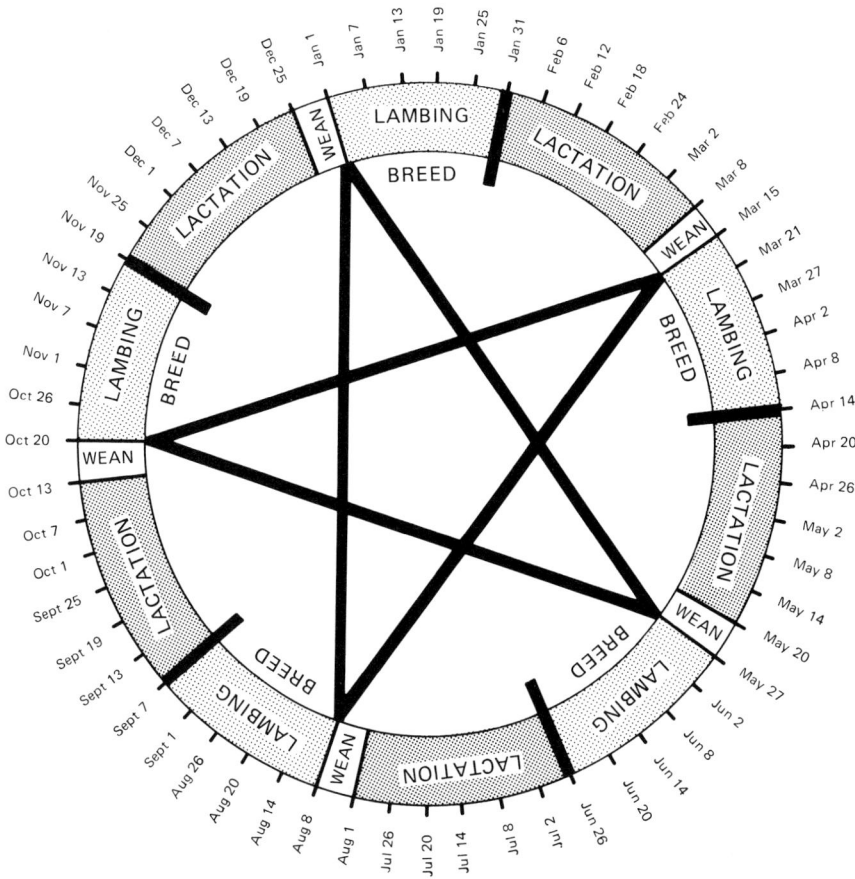

Figure 6.1 STAR Accelerated Lambing System, Cornell University, November 1983. (Copyright ©
1984 Cornell Research Foundation, reproduced with permission)

The schedule is depicted as follows over 2 years:

Month	Year 1												Year 2											
	J	F	M	A	M	J	J	A	S	O	N	D	J	F	M	A	M	J	J	A	S	O	N	D
Flock A	L			B					L			B				L				B				
Flock B		L				B				L				B				L			B			
Flock C			L			L		B					L		B					L				B
Flock D		B				L			B					L			B					L		
Monthly activity	L	B	L	B	L		B	L	B	L	B	L	B	L	B	L	B	L		B	L	B	L	B	L	B

It is apparent from the schedule that if all flocks are simply combined, every
alternate month is a breeding month and all interceding months are lambing
months, hence the name CAMAL. If rams are put with all ewes (including those

lactating), each ewe has the opportunity to lamb after an interval of 6, 8, 10 or 12 months. This system, therefore, combines the advantages of twice-a-year lambing, and three lambings in 2 years plus the possibility of a 10 month interval.

Dorset ewes at Cornell were managed on the CAMAL system for about 5 years, but management was not simple. Ewes actually do not have a 5 month gestation period but rather one of 146 days. Therefore, the alternate-month schedule was not exact and considerably more attention to detail was necessary than is desirable. Few ewes lambed after a 6 month interval and these were usually actually a 7 month interval as they were ewes that lambed early in the lambing month and bred late the following month. Also, lambing and breeding were not coincident which further complicated prelambing and prebreeding feeding, etc.

STAR system

The STAR system (*Figure 6.1*) was developed at Cornell (by Brian Magee, sheep farm manager) as an intended improvement on the CAMAL system (see below). It was realized that a successful frequent or accelerated lambing system should (1) not violate sheep biology, and (2) fit into a calendar year. It was also apparent that if breeding and lambing could be coincident, overall management would be simplified. The STAR programme solves the above dilemmas. Pregnancy in the sheep is 146 days, one-half pregnancy (146/2) equals 73 days which is exactly one-fifth of a year. The calendar year contains exactly five half sheep pregnancies as depicted on the attached figure. The annual calendar is depicted in circular fashion and divided into five equal 73 day periods. By connecting the first day of alternate periods, a perfect STAR is formed. By breeding or lambing the first 30 days of a period, the remaining management tasks are immediately apparent. If a ewe lambs in period 1, she is bred in period 2 to lamb in period 4, to be bred in period 5, etc. Each ewe can lamb five times in 3 years and complete the STAR. If she has five consecutive lambings with 7.2 month intervals, she is referred to as a STAR ewe. If she has twins at each lambing, she is referred to as an ALL-STAR ewe. The STAR system then:

(1) does not violate sheep biology;
(2) fits into the calendar year; and
(3) simplifies management, especially by making breeding and lambing dates exactly coincident.

The five lambing periods per year then:

(1) allows ewes to lamb five times in 3 years which equals 1.67 lambings/ewe/year;
(2) allows a uniform annual production of market lambs for a uniform supply and cash flow;
(3) increases efficiency of utilization of facilities such as lambing areas, etc.; and
(4) allows for the identification of ewes that not only will breed out of season but will breed following lambing in any season.

The Cornell Dorset Horn (Polled) flock has been managed on the STAR system for 4½ years. Lambing data are not fully analysed, but each year more ewes lamb in the autumn (periods 4 and 5), and the average lambing interval is decreasing. Of course, the flock represents selected sheep because long interval ewes are being culled. Sufficient STAR ewes have been identified to produce all breeding rams. Therefore, in the future, all lambs will be sired by rams from STAR ewes. These rams used as sires are also all twins or triplets.

Breeding and lambing records on these sheep are also establishing a database to allow study of genetic parameters and to develop optimal selection indices for the further identification of outstanding animals to be used in aseasonal frequent lambing systems.

Chapter 7

Advancement of puberty in male and female sheep

Ó.R. Dýrmundsson

Summary

Sexual development in the lamb is a gradual process, beginning early in fetal life, and is controlled by a complex interaction between the hypothalamus, anterior pituitary and the gonads. The considerable variation which exists, both between and within breeds of sheep, in the age and the body weight at puberty can be attributed to both genetic effects and environmental factors, particularly to nutrition in both sexes and to photoperiod in the female. Thus, for example, due to the close association between general body growth and sexual development, puberty may be advanced substantially by raising the plane of nutrition during rearing. A survey of the scientific literature clearly indicates that certain improvements may be achieved in the reproductive performance of both ewe and ram lambs. Moreover, early breeding is increasingly accepted as a means of enhancing flock productivity under a wide range of conditions.

Introduction

The inclusion of pubertal male and female lambs in the breeding flock is a means of increasing their lifetime productivity and hence the profitability of sheep farming. This particularly applies to ewe lambs in systems of intensified production. The main purpose of this paper is to review briefly various methods of advancing puberty paying special attention to practical aspects. In considering the reproductive potential of young sheep, it is pertinent to underline the fact that puberty precedes sexual maturity by variable lengths of time, both in the male (Colas, 1983) and in the female (Dýrmundsson, 1983).

Sexual maturation and the attainment of puberty in the sheep

The development of reproductive function in the lamb is a gradual process beginning early in fetal life (Mauleon, 1978). Much new information has been published in the past 10–15 years on the physiology of puberty in the sheep, underlining the complex interaction between the hypothalamus, anterior pituitary and the gonads (Pelletier, Carrez-Camous and Thiery, 1981; Quirke, Adams and Hanrahan, 1983; Foster, Yellon and Olster, 1985; Hochereau-de Reviers *et al.*, 1985). Special attention has been paid to the hormonal control of pubertal development, for example, studies on oestradiol feedback (Foster and Ryan, 1981), and more recently on the role of the pineal gland and its hormone melatonin (Kennaway, Gilmore and Dunstan, 1985; Rodway *et al.*, 1985).

It is well established that ewe lambs which have attained puberty, defined most commonly as the occurrence of first oestrus, have smaller ovaries, lower ovulation rates, shorter and less intense oestrus and both fewer and less regular oestrus cycles during the breeding season than mature ewes (Dýrmundsson, 1978a; 1981). One or two ovulations may precede the first oestrus (Hare and Bryant, 1982; Quirke, Stabenfeldt and Bradford, 1985), and oestrus without ovulation may also occur in ewe lambs (Edey, Kilgour and Bremner, 1978; Downing, 1980). Moreover, fertilized ova from ewe lambs are found to have lower survival rates to term than those of adult ewes, suggesting a quality difference which may partly account for disappointing conception and lambing rates often reported in mated ewe lambs (Quirke, Adams and Hanrahan, 1983; McMillan and McDonald, 1985). Fertility in ewe lambs may be improved slightly by delaying mating until the second oestrus (Hare and Bryant, 1985).

Puberty in the male is closely associated with testicular growth, but spermatozoa may be released by the time the testis has grown to approximately half its mature size (Courot, Hochereau and Ortavant, 1970; Dýrmundsson, 1978b). The completion of spermatogenesis is, however, preceded by certain anatomical developments such as by the descent of the testes into the scrotum and the detachment of preputial adhesions making the penis freely movable in the sheath. Physiological puberty, marked by the release of spermatozoa, normally precedes complete mating ability, even by several weeks (Dýrmundsson, 1972; Lees, 1978). In fact the attainment of puberty in the male cannot be measured as accurately as in the female.

When comparing information in the scientific literature, care must be taken in generalizing about the minimum pubertal age and body weight in either sex, since the criteria may range from the monitoring of endocrine parameters to the recording of behavioural events such as the first oestrus or the first copulation. Furthermore, sexual development is affected by both genetic and environmental factors and the interaction between these.

Improved reproductive performance of ewe lambs

Genetic selection and crossbreeding

Reviews of a large number of reports clearly show breed and strain differences in the incidence, and in the age and body weight at first oestrus (Hafez, 1952; Dýrmundsson, 1973a). Thus the incidence of oestrus in female sheep in their first year of life virtually ranges from 0 to 100% in the respective groups reported on, and most of the mean values cited for age at puberty are within the range of 6–18 months. There may be considerable genetic variability among individuals of the same breed. Genetic effects on puberty are, however, obscured to a variable extent by environmental factors, such as the plane of nutrition and the season of birth. Furthermore, published results on early breeding may not fully indicate the breed potential in each case.

Crossbred ewe lambs tend to have better reproductive performance than purebreds and heterosis may contribute to earlier sexual development (Jakubec, 1977; Land, 1978; McMillan and McDonald, 1983; Lahlou-Kassi and Marie, 1985; Young, Dickerson and Fogarty, 1985). Much of the information accumulated in recent years highlights the performance of highly precocious and prolific breeds and their crosses, such as the Finnish Landrace, the Romanov and the Booroola (Maijala and Österberg, 1977; Ricordeau et al., 1978; Smith et al., 1979; Hulet, Ercanbrack and Knight, 1984; Aboul-Naga, 1985; Montgomery et al., 1985; Quirke, Stabenfeldt and Bradford, 1985; Young, Dickerson and Fogarty, 1985).

The results certainly indicate that such breeds transmit sexual precocity and improved reproductive efficiency to their female progeny. Puberty may also be advanced by selection within breeds, but at a much slower rate than by crossbreeding. For example, there is a small positive association of ewe lamb oestrous activity and subsequent lamb production (Levine et al., 1978; Baker, Clarke and Diprose, 1981), and there are indications of a positive genetic relationship between lambing rate at 1 year of age and at older ages (Eikje, 1975; Jónmundsson, 1982). Relatively high ovulation and lambing rates have been recorded in ewe lambs carrying single high fecundity genes but apparently puberty was not advanced in these lambs (Jónmundsson and Adalsteinsson, 1985; Montgomery et al., 1985). Bodin et al. (1986) reporting on a considerable genetic variation in plasma FSH levels in young ewe lambs, have suggested that this trait may be a suitable early criterion of selection for ewe fertility.

Nutrition and growth promoters

There is a close association between general body growth and the growth and development of the reproductive organs (Pálsson and Vergés, 1952). Furthermore, it has been well established by several workers—for example, by Allen and Lamming (1961), Burfening et al. (1971) and Aboul-Naga, El-Shobokshy and Gabr (1980)—that raising the level of nutrition will advance puberty and that ewe lambs growing at faster rates will exhibit their first oestrus and are more likely to conceive at a lower age and heavier body weight than ewe lambs growing at slower rates. Foster and Olster (1985) have suggested that LH secretion may be impaired in undernourished ewe lambs with delayed puberty. Body composition traits such as the amount of carcass fat and protein do not seem to be important indicators of sexual development in the ewe lamb (Moore et al., 1985). In the work of Hugo, Roelofse and Els (1980) there was no evidence of a relationship between wool production in ewe lambs and age at puberty.

The early growth pattern of ewes can affect their reproductive potential (Gunn, 1977), and there is ample evidence that special nutritional allowances must be made, both in the first and in subsequent years, where early breeding is practised (Dýrmundsson, 1983). This stresses the need to give due consideration to the level of feeding, since in many locations nutrition would seem to be the environmental factor most limiting to successfully exploiting the breeding potential of ewe lambs. Although a high level of nutrition will, as a rule, advance puberty and even increase ovulation rate in ewe lambs (Downing, 1980; Hamra and Bryant, 1982), the feeding of very high-energy diets prior to breeding may be associated with an increased incidence of barrenness in ewe lambs, possibly due to overfatness (Stoerger et al., 1976). Results obtained by Robinson et al. (1971) and Quirke, Sheehan and Lawlor (1978) point to differences in the pattern of nutrient utilization of pregnant ewe lambs compared with that of adult ewes. Whilst maternal body weight gain was found to increase, lamb birthweight tended to decrease with increasing protein intake. Johnsson and Hart (1985) and Umberger *et al.* (1985) have presented evidence of high plane prepubertal feeding impairing mammary gland development and subsequent first lactation milk yield, again suggesting that nutrient requirements of ewe lambs need to be more precisely defined.

There is very limited information available on the effects of growth promoters on sexual development in ewe lambs. In recent trials anabolic steroids were found to have marked retarding effects on ovarian weight and other reproductive traits, while generally increasing body growth rate in ewe lambs (Fitzsimons and Crosby, 1985). Previously, Cooper (1981a, 1981b) had recorded both reduced ovulation rates and delayed puberty in ewe lambs treated with the oestrogenic compound zeranol.

Exogenous hormones and immunogens

It is well established that young female lambs can be induced to ovulate and display oestrus by the use of exogenous gonadotrophic and gonadal hormones (Dýrmundsson, 1973a; Quirke, 1981). Some advancement in puberty may be achieved but it is universally accepted that hormone therapy should not be applied to poorly grown animals. Thus it has been suggested that under practical conditions, only lambs which are heavier than 60% of their mature weight and more than 7 months of age should be considered for treatment (Gordon, 1967; Thimonier *et al.*, 1968; Quirke, Adams and Hanrahan, 1983). Most commonly a combined treatment of intravaginal progestagen sponges for 14 days and an injection of PMSG at pessary withdrawal is applied. Special care should be taken when the sponges are inserted and removed in order to avoid damage to the vaginal tissues. High levels of PMSG leading to an increase in ovulation rate should be avoided since this may result in the incidence of undesirable multiple births. The optimum level of PMSG is the minimum necessary for a good oestrous response and a normal ovulation rate, usually in the range of 250–500 IU varying according to breed (Quirke, Adams and Hanrahan, 1983; Cornu and Cognié, 1985). Although a progestagen–PMSG treatment can effectively induce and synchronize oestrus, it is unlikely to improve the fertility of ewe lambs beyond that which can be obtained under natural conditions during the breeding season (Thimonier *et al.*, 1968; Keane, 1974; Al-Wahab and Bryant, 1978a).

Melatonin administration may advance puberty in ewe lambs (Moore *et al.*, 1984; Fitzsimons *et al.*, 1985; Nowak and Rodway, 1985) and immunization of pubertal lambs against steroid hormones has been shown to increase ovulation rate (Fitzsimons *et al.*, 1985; Smith, 1985) but the practical significance of these new techniques remains to be evaluated. In fact, hormonal manipulation of breeding is not as widely applied in ewe lambs as in mature ewes.

Seasonality and photostimulation

Since Hammond (1944) first demonstrated a relationship between date of birth and that of first oestrus in ewe lambs, several reports from various locations have indicated that seasonality is an important factor in the attainment of puberty in the ewe. Thus in both northern and southern hemispheres, first oestrus is mainly confined to the autumn and winter months irrespective of time of birth (Joubert,

1963; Dyrmundsson, 1973a). For this reason puberty may be delayed by nearly a year if the lambs are late born or stunted. In subtropical breeds, however, ewe lambs experience a less well-defined seasonal onset of first oestrus than ewe lambs at higher latitudes (Mounib, Ahmed and Hamada, 1956; Younis *et al.*, 1978; Lahlou-Kassi and Marie, 1985).

There is mounting evidence that the photoperiod experienced by the lamb during rearing, interacting with nutrition and growth, is the most critical factor in the timing of puberty. As a matter of fact, considerable new knowledge has accrued in recent years on photostimulation in ewe lambs modifying the pubertal process to a variable extent (Al-Wahab and Bryant, 1978b; Foster and Ryan, 1981; Foster, Yellon and Olster, 1985; Foster and Yellon, 1985; Yellon and Foster, 1985). Nevertheless, the application of artificial photoperiods to advance puberty in the ewe does not as yet seem to be of any practical importance.

Temperature effects

There appears to be no information in the scientific literature on the direct effects of temperature on sexual development in ewe lambs. Sadleir (1969) has pointed out that the effects of temperature changes on puberty in mammals are closely related to the effect of temperature on body growth. This may be of special importance in subtropical and tropical regions. Puberty in ewe lambs may, for example, be delayed by the indirect effect of high temperature on the endocrine balance (Fitzgerald, 1981). Autumn shearing treatments did not have any clear effect on the onset of first oestrus and cyclic activity in ewe lambs in Wales (Dýrmundsson and Lees, 1972a), while premating shearing of winter-housed ewe lambs in Iceland (Adalsteinsson, 1972) and Norway (Nedkvitne, 1979) has been associated with better fertility and increased birthweight of the progeny.

Ram effect and mating management

There is a paucity of information in the scientific literature regarding the so-called 'ram effect' on the attainment of puberty. Dýrmundsson and Lees (1972b) reported that the sudden introduction of rams to ewe lambs in the transition from non-breeding to breeding activity resulted in a high degree of synchronization of first mating. Results obtained by Madani, Yagoub and Zawia (1985) support these findings. There are also indications that puberty may be advanced by the 'ram effect' (Oldham and Gray, 1984; Lopez Sebastian, Alonzo De Miguel and Gomez Brunet, 1985). However, Moore and McMillan (1981) failed to detect any effect of ram introduction on ovulation in pubertal ewe lambs. Thus the evidence of a male effect on puberty in female sheep remains equivocal.

It is well recognized that the behavioural signs of oestrus in ewe lambs are usually weak and the duration of oestrus is shorter than in mature ewes (Edey, Kilgour and Bremner, 1978; Lees, 1978). Thus ewe lambs in heat make little or no effort to approach the ram and may be reluctant to accept service when the latter makes his sexual advances. Rams show preference for mature ewes (Keane, 1976; Downing, 1980), and it is sound management to mate ewe lambs separately from older ewes in a confined mating area, to use active rams of small to medium size, and to maintain a high ram-to-ewe lamb ratio, especially when oestrus synchronization is applied. Colas (1979) and King *et al.* (1982) have successfully employed artificial insemination in well-developed ewe lambs at a synchronized oestrus. There is no evidence, however, to suggest that artificial insemination will improve conception in the ewe lamb and a controlled comparison between artificial insemination and natural mating is lacking.

Improved reproductive performance of ram lambs

Genetic effects

The marked differences in pubertal age and body weight between ram lambs of various breeds indicate that genetic factors may account for some of this variation. Within-breed variation may also be partly heritable (Dýrmundsson, 1973b). There is indeed considerable evidence of genetic variation, both between and within breeds, in traits associated with the attainment of puberty such as hormone levels, testicular growth, sperm production and sexual activity (Louda et al., 1981; Hochereau-de Reviers et al., 1985; Jónmundsson and Adalsteinsson, 1985; Kilgour et al., 1985; Notter et al., 1985). It appears, however, that no direct attempts have been made to select for advanced puberty in ram lambs. On the other hand, the relationship between testis size and both FSH and LH levels in ram lambs, and reproductive performance in related females, has been studied extensively with inconclusive results (Ricordeau, Blanc and Bodin, 1984; Lee and Land, 1985).

Nutrition and growth promoters

The various stages of sexual maturation in the male are strongly influenced by nutrition and growth. Thus lambs reared on high planes of nutrition normally attain puberty at lower ages and heavier body weights than those on lower feeding levels (Ragab, Sharafeldin and Khalil, 1966; Pretorius and Marincowitz, 1968). Although most ram lambs attain puberty within the age range of 4–8 months, retarded sexual development due to nutritive deficiency has been documented (Dýrmundsson, 1973b; Ketut Sutama and Edey, 1985). Improving nutritional conditions during rearing is clearly a most effective means of advancing puberty in the ram, by several weeks or even by a few months. In this context, management practices such as effective parasite control should not be neglected (Sykes, 1983).

Anabolic steroids promoting growth may adversely affect gonadotrophin synthesis and testis growth in ram lambs (Riesen et al., 1977; Fitzsimons and Crosby, 1985). Presumably, such products delay puberty and should not be used in ram lambs which are to be retained for breeding.

Photoperiod and temperature

Long photoperiods and high temperatures may have deleterious effects on sexual performance of adult rams (Ortavant, Courot and Hochereau, 1969; Colas, 1983). In ram lambs, there does not appear to be any direct relationship between the daylight environment and pubertal development but limited evidence suggests, however, that such a relationship might exist at the higher latitudes (Skinner and Rowson, 1968; Dýrmundsson and Lees, 1972c, 1972d). Exposure of ram lambs to long photoperiods from 10 to 21 weeks of age has been shown to increase growth rate and carcass weight without affecting concentrations of LH and testosterone (Schanbacher and Crouse, 1981). Treating ram lambs with melatonin, associated with photoperiodic control, has subtle effects on pubertal development according to Kennaway and Gilmore (1985). It would seem unlikely that the use of ram lambs for breeding can be advanced by photostimulation.

Although the detrimental effect of high temperature on the reproductive performance of the adult ram is well established (Colas, 1983), there does not seem to be any evidence of temperature affecting puberty in the ram lamb. However, the possibility of such a relationship should not be ruled out, perhaps in association with growth rate.

Behaviour and breeding practices

When first introduced to ewes in oestrus, pubertal ram lambs exhibit varying degrees of libido and mating dexterity, ranging from a complete lack of sexual interest to intense mounting and copulation. Ram lambs, being relatively smaller than adult ewes of their own breed, may sometimes experience difficulties in mating with them. Thus normally a smaller number of ewes is allocated to each ram lamb than to a mature ram during the breeding season (Dýrmundsson and Lees, 1972d; Dýrmundsson, 1973b). Due to the dominance of older rams, it is sound management to separate ram lambs and mature rams with their respective ewe groups, and in some instances hand-mating may improve the mating efficiency of ram lambs. Artificial insemination with ram lamb semen is not widely practised, due to its relatively poor quality and fertilizing capacity (Colas, 1983) but may, however, be applied in certain breeding programmes.

Conclusions

It is clear from the evidence reviewed that a considerable advancement of puberty may be achieved in both male and female sheep. Improved lamb growth rate by raising the level of nutrition and by effectively controlling parasitism would seem to be the most important factor to consider in many localities, in some cases in conjunction with crossbreeding and genetic selection. Exogenous hormones may be used with some practical advantage in ewe lambs. Although earlier breeding is less important economically in the male than in the female, the use of ram lambs for mating has several significant advantages. In considering the potential contribution of earlier sexual maturation and breeding, relative to other methods of intensifying sheep production, special attention should be paid to economic aspects under a range of environmental conditions.

References

ABOUL-NAGA, A.M. (1985). Crossbreeding for fecundity in subtropical sheep. In *Genetics of Reproduction in Sheep*, edited by R.B. Land and D.W. Robinson, pp. 55–62. London: Butterworths

ABOUL-NAGA, A.M., EL-SHOBOKSHY, A.S. and GABR, M.G. (1980). Using Suffolk sheep for improving lamb production from subtropical Egyptian sheep. *Journal of Agricultural Science, Cambridge*, **95**, 333–337

ADALSTEINSSON, S. (1972). Experiments on winter shearing of sheep in Iceland. *Acta Agriculturae Scandinavica*, **22**, 93–96

ALLEN, D.M. and LAMMING, G.E. (1961). Some effects of nutrition on the growth and sexual development of ewe lambs. *Journal of Agricultural Science, Cambridge*, **57**, 87–95

AL-WAHAB, R.M.H. and BRYANT, M.J. (1978a). Reproduction in young female sheep induced to breed at various ages. *Animal Production*, **26**, 309–316

AL-WAHAB, R.M.H. and BRYANT, M.J. (1978b). The effect of reduction in daylength, level of feeding and age on the reproduction of young female sheep mated at an induced ovulation. *Animal Production*, **26**, 317–324

BAKER, R.L., CLARKE, J.N. and DIPROSE, G.D. (1981). Effect of mating Romney ewe hoggets on lifetime production—preliminary results. *Proceedings of the New Zealand Society of Animal Production*, **41**, 198–203

BODIN, L., BIBE, B., BLANC, M. and RICORDEAU, G. (1986). Parametres génétiques de la concentration plasmatique en FSH des agnelles Lacaune viande. *Génétique, Sélection, Evolution*, **16**

BURFENING, P.J. HOVERSLAND, A.S., DRUMMOND, J. and VAN HORN, J.L. (1971). Supplementation for wintering range ewe lambs; effect on growth and estrus as ewe lambs. *Journal of Animal Science*, **33**, 711–714

COLAS, G. (1979). Fertility in the ewe after artificial insemination with fresh and frozen semen at the induced oestrus, and influence of the photoperiod on the semen quality of the ram. *Livestock Production Science*, **6**, 153–166

COLAS, G. (1983). Factors affecting the quality of ram semen. In *Sheep Production*, edited by W. Haresign, pp. 453–465. London: Butterworths

COOPER, R.A. (1981a). Some aspects of the use of the growth promoter zeranol in ewe lambs retained for breeding. 1. Effect on liveweight gain and puberty. *British Veterinary Journal*, **137**, 513–519

COOPER, R.A. (1981b). Some aspects of the use of the growth promoter zeranol in ewe lambs retained for breeding. II. Effects on reproductive tract, pituitary gland and gonadotrophin levels. *British Veterinary Journal*, **137**, 621–625

CORNU, C. and COGNIÉ, Y. (1985). The utilization of Romanov sheep in a system of integrated husbandry. In *Genetics of Reproduction in Sheep*, edited by R.B. Land and D.W. Robinson, pp. 383–389. London: Butterworths

COUROT, M., HOCHEREAU, M.T. and ORTAVANT, R. (1970). Spermatogenesis. In *The Testis*, Vol. I, edited by A.D. Johnson, W.R. Gomes and N.L. van Demark, pp. 339–432. London: Academic Press

DOWNING, J.M. (1980) Studies on the effects of date of birth and plane of nutrition on attainment of puberty and reproductive performance in Clun Forest ewe lambs. *PhD thesis*, University of Wales, UK

DÝRMUNDSSON, Ó.R. (1972). Studies on the attainment of puberty and reproductive performance in Clun Forest ewe and ram lambs. *PhD thesis*, University of Wales, UK

DÝRMUNDSSON, Ó.R. (1973a). Puberty and early reproductive performance in sheep. I. Ewe lambs. *Animal Breeding Abstracts*, **41**, 273–289

DÝRMUNDSSON, Ó.R. (1973b). Puberty and early reproductive performance in sheep. II. Ram lambs. *Animal Breeding Abstracts*, **41**, 419–430

DÝRMUNDSSON, Ó.R. (1978a). Studies on the breeding season of Icelandic ewes and ewe lambs. *Journal of Agricultural Science, Cambridge*, **90**, 275–281

DÝRMUNDSSON, Ó.R. (1978b). A note on sexual development of Icelandic rams. *Animal Production*, **26**, 335–338

DÝRMUNDSSON, Ó.R. (1981). Natural factors affecting puberty and reproductive performance in ewe lambs: a review. *Livestock Production Science*, **8**, 55–65

DÝRMUNDSSON, Ó.R. (1983). The influence of environmental factors on the attainment of puberty in ewe lambs. In *Sheep Production*, edited by W. Haresign, pp. 393–408. London: Butterworths

DÝRMUNDSSON, Ó.R. and LEES, J.L. (1972a). Effect of autumn shearing on breeding activity in Clun Forest ewe lambs. *Journal of Agricultural Science, Cambridge*, **79**, 431–433

DÝRMUNDSSON, Ó.R. and LEES, J.L. (1972b). Effect of rams on the onset of breeding activity in Clun Forest ewe lambs. *Journal of Agricultural Science, Cambridge*, **79**, 269–271

DÝRMUNDSSON, Ó.R. and LEES, J.L. (1972c). Puberal development of Clun Forest ram lambs in relation to time of birth. *Journal of Agricultural Science, Cambridge*, **79**, 83–89

DÝRMUNDSSON, Ó.R. and LEES, J.L. (1972d). A note on mating ability in Clun Forest ram lambs. *Animal Production*, **14**, 259–262

EDEY, T.N., KILGOUR, R. and BREMNER, K. (1978). Sexual behaviour and reproductive performance of ewe lambs at and after puberty. *Journal of Agricultural Science, Cambridge*, **90**, 83–91

EIKJE, E.D. (1975). Studies on sheep production records. VII. Genetic, phenotypic and environmental parameters for productivity traits of ewes. *Acta Agriculturae Scandinavica*, **25**, 242–252

FITZGERALD, J.A. (1981). The influence of artificial photoperiod, diet and temperature upon endocrine aspects of sexual maturity in the ewe lamb: changes in prolactin, LH and T4. *Dissertation Abstracts International, B*, **41**, 4403

FITZSIMONS, J.M. and CROSBY, T.F. (1985). Growth and reproductive traits in lambs treated with oestradiol/progesterone and trenbolone acetate. *Irish Journal of Agricultural Research*, **24**, 195–200

FITZSIMONS, J.M., HANRAHAN, J.P., EL-NAKLA, S. and ROCHE, J.F. (1985). Effect of melatonin and immunisation against androstenedione on puberty in ewe lambs. *Research Report 1985, Animal Production*. Dublin: An Foras Talúntais

FOSTER, D.L. and OLSTER, D.H. (1985). Effect of restricted nutrition on puberty in the lamb: patterns of tonic luteinizing hormone (LH) secretion and competency of the LH surge system. *Endocrinology*, **116**, 375–381

FOSTER, D.L. and RYAN, K.D. (1981). Endocrine mechanisms governing transition into adulthood in female sheep. *Journal of Reproduction and Fertility*, Supplement, **30**, 75–90

FOSTER, D.L. and YELLON, S.M. (1985). Photoperiodic time measurement is maintained in undernourished lambs with delayed puberty. *Journal of Reproduction and Fertility*, **75**, 203–208

FOSTER, D.L., YELLON, S.M. and OLSTER, D.H. (1985). Internal and external determinants of the timing of puberty in the female. *Journal of Reproduction and Fertility*, **75**, 327–344

GORDON, I. (1967). Progesterone-PMS therapy in the induction of pregnancy in anoestrous ewes and ewe lambs. *Journal of the Department of Agriculture, Republic of Ireland*, **64**, 38–50

GUNN, R.B. (1977). The effects of two nutritional environments from 6 weeks prepartum to 12 months of age on lifetime performance and reproductive potential of Scottish Blackface ewes in two adult environments. *Animal Production*, **25**, 155–164

HAFEZ, E.S.E. (1952). Studies on the breeding season and reproduction of the ewe. *Journal of Agricultural Science, Cambridge*, **42**, 189–265

HAMMOND, J. Jr (1944). On the breeding season in the sheep. *Journal of Agricultural Science, Cambridge*, **34**, 97–105

HAMRA, A.M. and BRYANT, M.J. (1982). The effects of level of feeding during rearing and early pregnancy upon reproduction in young female sheep. *Animal Production*, **34**, 41–48

HARE, L. and BRYANT, M.J. (1982). Characteristics of oestrous cycles and plasma progesterone profiles of young female sheep during their first breeding season. *Animal Production*, **35**, 1–7

HARE, L. and BRYANT, M.J. (1985). Ovulation rate and embryo survival in young ewes mated either at puberty or at the second or third oestrus. *Animal Reproduction Science*, **8**, 41–52

HOCHEREAU-DE REVIERS, M.T., BLANC, M.R., COLAS, G. and PELLETIER, J. (1985). Parameters of male fertility and their genetic variation in sheep. In *Genetics of Reproduction in Sheep*, edited by R.B. Land and D.W. Robinson, pp. 301–314. London: Butterworths

HUGO, W.J., ROELOFSE, C.S.M.B. and ELS, D.L. (1980). Studies on the sexual activity of Merino ewes. 1. Puberty in relation to feeding level and intensity of wool production. *Agroanimalia*, **12**, 45–49

HULET, C.V., ERCANBRACK, S.K. and KNIGHT (1984). Development of the Polypay breed of sheep. *Journal of Animal Science*, **58**, 15–24

JAKUBEC, V. (1977). Productivity of crosses based on prolific breeds of sheep. *Livestock Production Science*, **4**, 379–392

JOHNSSON, I.D. and HART, I.C. (1985). Pre-pubertal mammogenesis in the sheep. I. The effects of level of nutrition on growth and mammary development in female lambs. *Animal Production*, **41**, 323–332

JÓNMUNDSSON, J.V. (1982). Segir frjósemi gemlingsárid eitthvad um frjósemi ánna sídar? *Freyr*, **78**, 578–579 (paper in Icelandic)

JÓNMUNDSSON, J.V. and ADALSTEINSSON, S. (1985). Single genes for fecundity in Icelandic sheep. In *Genetics of Reproduction in Sheep*, edited by R.B. Land and D.W. Robinson, pp. 159–168. London: Butterworths

JOUBERT, D.M. (1963). Puberty in female farm animals. *Animal Breeding Abstracts*, **31**, 295–306

KEANE, M.G. (1974). Effect of progestagen-PMS hormone treatment on reproduction in ewe lambs. *Irish Journal of Agricultural Research*, **13**, 39–48

KEANE, M.G. (1976). Factors affecting reproduction in ewe lambs. *Proceedings of the 8th International Congress on Animal Reproduction and Artificial Insemination, Cracow*, pp. 473–476

KENNAWAY, D.J. and GILMORE, T.A. (1985). Effects of melatonin implants in ram lambs. *Journal of Reproduction and Fertility*, **73**, 85–91

KENNAWAY, D.J., GILMORE, T.A. and DUNSTAN, E.A. (1985). Pinealectomy delays puberty in ewe lambs. *Journal of Reproduction and Fertility*, **74**, 119–125

KETUT SUTAMA, I. and EDEY, T.N. (1985). Reproductive development during winter and spring of Merino ram lambs grown at three different rates. *Australian Journal of Agricultural Research*, **36**, 461–467

KILGOUR, R.J., PURVIS, I.W., PIPER, L.R. and ATKINS, K.D. (1985). Heritabilities of testis size and sexual behaviour in males and their genetic correlations with measures of female reproduction. In *Genetics of Reproduction in Sheep*, edited by R.B. Land and D.W. Robinson, pp. 343–345. London: Butterworths

KING, M.E., FRASER, C., DINGWELL, W.S. and ROBINSON, J.J. (1982). Artificial insemination versus service in ewes: a comparison of conception rates and numbers of lambs born. *Animal Production*, **34**, 369–370

LAHLOU-KASSI, A. and MARIE, M. (1985). Sexual and ovarian function of the D'man ewe. In *Genetics of Reproduction in Sheep*, edited by R.B. Land and D.W. Robinson, pp. 245–260. London: Butterworths

LAND, R.B. (1978). Reproduction in young sheep: some genetic and environmental sources of variation. *Journal of Reproduction and Fertility*, **52**, 427–436

LEE, G.J. and LAND, R.B. (1985). Testis size and LH response to LH-RH as male criteria of female reproductive performance. In *Genetics of Reproduction in Sheep*, edited by R.B. Land and D.W. Robinson, pp. 333–341. London: Butterworths

LEES, J.L. (1978). Factors affecting puberty and mating behaviour in sheep. In *The Management and Diseases of Sheep*, edited by the British Council, pp. 124–151. London: British Council and Commonwealth Agricultural Bureaux

LEVINE, J.M., VAVRA, M., PHILLIPS, R. and HOHENBOKEN, W. (1978). Ewe lamb conception as an indicator of future production in farm flock Columbia and Targhee ewes. *Journal of Animal Science*, **46**, 19–25

LOPEZ SEBASTIAN, A., ALONSO DE MIGUEL, M. and GOMEZ BRUNET, A. (1985). Caracteristicas del comienzo de la pubertad en corderas Manchegas mediante la estimulacion por machos en estacion desfavorable. *Anales del Instituto Nacional de Investigaciones Agararias*, **22**, 167–181

LOUDA, F., DONEY, J.M., ŠTOLC, L., KŘIŽEK, J. and ŠMERHA, J. (1981). The development of sexual activity and semen production in ram lambs of two prolific breeds: Romanov and the Finnish Landrace. *Animal Production*, **33**, 143–148

MADANI, M.O.K., YAGOUB, B.A. and ZAWIA, M.T. (1985). The mating of Libyan fat-tailed ewe lambs. *British Veterinary Journal*, **141**, 401–408

MAIJALA, K. and ÖSTERBERG, S. (1977). Productivity of pure Finnsheep in Finland and abroad. *Livestock Production Science*, **4**, 355–377

MAULEON, P. (1978). Ovarian development in young mammals. In *Control of Ovulation*, edited by D.B. Crighton, N.B. Haynes, G.R. Foxcroft and G.E. Lamming, pp. 141–158. London: Butterworths

McMILLAN, W.H. and McDONALD, M.F. (1983). Reproduction in ewe lambs and its effect on 2-year-old performance. *New Zealand Journal of Agricultural Research*, **26**, 437–442

McMILLAN, W.H. and McDONALD, M.F. (1985). Survival of fertilized ova from ewe lambs and adult ewes in the uteri of ewe lambs. *Animal Reproduction Science*, **8**, 235–240

MONTGOMERY, G.W., KELLY, R.W., DAVIS, G.H. and ALLISON, A.J. (1985). Ovulation rate and oestrus in Booroola genotypes: some effects on age, season and nutrition. In *Genetics of Reproduction in Sheep*, edited by R.B. Land and D.W. Robinson, pp. 237–243. London: Butterworths

MOORE, R.W. and McMILLAN, W.H. (1981). The effect of nutrition and the time of ram introduction on the onset of puberty in Romney ewe lambs. *Annual Report 1980/81*, Whatawhata Hill Country Research Station

MOORE, R.W., MILLER, C.M., LYNCH, P.R., WELCH, R.A.S., BARNES, D.R. and HOCKEY, H-U.P. (1984). The effect of melatonin on the onset of first oestrus in Romney ewe lambs. *Proceedings of the New Zealand Society of Animal Production*, **44**, 21–23

MOORE, R.W., BASS, J.J., WINN, G.W. and HOCKEY, H-U.P. (1985). Relationship between carcass composition and first oestrus in Romney ewe lambs. *Journal of Reproduction and Fertility*, **74**, 433–438

MOUNIB, M.S., AHMED, I.A. and HAMADA, M.K.O. (1956). A study of sexual behaviour of the female Rahmany sheep. *Alexandria Journal of Agricultural Research*, **4**, 85–108

NEDKVITNE, J.J. (1979). Effect of nutrition and of shearing before the mating season on the breeding activity of female lambs. *Proceedings of the EAAP, Sheep and Goat Commission, Harrogate*. British Society of Animal Production

NOTTER, D.R., LUCAS, J.R. McCLAUGHERTY, F.S. and COPENHAVER, J.S. (1985). Breed group differences in testicular growth patterns in spring-born ram lambs. *Journal of Animal Science*, **60**, 622–631

NOWAK, R. and RODWAY, R.G. (1985). Effect of intravaginal implants of melatonin on the onset of ovarian activity in adult and prepubertal ewes. *Journal of Reproduction and Fertility*, **74**, 287–293

OLDHAM, C.M. and GRAY, S.J. (1984). The 'ram effect' will advance puberty in 9 to 10 month old Merino ewes independent of their season of birth. *Proceedings of the Australian Society of Animal Production*, **15**, 727

ORTAVANT, R., COUROT, M. and HOCHEREAU, M.T. (1969). Spermatogenesis and morphology of the spermatozoon. In *Reproduction in Domestic Animals*, edited by H.H. Cole and T.T. Cupps, pp. 251–267. London: Academic Press

PÁLSSON, H. and VERGÉS, J.B. (1952). Effects of plane of nutrition on growth and development of carcase quality in lambs. *Journal of Agricultural Science, Cambridge*, **42**, 1–149

PELLETIER, J., CARREZ-CAMOUS, S. and THIERY, J.C. (1981). Basic neuroendocrine events before puberty in cattle, sheep and pigs. *Journal of Reproduction and Fertility*, Supplement, **30**, 91–102

PRETORIUS, P.S. and MARINCOWITZ, G. (1968).Post-natal penis development, testes descent and puberty in Merino ram lambs on different planes of nutrition. *South African Journal of Agricultural Science*, **11**, 319–334

QUIRKE, J.F. (1981). Regulation of puberty and reproduction in female lambs: a review. *Livestock Production Science*, **8**, 37–53

QUIRKE, J.F., ADAMS, T.E. and HANRAHAN, J.P. (1983). Artificial induction of puberty in ewe lambs. In *Sheep Production*, edited by W. Haresign, pp. 409–429. London: Butterworths

QUIRKE, J.F., SHEEHAN, W. and LAWLOR, M.J. (1978). The growth of pregnant female lambs and their progeny in relation to dietary protein and energy during pregnancy. *Irish Journal of Agricultural Research*, **17**, 33–42

QUIRKE, J.F., STABENFELDT, G.H. and BRADFORD, G.E. (1985). Onset of puberty and duration of the breeding season in Suffolk, Rambouillet, Finnish Landrace, Dorset and Finn–Dorset ewe lambs. *Journal of Animal Science*, **60**, 1463–1471

RAGAB, M.T., SHARAFELDIN, M.A. and KHALIL, I.A. (1966). Sexual behaviour of male lambs as affected by the plane of nutrition. *Journal of Animal Production of the United Arab Republic*, **6**, 89–94

RICORDEAU, G., BLANC, M.R. and BODIN, L. (1984). Teneurs plasmatiques en FSH et LH des agneaux mâles et femelles issus de béliers Lacaune prolifiques et non prolifiques. *Génétique, Sélection, Evolution*, **16**, 195–210

RICORDEAU, G., TCHAMITCHIAN, L., THIMONIER, J., FLAMANT, J.C. and THERIEZ, M. (1978). First survey of results obtained in France on reproductive and maternal performance in sheep, with particular reference to the Romanov breed and crosses with it. *Livestock Production Science*, **5**, 181–201

RIESEN, J.W., BEELER, B.J., ABENES, F.G. and WOODY, C.D. (1977). Effects of zeranol on the reproductive system of lambs. *Journal of Animal Science*, **45**, 293–298

ROBINSON, J.J., FRASER, C., CORSE, E.L. and GILL, J.C. (1971). Reproductive performance and protein utilization in pregnancy of sheep conceiving at eight months of age. *Animal Production*, **13**, 653–660

RODWAY, R.G., SWIFT, A.D., NOWAK, R., SMITH, J.A. and PADWICK, D. (1985). Plasma concentrations of melatonin and the onset of puberty in female lambs. *Animal Reproduction Science*, **8**, 241–246

SADLEIR, R.M.F.S. (1969). *The Ecology of Reproduction in Wild and Domestic Mammals*. London: Methuen

SCHANBACHER, B.D. and CROUSE, J.D. (1981). Photoperiodic regulation of growth: a photosensitive phase during light-dark cycle. *American Journal of Physiology*, **241**, E1–E5

SKINNER, J.D. and ROWSON, L.E.A. (1968). Puberty in Suffolk and cross-breed rams. *Journal of Reproduction and Fertility*, **16**, 479–488

SMITH, C., KING, J.W.B., NICHOLSON, D., WOLF, B.T. and BAMPTON, P.R. (1979). Performance of crossbred sheep from a synthetic dam line. *Animal Production*, **29**, 1–9

SMITH, J.F. (1985). Immunization of ewes against steroids: a review. *Proceedings of the New Zealand Society of Animal Production*, **45**, 171–177

STOERGER, M.F., HINDS, F.G., LEWIS, J.M., WALLACE, M. and DZUIK, P.J. (1976). Influence of dietary roughage level on reproductive rate in ewe lambs. *Journal of Animal Science*, **43**, 952–958

SYKES, A.R. (1983). Effects of parasitism on metabolism in the sheep. In *Sheep Production*, edited by W. Haresign, pp. 317–334. London: Butterworths

THIMONIER, J., MAULÉON, P., COGNIÉ, Y. and ORTAVANT R. (1968). Déclenchement de l'oestrus et obtention précoce de gestations chez des agnelles à l'aide d'éponges vaginales imprégnées d'acétate des fluorogestone. *Annales de Zootechnie*, **17**, 275–288

UMBERGER, S.H., GODE, L., CARUOLO, E.V., HARVEY, R.W., BRITT, J.H. and LINNERUD, A.C. (1985). Effects of accelerated growth during rearing on reproduction and lactation in ewes lambing at 13 to 15 months of age. *Theriogenology*, **23**, 555–564

YELLON, S.M. and FOSTER, D.L. (1985). Alternate photoperiods time puberty in the female lamb. *Endocrinology*, **116**, 2090–2097

YOUNG, L.D., DICKERSON, G.E. and FOGARTY, N.M. (1985). Evaluation and utilization of Finn sheep. In *Genetics of Reproduction in Sheep*, edited by R.B. Land and D.W. Robinson, pp. 25–38. London: Butterworths

YOUNIS, A.A., EL-GABOORY, I.A., EL-TAWIL, E.A. and EL-SHOBOKSHY, A.S. (1978). Age at puberty and possibility of early breeding in Awassi ewes. *Journal of Agricultural Science, Cambridge,* **90**, 255–260

Part III

Reproduction methods

With the advance of reproductive biology new techniques for reproduction have become available, particularly artificial insemination, to replace natural mating and embryo transfer to allow the augmentation of the breeding capacity of superior ewes. These techniques have been available for some time and artificial insemination for example has become widely used in the bovine species.

In this section the present stage of the application of these techniques to sheep is discussed and some of the improvements that may make these methods more widely usable in practice outlined.

A major limitation to the more widespread use of sheep artificial insemination has been the difficulty of efficient semen storage and this problem rightly deserves attention in the present section. The possibilities of sheep artificial insemination will reach a different dimension when frozen semen can be efficiently used.

Embryo transfer using surgical methods has already made some contribution to sheep research, but improvement of the technique could open further avenues of exploration, particularly in our understanding of the factors influencing embryo mortality.

Whether non-surgical methods, amenable to use on normal farms, can be developed is a question that hangs over its future as a technique for sheep breeding.

Chapter 8

Embryo transfer

I. Wilmut

Summary

The procedure of embryo transfer now in use is described. In the donors, hormones are used to control the time of oestrus and induce superovulation, the preparations used, the methods employed and the phase of treatment are mentioned. In the recipient, synchrony of the oestrus is essential for successful embryo transfer and is achieved by progestagen treatment. The time of the onset of oestrus should be determined by use of a vasectomized ram, at least twice daily. Embryo recovery is conducted either by flushing the oviduct or the uterus. Flushing the oviduct is appropriate for all stages of embryo development up to day 7 when the embryos are blastocysts in the zona pellucida. Flushing the uterus is successful only with embryos recovered on day 4 or later. The two methods are of similar efficiency. The site to which embryos are transferred is determined by the stage of embryo development. Embryos collected before day 4 should be transferred to the oviduct. Those collected after day 4 should be transferred to the uterus, while those collected on day 4 can be transferred to either site.

The *in vitro* maturation and fertilization of the mature oocytes and culture and storage of embryos is discussed.

Splitting of sheep embryos into halves or quarters could be practised to increase the number of offspring from selected donors as could transferring nuclei from later stage embryos to an oocyte.

Introduction

Embryo transfer between ewes has been carried out routinely for 30 years (Hunter, Adams and Rowson, 1955). During that time the methods have been improved considerably and a number of techniques have been developed for the manipulation and storage of embryos. The history of embryo transfer has been described by others (Adams, 1982a; Moore, 1982), and there are earlier reviews of the techniques (Moore, 1977; 1982). The purpose in this chapter is to describe procedures in use at the present time before considering their practical value.

Procedures for embryo transfer

Treatment of donors

Hormones are used to control the time of oestrus and induce superovulation in donor ewes. The gonadotrophin preparations that have been used most frequently are pregnant mares' serum gonadotrophin (PMSG) and pituitary follicle stimulating hormone (FSH), usually prepared from the horse pituitary. In cyclic ewes, PMSG is given as a single injection, either intramuscularly or subcutaneously, on day 12 or 13 of the oestrous cycle (day 0 = onset of oestrus). By contrast, pituitary FSH has been given in two to six injections beginning at the same stage of the cycle. In each case, promotion of follicle growth leads to a reduction in oestrous cycle length by approximately 24 h.

The availability of hormone treatments to control the time of oestrus permits precise planning of embryo transfer. Usually oestrus is controlled by treatment for 12–14 days with an intravaginal sponge containing progestagen. The injection of PMSG is given 24–36 h before removal of the sponge and oestrus occurs typically some 24–36 h after sponge withdrawal. Treatment with FSH should begin 48–24 h before sponge withdrawal. The interval from the end of treatment to the onset of oestrus varies with breed of ewe (I. Wilmut, unpublished observations) and with the progestagen (Robinson, 1967).

Injections of prostaglandin F_2 can be used to induce luteolysis and so control the time of oestrus. Administration of two injections, 8 days apart, controls oestrus; however, the yield of normal embryos is reduced in ewes that were at early stages of the cycle when treatment commences (I. Wilmut and D.I. Sales, unpublished observations). However, a single injection of prostaglandin can be used to improve the synchronization of oestrus in a group of ewes whose cycles are loosely grouped. An injection of PMSG should be given 48 h before that of prostaglandin and oestrus can be expected in most ewes 24–36 h after the injection of prostaglandin.

The choice of dose of gonadotrophin is critical. As the amount being given is increased beyond the optimum level, the number of follicles being stimulated reaches a threshold beyond which follicle growth and ovulation are disturbed and the number of ovulations decreases. A greater yield of normal embryos is probably obtained with FSH preparations than with PMSG.

In order to ensure that the majority of eggs are fertilized, it is essential either to supervise mating or to perform surgical insemination. This is particularly important if a number of the donors, being treated at any one time, are to be mated by the same ram. Observation to determine the time of the onset of oestrus should be carried out at least twice daily at approximately 12 h intervals (*see* p.83) and each ewe should be allowed to mate at least once at both the first and the second period of observation when they are in oestrus. Alternatively, 20 µl of semen should be deposited into the lumen of each uterine horn, near to the uterotubal junction, soon after the onset of oestrus (Trounson and Moore, 1974a, 1974b). Such surgical insemination may reduce the proportion of embryos recovered, but the yield of normal embryos is increased because the proportion of eggs fertilized is very much greater after surgical insemination (Trounson and Moore, 1974a, 1974b).

Treatment of recipients

Successful embryo transfer depends upon the occurrence of oestrus at the same time in donor and recipient animals (*see below*, p. 83). This requirement for 'synchrony' can be met by use of progestagen treatment to control the time of oestrus in recipient ewes. Gonadotrophin injections are not given, as induction of superovulation is unnecessary and might be harmful if a number of unovulated follicles are formed. Typically, sponge withdrawal should be 24 h earlier in recipients than in donors, as the interval from sponge removal until the onset of oestrus is shorter in donor ewes given gonadotrophin. The time of the onset of oestrus should be determined by use of a vasectomized ram, at least twice daily.

Embryo recovery

Immediately after ovulation, oocytes and their enveloping cumulus cells pass along the oviduct to the ampullary-isthmus junction. The isthmus retains the eggs in this position for approximately 60 h, by which time typical embryos have four cells (Holst, 1974). Most embryos enter the uterus during the following 12 h.

There are two methods of recovering embryos: to flush the uterus or the oviduct. When flushing the oviduct, medium is passed from the uterine lumen through the uterotubal junction and along the oviduct before it is collected into a dish through a cannula. This approach is appropriate for all stages of embryo development up to day 7 when the embryos are blastocysts in the zona pellucida. When flushing only the uterus, medium is introduced into the uterine lumen near to the uterotubal junction and recovered through a catheter inserted into the uterus near the external bifurcation. This procedure is successful only with embryos recovered on day 4 or later, by which time the embryos have passed into the uterus. Efficiency of embryo recovery is similar with these methods, but it is probable that fewer adhesions are caused by flushing the uterus and so is more appropriate in circumstances in which repeated recovery is to be attempted.

Success in embryo recovery from the oviduct requires careful insertion of the catheter into the oviduct and application of steady, but gentle, pressure on the uterus to force medium into the oviduct. A total of 10–20 ml medium is introduced into the uterine lumen through a blunt-ended needle (18 FG) inserted into the uterus approximately 4–6 cm below the uterotubal junction. Location of the lumen is much easier with a blunt needle, as a sharp-tip passes through tissue too readily. By grasping the uterus around the needle, the surgeon is able to prevent passage of fluid down the uterine horn toward the cervix. The uterus becomes distended as fluid enters the uterus (after approximately 5 ml), and it is important to maintain pressure within the uterus without causing rupture of the uterine wall.

A Foley catheter (10 FG, 5 ml balloon) is required for a uterine flush. It is inserted into the uterine lumen just above the external bifurcation, through a hole made by forcing the tip of a pair of artery forceps through the uterine wall. The catheter is passed 4–6 cm along the uterine horn toward the oviduct before the balloon is inflated with air (4–10 ml depending upon size of the uterus). The balloon should be sufficiently firm to prevent movement if the catheter is pulled gently. Medium is introduced into the uterine lumen to rinse the tip of the uterine horn, and allowed to flow to the tip of the uterine horn, but the surgical assistant prevents it from entering the oviduct by holding the uterotubal junction between a finger and thumb. Each uterine horn is rinsed with 20 ml medium. Efficient recovery of embryos is assumed to depend upon a free flow of medium able to rinse the uterine lumen thoroughly and this is achieved by gentle massage of the uterine horn. The hole made to allow access for the Foley catheter is closed with a suture.

Uterine flushing of this kind was first carried out during a midventral laparotomy (Wilmut and Sales, 1981), but has recently been performed by laparoscopy (McKelvey and Robinson, 1986). Essentially the same procedures are followed except that all manipulations of the reproductive tract are performed with the aid of grasping forceps. At laparoscopy a greater volume of medium (60 ml) was used to flush each uterine horn. Embryo recovery was 35%, 76% and 66% in three trials. No direct comparison with routine procedures was made, but it is probable that these recovery rates are slightly lower than those obtained during a midventral laparotomy. The main advantages of this approach are that ewes are unlikely to develop adhesions and that the stress on the animal is much less.

Non-surgical collection has been described, but eggs were only obtained from 11 of 26 (42%) ewes (Coonrod et al., 1986). A number of factors impose severe limitations on this approach. The cervical canal is extremely narrow and convoluted making passage of a catheter difficult, and perhaps impossible in some females. In addition, the inability to inject medium into the uterine lumen at the tip of the uterine horn or to massage the uterus, lead to the use of large volumes of fluid. In turn this takes longer for the flushing procedure. Finally, it seems improbable that both uterine horns will be flushed efficiently.

Embryo transfer

Stage of embryo development determines the site to which the embryos should be transferred (Moore, 1982). Embryos collected before day 4 should be transferred to the oviduct. Those collected after day 4 should be transferred to the uterus, while those collected on day 4 can be transferred to either site.

Transfer to the oviduct is achieved by passing a fine Pasteur pipette into the ampulla. The tip of the pipette should be smooth and approximately 0.5 mm outside diameter. Movement of fluid is controlled by a mouthpiece. The fimbria of the oviduct is held with fingers to expose the lumen of the oviduct and in such a manner that up to 5 cm of the ampulla is straightened. After passing the pipette down this section of ampulla the fluid containing the embryo is expelled gently.

Embryos are transferred to the uterus using a larger Pasteur pipette fitted to a syringe. A blunted 18 FG needle is passed into the uterine lumen some 10–20 cm from the uterotubal junction and directed towards it. The pipette, with a flame-polished tip (outside diameter 1–2 mm), is passed into the lumen for a distance of 5–10 cm before the fluid is expelled. If fluid is drawn into the pipette before the embryo, and is separated from the fluid containing the egg by air bubbles, expulsion of fluid and embryo from the pipette can be monitored by watching the movement of the additional fluid. The latter is not expelled into the uterus.

Transfer to the uterus has been achieved reliably by means of laparoscopy (McKelvey, Robinson and Aitken, 1985). Essentially the procedure was that described above. The conception rate was comparable to that described by those using midventral laparotomy. This approach halved the time taken for each recipient and has the additional advantage of imposing less stress on the animals. In future, when transferring embryos to the uterus it would be appropriate to use laparoscopy. It is likely that experience gained during transfer would also lead to improved rates of recovery by laparoscopy.

Factors influencing the success of transfer

The requirement that embryos be transferred to recipients that are suitable for their stage of development is the most important factor to influence the outcome of embryo transfer. Initially, this requirement for 'synchrony' was defined in relation to the interval from the onset of oestrus to surgery in donor and recipient ewes (Moore and Shelton, 1964; Rowson and Moor, 1966). Investigations since then have established that the fundamental characteristic in donors is the stage of embryo development. In recipients, it is the time when peripheral plasma concentrations of progesterone typical of the luteal phase initiate changes in uterine function (Wilmut, Sales and Ashworth, 1985a, 1985b). In practice, it is still necessary to use determination of the time of onset of oestrus as the criterion to establish synchrony, although it may be possible in the future to develop methods of administering hormones to control the time of changes in recipients.

Precise estimates of the effect of asynchrony on embryo survival are difficult to provide and there are important implications for labour requirements of a recommendation that exact synchrony be established. In the two major research papers on the subject, increasing degrees of asynchrony led to a reduced proportion of pregnancies only when increased beyond 12 and 48 h respectively (Moore and Shelton, 1964; Rowson and Moor, 1966). In other reports, there is evidence of effects of as little as 8 h asynchrony (McKelvey, Robinson and Aitken, 1985), a comment that depends upon observations at least every 4 h. A group would be well advised to observe for oestrus at least twice daily and then examine their results. Evidence of a poor overall result or of markedly smaller proportion of pregnancies when synchrony varied by 12 or 24 h would justify more frequent observation and attempts at greater synchrony.

Survival of very early embryos may be less than that of later stages of development. As embryo stage advanced from two to at least eight cells in eggs recovered 48–84 h after the onset of oestrus, the proportion of eggs represented by live lambs increased from 44% to 78% (Moore and Shelton, 1964).

The proportion of embryos surviving to term is influenced by the number transferred and there are differences between breeds in the number of lambs they can carry (Land and Wilmut, 1977). It is appropriate to transfer a number of eggs equal to the typical number of ovulations for the breed. The proportion of pregnancies is not affected by small variations in the number of embryos transferred suggesting that in some cases failure to conceive reflects the influence of an abnormality in the recipient. If only one embryo is transferred, it should be placed in the horn ipsilateral to the ovulation. Larger numbers of embryos should be divided between the horns.

Subjective assessments of embryo development during transfer of cattle embryos have revealed that even quite markedly atypical embryos do occasionally establish a normal pregnancy. However, the proportion of eggs surviving to term decreases as the extent to which the embryo is atypical increases. A decision on whether or not to transfer embryos depends upon the objectives of the transfer and upon which resource is limiting. All embryos should be transferred if the objective is to obtain as many lambs as possible from a particular donor, and if the supply of recipients is unlimited. By contrast, when the number of recipients is limited, a choice among embryos should be made. The causes of atypical development are not known, but probably include the influence of abnormal hormone patterns on follicle development and maturation and upon functioning of the reproductive tract. The presence of unovulated follicles is commonly associated with recovery of atypical embryos.

Factors such as nutrition, body condition, stress and disease that can prejudice survival of embryos in mated ewes, would be expected to have similar effects after transfer. Many studies have employed older ewes of proven fertility as recipients. These are preferable to maiden animals as the broad ligament has been stretched, but it is essential that they are healthy animals.

Finally, success depends upon practice and upon the availability of sterile equipment. There are two published accounts of improvement in results with greater experience, one in mice the other in cattle (McLaren and Michie, 1959; Church and Shea, 1976). There are many aspects of the technique that can be modified to produce optimum performance including manual dexterity during transfer. Evidence that infection arising during transfer causes loss of embryos is circumstantial (Adams, 1982b), but it seems prudent to perform the surgery in a clean area and to carry out embryo manipulation in a room away from the ewes.

The yield of embryos from a group of ewes can be expected to be two to six/donor/recovery depending upon the breed. The number recovered from an individual donor may vary from 0 to 40. After transfer, 50–70% of embryos are expected to survive to term. As a result, each donor might have one to four additional lambs per recovery.

In vitro maturation of oocytes

Those who carry out embryo transfer have long been tempted by the possibility of harvesting some of the thousands of oocytes present in each ovary. Oocytes recovered from antral follicles resume meiosis in culture, but exhibit severe abnormalities during early development. By contrast, oocytes matured within follicles are able to develop normally provided that gonadotrophins and oestradiol-17β are present in the medium (Moor and Trounson, 1977). Normal maturation of oocytes outside the follicle has been achieved by gentle agitation of oocytes enclosed in cumulus cells cultured in the presence of hormones (Moor, Kruip and Green, 1984). Under optimal conditions, the proportion of oocytes able to develop to blastocysts after fertilization in vitro was 55%, compared to 65% if oocytes were allowed to mature in vivo (Moor, Kruip and Green, 1984). Despite these advances in understanding, this system cannot yet be recommended for routine use.

In vitro fertilization

Fertilization of in vitro matured oocytes can be achieved in the oviduct of a recipient animal (Moor and Trounson, 1977). Direct transfer to a recipient is appropriate when the only objective is to produce offspring from a particular donor. By contrast, in vitro techniques have the potential of providing embryos at particular stages of development for subsequent micromanipulation. Development of techniques of in vitro fertilization in farm animals is taking place rapidly (see Wright and Bondioli, 1981; Brackett, 1983), but routine application of the procedures will remain inappropriate until the frequency of live-birth approaches that obtained following recovery of early embryos.

Culture and storage of embryos

The medium (PBS) used for recovery and manipulation of embryos is Dulbecco's phosphate buffered saline supplemented with energy sources, protein and antibiotic (Whittingham, 1971). Fetal bovine or heterologous sheep serum are added (10–20% v/v) by some groups during routine storage and manipulation.

The choice of medium for culture depends upon the stage of development of the embryo when culture begins. There is no medium which will support development from one cell to blastocysts with a normal potential to establish pregnancy, although some embryos will reach the blastocyst stage. Eggs at any stage from oocyte to eight to 16 celled stages develop for several days in bicarbonate buffered medium supplemented with bovine serum albumin (Tervit, Whittingham and Rowson, 1972; Wright et al., 1976). By contrast, morulae and blastocysts benefit from having serum in the medium, and will develop in medium buffered with either bicarbonate or phosphate (Moor and Cragle, 1971; Moore and Bilton, 1973; Trounson and Moore, 1974a, 1974b).

Caution should be exercised in planning the use of culture systems. Very few of the published experiments have assessed viability of cultured embryos by transfer, and it is probable that after 2 or 3 days' culture some embryos are abnormal. Systems of co-culture with cells from the oviduct may create new opportunities (Rexroad and Powell, 1986).

Storage of embryos can be carried out by freezing, or by holding them in a refrigerator or at room temperature. Storage for several hours is routine at a temperature of 20°C in phosphate buffered saline (PBS). Storage at low temperatures has not been developed as thoroughly. There is evidence of a considerable proportion of morulae and blastocysts surviving storage for 48 h at 5–10°C (Kardymowicz, Kardymowicz and Grochowalski, 1964, 1966; Moore and Bilton, 1973).

Longer-term storage can be achieved by freezing and thawing (Willadsen, 1977). Survival after freezing and thawing depends upon control of several factors. Medium containing cryoprotectant (glycerol or dimethylsulphoxide, DMSO) is added at room temperature (around 20°C) usually to a final concentration of 10% (v/v). After transfer to a suitable container, such as a straw or ampoule, the embryos are cooled to −7°C. At this temperature, ice formation is induced in the medium by holding a pair of forceps that have been cooled in liquid nitrogen against the outside of the container. This process of 'seeding' is essential. Embryos should then be cooled very slowly, usually at 0.2–0.3°C/min. Storage is in liquid nitrogen (−196°C). Transfer to liquid nitrogen can be from a temperature of −35°C or −60°C. This choice also determines the method of thawing. A rapid procedure (for example, shaking in water at 37°C) is required if samples are plunged from −35°C. By contrast, slow warming, in air at room temperature is essential for samples plunged from −60°C. Survival is probably similar with the two approaches (Willadsen, 1977), but the procedure is much quicker when samples are plunged from the higher temperature. Overall, approximately 35% of frozen and thawed embryos can be expected to develop to term. Survival could probably be increased by refinement of the method.

Splitting of embryos

Sheep embryos can be divided into halves or quarters to increase the number of offspring from selected donors (Willadsen, 1982). Embryos with four or eight cells can be divided into quarters, but must then be transferred to the oviduct of a recipient ewe for several days to develop to the blastocyst stage. In order to prevent phagocytosis of the cells, the zona is embedded in a cylinder of agar. Blastocysts dissected free from agar can be transferred to recipient ewes. By contrast, it is impracticable to cut morulae or blastocysts into more than two sections, but they can be bisected within their own zona pellucida and then transferred direct to recipients (Willadsen and Godke, 1984).

The techniques have not been applied extensively so that it is difficult to estimate the yield of lambs in practice. It is not possible to double the number of lambs by bisecting embryos, but an increase of approximately 50% is likely.

Nuclear transfer

In principle, a more effective method for increasing the number of young from a particular donor would be to transfer nuclei from later stage embryos to an oocyte. This has recently been achieved (Willadsen, 1986). The methods are not yet routine, but rapid improvement can be expected.

Gene transfer

The technique of gene transfer provides the opportunity to induce specific modifications to farm animals. In mice, following the direct injection of several hundred copies of a gene into a pronucleus of an egg, the gene is incorporated into a chromosome in a proportion of the resulting offspring. In the great majority of cases it is inherited in a simple Mendelian manner and is expressed (Gordon, 1983). Early results in sheep suggest a less frequent occurrence of incorporation than in mice (Hammer et al., 1985; J.P. Simons, A.J. Clark and I. Wilmut, unpublished observations). However, it is probable that future developments will create the opportunity to apply the techniques in sheep.

Improvements in the design of the injected gene, the construct, are also to be expected. Genes contain elements that govern the tissue in the body that express the gene and also the stage in development when it is expressed. By choice of suitable control sequences, it will be possible to arrange for hormones and other proteins to be produced in a novel manner (*see below*, p.87).

Practical application

The extent to which embryo transfer is used commercially is limited by the costs involved. After analysing the potential economic return from carrying out embryo transfer in cattle, it was judged that a breeder could not justify applying the procedure simply to increase the yield of milk or meat from the herd (Wilmut and Hume, 1978). By contrast, it may be justifiable to produce additional selected young for sale. A similar situation applies in sheep and the technique has been used. Embryo transfer is useful for importation and perhaps to increase the number of animals of an unusual breed whenever it commands very high prices. It has been used during a selection programme to increase the rate of genetic improvement (Hanrahan and Quirke, 1980). Rams from the flock could then be used to spread the improvement throughout the national flock. The greatest difficulty in this approach is in agreeing upon a selection criterion. Once transfer methods are established it is probably useful to apply whatever methods are available for embryo splitting or nuclear transfer.

In time, it will be possible to modify many aspects of sheep production by gene transfer. Incorporation of a construct containing the growth hormone gene led to a significant increase in growth rate in mice (Palmiter *et al.*, 1982). In these cases, expression was controlled by sequences from a gene that is activated by heavy metal ions. Secretion of growth hormone was stimulated by treatment with zinc. By careful selection of control sequences, it may be possible to obtain expression of the gene at the appropriate time without the need for treatment. Alterations to the pattern of growth are also likely to lead to changes in carcass composition. It may be possible to improve a variety of other aspects of performance including wool production by modification of keratin genes (Frankham and Gillings, 1983), and reproduction by changing the feedback system between ovarian hormones and pituitary gonadotrophins.

Conclusions

Embryo transfer is an extremely powerful technique and plays a unique role in research (Wilmut, 1982). There has been relatively little commercial application of the technique. However, sheep production will probably benefit more from application of the associated procedure of gene transfer. As the number of animals involved will be small and the techniques are sophisticated they should be established in a few specialist centres.

References

ADAMS, C.E. (1982a). Egg transfer: histological aspects. In *Mammalian Egg Transfer*, edited by C.E. Adams, pp. 1–17. Baton Rouge, Florida: CRC Press

ADAMS, C.E. (1982b). Factors affecting the success of egg transfer. In *Mammalian Egg Transfer*, edited by C.E. Adams, pp. 175–183. Baton Rouge, Florida: CRC Press

BRACKETT, B.C. (1983). A review of bovine fertilization *in vitro*. *Theriogenology*, **19**, 1–15

CHURCH, R.B. and SHEA, B. (1976). Some aspects of bovine embryo transfer. In *Egg Transfer in Cattle*, edited by L.E.A. Rowson, pp. 73–91. Luxembourg: CEC

COONROD, S.A., COREN, B.R., McBRIDE, B.L., BOWEN, M.J. and KRAEMER, D.C. (1986). Successful non-surgical collection of ovine embryos. *Theriogenology*, **25**, 149

FRANKHAM, R. and GILLING, M.R. (1983). Molecular biology and its application to domestic animals. In *Animal Genetic Resources: Cryogenic Storage of Germplasm and Molecular Engineering*, pp. 89–105. Proceedings of the Joint FAO/UNEP Experimental Panel. Rome: FAO

GORDON, J.W. (1983). Transgenic mice: a new and powerful experimental tool in mammalian developmental genetics. *Developmental Genetics*, **4**, 1–20

HAMMER, R.E., PURSEL, V.G., REXROAD, C.E. Jr, WALL, R.J., BOLT, D.J., EBERT, K.M., PALMITER, R.D. and BRINSTER, R.L. (1985). Production of transgenic rabbits, sheep and pigs by microinjection. *Nature (London)*, **315**, 680–683

HANRAHAN, J.P. and QUIRKE, J.F. (1980). Selection on ovulation rate in sheep aided by the use of superovulation and egg transfer. *Proceedings of the World Conference on Sheep and Cattle Breeding*, Vol. 2, edited by R.H. Barton and W.C. Smith, pp. 329–335. Palmerston North, New Zealand: The Dunmore Press

HOLST, P.J. (1974). The time of entry of ova into the uterus of the ewe. *Journal of Reproduction and Fertility*, **36**, 427–428

HUNTER, G.L., ADAMS, C.E. and ROWSON, L.E.A. (1955). Inter-breed ovum transfer in sheep. *Journal of Agricultural Science, Cambridge*, **46**, 143

KARDYMOWICZ, M., KARDYMOWICZ, O. and GROCHOWALSKI, K. (1964). The influence of the storage of sheep ova in various temperatures on their implantation. *Acta Biologica Cracoviensia, Serié Zoologique*, **7**, 141–147

KARDYMOWICZ, M., KARDYMOWICZ, O. and GROCHOWALSKI, K. (1966). A study on the effect of cooling of sheep ova at 10 °C on their capability of further development. *Acta Biologica Cracoviensia, Serié Zoologique*, **9**, 113–116

LAND, R.B. and WILMUT, I. (1977). The survival of embryos transferred in large groups to sheep of breeds with different ovulation rates. *Animal Production*, **24**, 183–187

McKELVEY, W.A.C. and ROBINSON, J.J. (1986). Repeated recoveries of ovine ova by laparoscopy. *Theriogenology*, **25**, 171

McKELVEY, W.A.C., ROBINSON, J.J. and AITKEN, R.P. (1985). A simplified technique for the transfer of ovine embryos by laparoscopy. *Veterinary Record*, **117**, 492–494

McLAREN, A. and MICHIE, D. (1959). Studies on the transfer of fertilised mouse eggs to uterine foster-mothers. 2: The effect of transferring large numbers of eggs. *Journal of Experimental Biology*, **36**, 40–46

MOOR, R.M. and CRAGLE, R.G. (1971). The sheep egg: enzymatic removal of the zona pellucida and culture of eggs *in vitro*. *Journal of Reproduction and Fertility*, **27**, 401–409

MOOR, R.M., KRUIP, Th.A.M. and GREEN, D. (1984). Intraovarian control of folliculogenesis: limits to superovulation. *Theriogenology*, **21**, 103–116

MOOR, R.M. and TROUNSON, A.O. (1977). Hormonal and follicular factors affecting maturation of sheep oocytes *in vitro* and their subsequent developmental capacity. *Journal of Reproduction and Fertility*, **49**, 101–109

MOORE, N.W. (1977). A review of techniques and applications. In *Embryo Transfer in Farm Animals*, edited by K.J. Betteridge, pp. 38–40. Ottawa: Canada Department of Agriculture

MOORE, N.W. (1982). Egg transfer in the sheep and goat. In *Mammalian Egg Transfer*, edited by C.E. Adams. Baton Rouge, Florida: CRC Press

MOORE, N.W. and BILTON, R.J. (1973). The storage of fertilized sheep ova at 5 °C. *Australian Journal of Biological Science*, **26**, 1421–1427

MOORE, N.W. and SHELTON, J.N. (1964). Egg transfer in sheep. Effect of degree of synchronization between donor and recipient, age of egg, and site of transfer on the survival of transferred eggs. *Journal of Reproduction and Fertility*, **7**, 145–152

PALMITER, R.D., BRINSTER, R.L., HAMMER, R.E., TRUMBAUER, M.E., ROSENFELD, M.G., BIRNBERG, N.C. and EVANS, R.M. (1982). Dramatic growth of mice that develop from eggs microinjected with metallothionein-growth hormone fusion genes. *Nature (London)*, **300**, 611–615

REXROAD, C.E. Jr and POWELL, A.M. (1986). Co-culture of sheep ova and cells from sheep oviduct. *Theriogenology*, **25**, 187

ROBINSON, T.J. (1967). *The Control of the Ovarian Cycle in Sheep*. Sydney: University Press

ROWSON, L.E.A. and MOOR, R.M. (1966). Embryo transfer in the sheep: the significance of synchronizing oestrus in the donor and recipient animal. *Journal of Reproduction and Fertility*, **11**, 207–212

TERVIT, H.R., WHITTINGHAM, D.G. and ROWSON, L.E.A. (1972). Successful culture *in vitro* of sheep and cattle ova. *Journal of Reproduction and Fertility*, **30**, 493–497

TROUNSON, A.O. and MOORE, N.W. (1974a). The survival and development of sheep eggs following complete or partial removal of the zona pellucida. *Journal of Reproduction and Fertility*, **41**, 97–105

WHITTINGHAM, D.G. (1971). Survival of mouse embryos after freezing and thawing. *Nature (London)*, **233**, 125–126

WILLADSEN, S.M. (1977). Factors affecting the survival of sheep embryos during deep-freezing and thawing. In *The Freezing of Mammalian Embryos*, edited by K. Elliot and J. Whelan, pp. 52, 175–201, CIBA Foundation Symposium

WILLADSEN, S.M. (1982). Micromanipulation of embryos of the large domestic species. In *Mammalian Egg Transfer*, edited by C.E. Adams, pp. 185–210. Baton Rouge, Florida: CRC Press

WILLADSEN, S.M. (1986). Nuclear transplantation in sheep embryos. *Nature (London)*, **320**, 63–65

WILLADSEN, S.M. and GODKE, R.A. (1984). A simple procedure for the production of identical sheep twins. *Veterinary Record*, **114**, 240–243

WILMUT, I. (1982). Applications of egg transfer to animal production, breeding, and research. In *Mammalian Egg Transfer*, edited by C.E. Adams, pp. 211–230. Baton Rouge, Florida: CRC Press

WILMUT, I. and HUME, A. (1978). The value of embryo transfer to cattle breeding in Britain. *Veterinary Record*, **103**, 107–110

WILMUT, I. and SALES, D.I. (1981). Effect of an asynchronous environment on embryonic development in sheep. *Journal of Reproduction and Fertility*, **61**, 179–184

WILMUT, I., SALES, D.I. and ASHWORTH, C.J. (1985a). Physiological criteria for embryo mortality—is asynchrony between embryo and ewe a significant factor? In *The Genetics of*

Reproduction in Sheep, edited by R.B. Land and D.W. Robinson, pp. 275–289. London: Butterworths

WILMUT, I., SALES, D.I. and ASHWORTH, C.J. (1985b). The influence of variation in embryo stage and maternal hormone profiles on embryo survival in farm animals. *Theriogenology*, **23**, 107–119

WRIGHT, R.W. Jr and BONDIOLI, K.R. (1981). Aspects of *in vitro* fertilization and embryo culture in domestic animals. *Journal of Animal Science*, **53**, 702–729

WRIGHT, R.W. Jr, ANDERSON, G.B., CUPPS, P.T., DROST, M. and BRADFORD, G.E. (1976). *In vitro* culture of embryos from adult and prepubertal ewes. *Journal of Animal Science*, **42**, 912–917

Chapter 9

Artificial insemination

M.N. El-Gaafary, R.F.E. Axford and A.G. Chamberlain

Summary

The success of artificial insemination of sheep depends on various factors that act in a complex way. These include quality of semen, sperm cell count per dose, the timing of oestrus in sheep, number of inseminations, time of insemination and site of semen deposition.

Diluents proposed in the literature for the liquid storage of ram semen are either skimmed milk or egg yolk-based. The choice between them depends on the temperature and the period of storage. Fertilizing ability of spermatozoa is depressed following storage at $-196\,°C$, regardless of the freeze-thawing method employed and the type of packaging (pellets or straws).

It is not advisable to use less than 200 and 400 million spermatozoa in the insemination dose for fresh and frozen semen, respectively.

Fixed time insemination (56 h after sponge withdrawal) gives good fertility results, decreases the costs and labour, thus opening the way to much greater use of the procedure in sheep breeding programmes.

Intrauterine insemination by laparoscopy results in a considerably higher fertilization rate than cervical insemination.

Supplementation of ram semen with prostaglandins up to $600\,\mu g/ml$ increases sperm motility and fertility indicating that prostaglandins might improve the conception rate of frozen ram semen.

Introduction

The application of artificial insemination to sheep has been inhibited by the anatomy of the ewe's cervix. The cervical canal of the ewe is tortuous and narrow. For practical purposes this has limited the deposition of semen to its entrance and does not permit as easy sperm penetration as in many other species. As for cattle, the development of artificial insemination for sheep would allow greater use of the best sires for chosen traits (Colas and Courot, 1979). Many attempts therefore have been made to develop insemination techniques adapted to the anatomy of the ewe's genitalia.

The purpose of this chapter is to review some of the factors which affect fertility following artificial insemination, and discuss recent developments which are aimed at overcoming the difficulties associated with sperm transport in the genital tract of the ewe.

Some factors affecting fertility

Type of semen

Liquid semen
Ram semen has been successfully stored for short periods of time, without serious loss of its fertilizing ability, using diluents based on either skimmed milk or egg yolk. The choice between the two types of diluents depends on the temperature selected for storage. As shown in *Table 9.1*, at 15 °C those based on skimmed milk

TABLE 9.1 Lambing percentage following insemination with skimmed milk and egg yolk–citrate diluent stored at 15 °C (from Barlow, Pryce-Jones and Reed, 1974)

Diluent	Number of ewes inseminated	Number of ewes lambed	Per cent lambing
Reconstituted skim milk	483	285	59
Buffered egg yolk/citrate	478	201	42

TABLE 9.2 Fertility of ram semen diluted with skimmed milk or tris-based diluent and stored for 3 or 6 h at 5 °C (from El-Gaafary, unpublished data)

Type of diluent	Age of semen (h)	Number of ewes inseminated	Per cent lambing
Reconstituted skim milk	3	25	68.0
	6	24	41.7
Egg yolk-fructose-tris*	3	25	60.0
	6	25	44.0

*Tris (hydroxymethyl) aminomethane

preserve the fertilizing capacity of spermatozoa better than those based on egg yolk (Barlow, Pryce-Jones and Reed, 1974). However, milk diluent alone is not a suitable storage medium at 4 °C (Colas *et al.*, 1980). According to our results, storage of semen at 5 °C in skimmed milk for 3 h produced better lambing results than when egg-yolk based diluent was used. However, when storage time at this temperature was extended to 6 h, the egg yolk proved to be the better diluent (*Table 9.2*). The deleterious effect of the milk diluent at 5 °C is related to the rapid accumulation of lactic acid, resulting in a significant drop in pH and leading to a high mortality rate of spermatozoa (Tiwari, Srivastava and Sahni, 1977). The protection given by buffers against the detrimental effect of low temperature is ascribed to their protein fractions (Blackshaw, 1954). In egg yolk, lipoproteins provide protection against cold shock (Quinn, Chow and White, 1980). Generally, storage of ram semen as liquid semen, with maintenance of its capacity fertilizing ability is possible for a short duration of less than 24 h (Colas and Courot, 1976).

Recently, New Zealand workers have developed a new technique for storing ram (Tervit *et al.*, 1982) and bovine (Shannon, Curson and Pitt, 1982) spermatozoa for up to 3 days at room temperature. This involves inactivation of the sperm by centrifugation through a 7% Ficoll solution. A high proportion of motile sperm (80%) was recovered after 3 days storage. It was shown by Ca^{2+} flux measurements that immotile spermatozoa still had intact plasma membranes. Furthermore, these immotile spermatozoa were shown to be resistant to cold-shock (Jansen, Van Eerten and Forrester, 1982). However, fertility results obtained following artificial insemination with the reactivated spermatozoa were disappointing, for reasons as yet unknown. The addition of potential activators such as bovine serum albumin, seminal plasma, egg yolk and stimulators of cyclic AMP should be considered in further studies during the reactivation of the sperm.

Frozen semen
The fertilizing ability of ram sperm is commonly reduced following storage in liquid nitrogen at $-196\,°C$. Several studies have shown that semen quality after thawing is affected by the osmolality of the diluent, its nature, the concentration of the cryoprotective agent, and freezing and thawing rates (Fiser, Ainsworth and Langford, 1981; Fiser, Fairfull and Marcus, 1986). The protective action of cryoprotectants such as glycerol is largely attributed to their 'salt buffering capacity', which minimizes electrolytic damage as the water freezes out (Graham, 1976). It is generally accepted that the optimum level of glycerol in the diluent is 4% (Fiser, Fairfull and Marcus, 1986). The most common methods for preserving ram semen in liquid nitrogen are as pellets or in straws. Different techniques of freezing have been proposed. According to our laboratory work, Tris buffer was the best diluent for maintaining sperm motility post-thawing when one-step dilution and freezing the semen in pellets were used, while lactose egg yolk was the best when two-step dilution of semen followed by freezing in straws was used. The composition of the two diluents are as follows:

Tris diluent	*Lactose—egg yolk*
360 mM Tris	343 mM Lactose
33.3 mM Glucose	20% Egg yolk
113.7 mM Citric acid	4% Glycerol
18% Egg yolk	
6% Glycerol	

Sperm dose and number of inseminations

Adequate numbers of sperm in the insemination doses are needed for successful artificial insemination. The number of spermatozoa inseminated may be varied by altering either the volume of the inseminate or the concentration of sperm in it or by a combination of the two factors. Most reports claim a close relationship between lambing rate and the total number of spermatozoa inseminated whether oestrus is natural (Salamon and Lightfoot, 1970; Schindler and Amir, 1973), or induced (Allison and Robinson, 1971; Anderson, Aamdal and Fougner, 1973; Langford and Marcus, 1982; Armstrong and Evans, 1984), and whether the semen is fresh (Zlatarev, 1976; Tervit *et al.*, 1978) or frozen (Visser and Salamon, 1973; Salamon, 1977; Tervit *et al.*, 1984). In general, it is recommended that not less than 100×10^6 spermatozoa per female be used for ewes at naturally occurring oestrus and 200×10^6 for progestagen-treated ewes. If the semen has been frozen, the

TABLE 9.3 Lambing percentage following single and double inseminations with different sperm doses of fresh and chilled-stored semen (from Maxwell, 1978)

Age of semen (days)	Number of sperm/0.1 ml inseminate ($\times 10^6$)	Number of inseminations	Number of ewes inseminated	Number of ewes lambing (%)
0	150	1	40	22 (55.0)
(Fresh)		2	39	23 (59.0)
	300	1	38	23 (60.5)
		2	39	29 (74.4)
		Total and means	156	97 (62.2)
1	150	1	21	4 (19.0)
		2	21	12 (57.1)
	300	1	25	10 (40.0)
		2	23	13 (56.5)
		Total and means	90	39 (43.3)
2	150	1	21	6 (28.6)
		2	18	8 (44.4)
	300	1	20	5 (25.0)
		2	10	5 (50.0)
		Total and means	69	24 (34.8)
3	150	1	19	1 (5.3)
		2	31	14 (45.2)
	300	1	24	1 (4.2)
		2	22	8 (36.4)
		Total and means	96	24 (25.0)

TABLE 9.4 Fertility of thawed frozen semen in relation to number of spermatozoa (from Visser and Salamon, 1974)

Number of motile sperm/inseminate	Number of ewes inseminated	Number of ewes lambing (%)
90×10^6	75	19 (25.3)
180×10^6	69	25 (36.2)
360×10^6	66	37 (56.1)

minimum number of motile sperm required is about 400×10^6 per female in both circumstances.

The double insemination technique produced better fertility results than a single insemination when chilled semen (Maxwell, 1978) or frozen semen (Salamon and Lightfoot, 1970; Visser and Salamon, 1974) were used (*Tables 9.3* and *9.4*). The benefit of double insemination in these studies was ascribed to the fact that the number of motile sperm deposited in two inseminations was twice that deposited in a single insemination. Deposition of equal total numbers of frozen and thawed motile spermatozoa by single and double insemination within the same time range yielded similar lambing results (Salamon, 1977). Langford (1986) indicated that there was no advantage to be gained by using a double insemination technique when it was applied either to mature ewes or to ewe lambs. As a general rule, a single insemination with sufficient numbers of spermatozoa, close to the time of ovulation, will produce fertility equivalent to a double insemination with similar total sperm numbers.

Experiments showed that varying the inseminate volume had little effect on the percentage of ewes lambing (Allison and Robinson, 1971; Martin and Watson, 1976). However, in order to deposit high numbers of spermatozoa at insemination, concentrated semen is required. This may be achieved by using a low dilution rate for storage or by concentrating the spermatozoa by centrifugation before insemination.

Time of insemination

The efficiency of sperm transport in the female genital tract depends on the stage of heat at which the ewes are inseminated. Insemination in early or late oestrus usually results in low fertility. Insemination of ewes in late oestrus is unlikely to provide sufficient spermatozoa in the oviduct to achieve normal fertilization. Mattner and Braden (1969) and Killeen and Moore (1970) found that the increased resistance of cervical mucus to penetration by ram spermatozoa in late oestrus was responsible for the poor fertilization. Emmens and Robinson (1962) and Salamon (1971) demonstrated that the greatest probability of successful fertilization was obtained by inseminating at mid-oestrus and the least probability at late oestrus. Therefore their general recommendation was to use two inseminations given at 12–14 and 23–25 h after the onset of oestrus (Visser and Salamon, 1974).

Oestrus and ovulation can be successfully induced or synchronized in ewes in late anoestrus or during the breeding season by means of vaginal pessaries, impregnated with a suitable progestagen, followed by the injection of pregnant mare serum gonadotrophin (PMSG), at sponge withdrawal. However, there is evidence that impaired sperm transport and survival could be responsible for lowered fertility at the induced heat (Evans and Armstrong, 1984).

An important advance towards the development of a commercially acceptable artificial insemination technique came with the reduction from two inseminations to a single one. In the fixed-time insemination method, oestrus is not looked for, and the timing of insemination is based on the time of sponge withdrawal. Much work has been done to identify the optimal time for insemination following sponge withdrawal and PMSG treatment (Gordon, 1974; 1983). In Ireland, application of the fixed-time insemination method at 56 h after sponge withdrawal gave good results, considerably decreased the cost and labour involved in farm applications and opened the way to much greater use of the procedure in sheep breeding programmes (Gordon, 1983).

Site of insemination

Three sites have been used for insemination of ewes: (a) vaginal insemination, (b) cervical insemination which is commonly used, and (c) intrauterine insemination which is being developed to overcome the sperm losses associated with passage through the ewe's cervix.

Vaginal insemination

This is commonly termed as the shot in the dark (SID) or the blind insemination method. It has received little attention in the literature. However, satisfactory fertility results have been obtained after insemination with fresh semen by Tervit *et al.* (1984) and Maxwell and Hewitt (1986) as shown in *Tables 9.5* and *9.6*. This

TABLE 9.5 Effects of type of semen, number of inseminations, site of insemination, and total number of sperm inseminated on lambing percentages following artificial insemination (from Tervit et al., 1984)

Semen type	Position deposited	Number of inseminations	Total number of sperm inseminated $(\times 10^6)$	Ewes lambing (%)
Fresh	Cervix	1	225	55.3
	Blind	1	225	65.4
		1	450	77.4
		2	450	74.4
	Intrauterine	1	30	82.9
Frozen	Cervix	2	900	17.0
	Blind	2	900	22.7
	Intrauterine	1	60	37.8
Overall				54.7

TABLE 9.6 Pregnancy after cervical and vaginal insemination (from Maxwell and Hewitt, 1986)

Method of insemination	Type of semen	Dose of inseminate (ml)	Number of ewes Inseminated	Number of ewes Pregnant* (%)
Cervical	Fresh	0.1	50	30 (60.0)
	Frozen	0.1	49	9 (18.4)
Vaginal	Fresh	0.1	50	32 (64.0)
	Frozen	0.1	51	9 (17.6)
		0.6	52	12 (23.1)

*Determined by ultrasonic scanner 40 days after insemination

TABLE 9.7 Pregnancy after different methods of insemination with frozen-thawed semen (from Maxwell and Hewitt, 1986)

Method of insemination	Dose of inseminate (ml)	Number of ewes inseminated	Number of ewes pregnant* (%)
Cervical	0.1	48	9 (18.8)
	0.6	45	19 (42.2)
Vaginal	0.1	47	8 (17.0)
	0.6	46	8 (17.4)
Intrauterine	0.1	45	25 (55.6)
	0.6	48	31 (64.6)

*Determined at slaughter 50 days after insemination

TABLE 9.8 Effects of type of semen, site of insemination and number of spermatozoa deposited into the uterus on the fertility of ewes (from El-Gaafary, unpublished data)

Type of semen	Site of insemination	Number of ewes inseminated	Number of ewes lambing	Per cent lambing
1 Fresh semen	Cervix	28	13	46.4
2 Frozen semen	A. Cervical	32	4	12.5
	B. Vaginal	43	4	9.3
	C. Intrauterine			
	1. A. 50×10^6	35	18	51.4
	2. B. 200×10^6	30	13	43.3

method is certainly rapid, easy and practical when fresh semen is used, but if frozen semen is employed unsatisfactory results are obtained (*Tables 9.5, 9.6, 9.7* and *9.8*).

Cervical insemination

The cervix acts as a significant barrier to passage of spermatozoa. This could be easily detected from the marked improvement in fertility which results from surgical deposition of the semen directly into the uterine lumen, bypassing the cervix. For cervical insemination the cervix is viewed by means of a speculum and the semen deposited from a straw or pipette in the first fold of the os-cervix. The inseminate can be deposited deeply into the tortuous canal, but deep penetration may result in tissue damage and adversely affect the success of inseminations (Lightfoot and Salamon, 1970; Anderson, Aamdal and Fougner, 1973). Treatment of ewes with relaxin or oxytocin in attempts to permit deeper insemination and to improve sperm transport in the ewe genital tract have resulted in negative results and decreased fertility (Salamon and Lightfoot, 1970).

Intrauterine insemination

Intrauterine insemination has been shown to increase the fertilization rate using both fresh and frozen ram semen (Tervit *et al.*, 1984). Fertility was higher both in ewes which were in heat naturally and in those whose oestrus had been artificially induced (Anderson, Aamdal and Fougner, 1973; Fukui and Roberts, 1976; Armstrong and Evans, 1984).

The midventral surgical approach used by Lightfoot and Salamon (1970) and Smith *et al.* (1975) may not be commercially practical. Anderson, Aamdal and Fougner (1973) described a non-surgical technique to penetrate the cervix by retracting it with special sponge-holding forceps, and guiding the insemination pipette through it by digital palpation via the rectum of the ewe; penetration is however technically difficult and was only successfully accomplished in 62% of ewes. A similar non-surgical approach for intrauterine insemination involving penetration of the cervical canal by modified hypodermic needle has been used by Fukui and Roberts (1979), but the possibility of trauma to the cervix during the procedure and the failure to penetrate 55% of the ewe cervixes have prevented its general adoption. Killeen and Caffery (1982) described a technique for location of the uterus and intrauterine insemination by laparoscopy. Their results, together with those of other workers (Tervit *et al.*, 1984; Maxwell and Hewitt, 1986; El-Gaafary, unpublished data) are summarized in *Tables 9.5, 9.6, 9.7* and *9.8* and show that high numbers of fertilized eggs can be obtained in ewes following intrauterine insemination with fresh or frozen semen. Maxwell, Butler and Wilson (1984) stated that intrauterine insemination with thawed frozen semen by laparoscopy at 60 h following progestagen sponge removal and PMSG injection might be as effective as cervical insemination of fresh semen 55 h after synchronization treatments (76% versus 80% fertilized eggs, respectively).

The intrauterine insemination technique is relatively more efficient than cervical insemination because of the small sperm numbers (20 million) required to achieve an acceptable fertilization rate, which is about 10–20 times less than in the case of traditional insemination. Davis *et al.* (1984) found no differences in conception rates in ewes following intrauterine insemination with fresh sperm doses of 12.5, 25, 50 or 100×10^6 spermatozoa given in 0.1 ml. In general, intrauterine insemination by laparoscopy has proved a relatively simple and convenient means of achieving high fertilization rates and the technique has been put into practice in Australia and many other countries. The results obtained during the last 4 years are promising.

Prostaglandin supplementation

The supplementation of ram semen with prostaglandins has recently been investigated for its potential in promoting fertility through improving sperm transport and/or survival in the female genitalia.

Two prostaglandins, PGE and PGF, have been isolated from sheep's vesicular glands and crystallized (Eliasson, 1959). Their concentration in ram semen is much higher than in bull, boar and stallion semen. In human semen, the prostaglandin concentrations are low and some investigators have suggested that they are related to infertility (Collier, Flower and Stanton, 1975). Recently, it has been shown that the motility of ram semen can be influenced by the variety and concentration of prostaglandins (Schlegel et al., 1981; Memon et al., 1984). However, Mai and Kinsella (1981) reported no apparent correlation between prostaglandin E and sperm motility in bovine seminal fluid. The mechanism by which seminal prostaglandin stimulates sperm motility may involve an enhancement of adenylate cyclase with a consequent rise in cyclic AMP concentration (Aitken and Kelly, 1985). Prostaglandin supplementation increased the fertility of both fresh (Dimov and Georgiev, 1977) and frozen ram semen (Gustafsson et al., 1975).

The concentrations and combinations of prostaglandins for optimal sperm transport are not known. However, low concentrations of prostaglandin, irrespective of whether it was a single variety or a combination of two or several PGs, increased sperm motility and did not affect the acrosome morphology. Concentrations above 600 µg/ml frozen semen caused significant damage to the acrosomes (Memon et al., 1980, 1984).

TABLE 9.9 Effect of prostaglandin supplementation on the fertility of diluted ram semen (from Dimov and Georgiev, 1977)

Type and amount of prostaglandins added to one dose of semen for artificial insemination	Number of rams	Control group		Experimental group	
		Number of ewes inseminated	Per cent lambing	Number of ewes inseminated	Per cent lambing
$PGF_{2\alpha}$ (5 µg)	3	48	62.5	52	67.3
$PGE_{2\alpha}$ (50 µg)	3	41	58.5	51	72.5
$PGF_{2\alpha}$ (5 µg) + $PGE_{2\alpha}$ (50 µg)	3	91	68.1	35	82.9

TABLE 9.10 The fertility of deep-frozen ram semen supplemented or not with $PGF_{2\alpha}$ prior to freezing (from Gustafsson et al., 1975)

	$PGF_{2\alpha}$ added to the semen (300 µg/ml)	Controls
Number of inseminated ewes	10	10
Number of lambing ewes	7	3
Lambing rate (%)	70	30
Number of lambs born per pregnancy	1.6	1.3

In general, addition of prostaglandin may improve the fertility of both chilled and frozen ram semen (*Tables 9.8* and *9.9*). However, more work is needed to establish the levels and combinations which give the best sperm survival *in vitro* and *in vivo* and the maximum fertility results.

Conclusions

Conception rates in ewes obtained after artificial insemination with ram semen have been lower than after natural mating. This is due to the failure of artificial insemination to provide a sufficient number of active sperm in the oviduct at the time of fertilization. Contributing to this failure are the decreased sperm numbers provided in the insemination dose; the difficulties of detecting oestrus and of determining the best time for insemination; and inadequate sperm transport in the female genital tract. Approaches towards overcoming the problem include (1) increasing the numbers of active sperm given in the insemination dose; (2) establishing a routine for the fixed time insemination of ewes following progestagen PMSG treatments; (3) developing and applying convenient intrauterine insemination techniques; and (4) improving sperm transport through the ewe genitalia by the supplementation of ram semen with prostaglandins. Most of these approaches hold promise, but fully satisfactory procedures are not laid down.

References

AITKEN, R.J. and KELLY, R.W. (1985). Analysis of the direct effects of prostaglandins on human sperm function. *Journal of Reproduction and Fertility*, **73**, 139–146

ALLISON, A.J. and ROBINSON, T.J. (1971). Fertility of progestagen-treated ewes in relation to the numbers and concentrations of spermatozoa in the inseminate. *Australian Journal of Biological Science*, **24**, 1001–1008

ANDERSON, K., AAMDAL, J. and FOUGNER, J.A. (1973). Intrauterine and deep cervical insemination with frozen semen in sheep. *Zuchthygiene*, **8**, 113–118

ARMSTRONG, D.T. and EVANS, G. (1984). Intrauterine insemination enhances fertility of frozen semen in superovulated ewes. *Journal of Reproduction and Fertility*, **71**, 89–94

BARLOW, M., PRYCE-JONES, D. and REED, H.C.B. (1974). MLC Sheep AI field trials: a comparison of milk and egg yolk diluents. *Society for the Study of Animal Breeding, Veterinary Record*, **94**, 159–160

BLACKSHAW, A.W. (1954). The prevention of temperature shock of bull and ram semen. *Australian Journal of Biological Science*, **7**, 573–582

COLAS, G. and COUROT, M. (1976). Storage of ram semen. In *Sheep Breeding*, 2nd edition, edited by G.J. Tomes, D.E. Robertson and R.J. Lightfoot, revised by W. Haresign, pp. 521–532. London: Butterworths

COLAS, G., TRYER, M., GUERIN, Y. and AGUER, D. (1980). Survival and fertilizing ability of ram sperm stored in a liquid state during 24 hours. *Proceedings of the Ninth International Congress on Animal Reproduction and Artificial Insemination*, Madrid, **3**, 315

COLLIER, J.G., FLOWER, R.J. and STANTON, S.L. (1975). Seminal prostaglandins in infertile men. *Fertility and Sterility*, **26**, 868–871

DAVIS, I.F., KERTON, D.J., McPHEE, S.R., GRANT, I. and CAHILL, L.P. (1984). Effect of sperm dose on conception rate following uterine AI in ewes. *Proceedings of the Sixteenth Annual Conference of the Australian Society for Reproductive Biology*, Melbourne, 101

DIMOV, V. and GEORGIEV, G. (1977). Ram semen prostaglandin concentration and its effect on fertility. *Journal of Animal Science*, **44**, 1050–1054

ELIASSON, R. (1959). Studies on prostaglandin. Occurrence, formulation and biological actions. *Acta Physiologica Scandinavica*, **Supplement 46 (158)**, 1–72

EMMENS, C.W. and ROBINSON, T.J. (1962). Artificial insemination in the sheep. In *The Semen of Animals and Artificial Insemination*, edited by J.P. Maule, pp. 205–251. Farnham Royal, Bucks: Commonwealth Agricultural Bureaux

EVANS, G. and ARMSTRONG, D.T. (1984). Reduction of sperm transport in ewes by superovulation treatments. *Journal of Reproduction and Fertility*, **70**, 47–53

FISER, P.S., AINSWORTH, L. and LANGFORD, G.A. (1981). Effect of osmolality of skim milk diluents and thawing rate on crysosurvival of ram spermatozoa. *Cryobiology*, **18**, 399–403

FISER, P.S., FAIRFULL, R.W. and MARCUS, G.J. (1986). The effect of thawing velocity on survival and acrosomol integrity of ram spermatozoa frozen at optimal rates in straws. *Cryobiology*, **23**, 141–149

FUKUI, Y. and ROBERTS, E.M. (1976). Fertility after non-surgical intra-uterine insemination with frozen-pelleted semen in ewes treated with prostaglandin $F_{2\alpha}$. In *Sheep Breeding*, 2nd edition, edited by G.J. Tomes, D.E. Robertson and R.J. Lightfoot, revised by W. Haresign, pp. 533–545. London: Butterworths

GORDON, I. (1974). Controlled breeding in sheep. *Irish Veterinary Journal*, **28**, 118–126

GORDON, I. (1983). Control and manipulation of reproduction in sheep. In *Controlled Breeding in Farm Animals*. Oxford: Pergamon Press

GRAHAM, E.F. (1976). Fundamentals of the preservation of spermatozoa. *The Integrity of Frozen Spermatozoa*, pp. 4–44. Washington DC: National Academy of Sciences

GUSTAFSSON, B., EDQVIST, S., EINARSSON, S. and LINGE, F. (1975). The fertility of deep frozen ram semen supplemented with $PGF_{2\alpha}$. *Acta Veterinaria Scandinavica*, **16**, 468–470

JANSEN, G.J., VAN EERTEN, M.T.W. and FORRESTER, I.T. (1982). Biochemical events associated with Ficoll washing of ram spermatozoa. In *Proceedings of the New Zealand Society of Animal Production*, **42**, 95–97

KILLEEN, I.D. and CAFFERY, G.J. (1982). Uterine insemination of ewes with the aid of a laparoscope. *Australian Veterinary Journal*, **59**, 95

KILLEEN, I.D. and MOORE, N.W. (1970). Transport of spermatozoa and fertilization in the ewe following cervical and uterine insemination early and late in oestrus. *Australian Journal of Biological Science*, **23**, 1271–1277

LANGFORD, G.A. (1986). Influence of body weight and number of inseminations on fertility of progestagen-treated ewe lambs raised in controlled environments. *Journal of Animal Science*, **62**, 1058–1062

LANGFORD, G.A. and MARCUS, G.J. (1982). Influence of sperm number and seminal plasma on fertility of progestagen-treated sheep in confinement. *Journal of Reproduction and Fertility*, **65**, 325–329

LIGHTFOOT, R.J. and SALAMON, S. (1970). Fertility of ram spermatozoa frozen by the pellet method. II. The effects of method of insemination on fertilization and embryonic mortality. *Journal of Reproduction and Fertility*, **22**, 399–408

MAI, J. and KINSELLA, J.E. (1981). Prostaglandin E_1 and E_2 in bovine semen: quantification of gas chromatography. *Prostaglandins*, **21**, 153–163

MARTIN, I.C.A. and WATSON, P.F. (1976). Artificial insemination of sheep: effects on fertility of number of spermatozoa inseminated and of storage of diluted semen for up to 18 hours at 5°C. *Theriogenology*, **5**, 29–35

MATTNER, P.E. and BRADEN, A.W.H. (1969). Effect of time of insemination on the distribution of spermatozoa in the genital tract in ewes. *Australian Journal of Biological Science*, **22**, 1283–1286

MAXWELL, W.M.C. (1978). Studies on the survival and fertility of chilled-stored ram spermatozoa and frozen-stored boar spermatozoa. *PhD thesis*, University of Sudney, Australia

MAXWELL, W.M.C. and HEWITT, L.J. (1986). A comparison of vaginal, cervical and intrauterine insemination in sheep. *Journal of Agricultural Science, Cambridge*, **106**, 191–193

MAXWELL, W.M.C., BUTLER, L.G. and WILSON, H.R. (1984). Intrauterine insemination of ewes with frozen semen. *Journal of Agricultural Science, Cambridge*, **102**, 233–235

MEMON, M.A., GUSTAFSSON, B.K., GRAHAM, E.F. and CRABO, B.G. (1980). Influence of prostaglandins on acrosome morphology of ram spermatozoa. *Proceedings of the Ninth International Congress of Animal Reproduction and Artificial Insemination*, Madrid, **3**, 148

MEMON, M.A., GUSTAFSSON, B.K., GRAHAM, E.F. and CRABO, B.G. (1984). Effect of prostaglandin supplementation on frozen thawed ram spermatozoa. *Proceedings of the 10th International Congress of Animal Reproduction and Artificial Insemination*, University of Illinois, **3**, 201–203

QUINN, P.J., CHOW, P.Y.W. and WHITE, I.G. (1980). Evidence that phospholipid protects ram spermatozoa from cold shock at a plasma membrane site. *Journal of Reproduction and Fertility*, **60**, 403–407

SALAMON, S. (1971). Fertility of ram spermatozoa following pellet freezing on dry ice at −79°C and −140°C. *Australian Journal of Biological Science*, **24**, 183–185

SALAMON, S. (1977). Fertility following deposition of equal numbers of frozen-thawed ram spermatozoa by single and double insemination. *Australian Journal of Agricultural Research*, **28**, 477–479

SALAMON, S. and LIGHTFOOT, R.J. (1970). Fertility of ram spermatozoa frozen by the pellet method. III. The effects of insemination technique, oxytocin and relaxin on lambing. *Journal of Reproduction and Fertility*, **22**, 409–423

SCHINDLER, H. and AMIR, D. (1973). The conception rate of ewes in relation to sperm dose and time of insemination. *Journal of Reproduction and Fertility*, **34**, 191–196

SCHLEGEL, W., ROTERMUND, S., FARBER, G. and NIESCHLAG, E. (1981). The influence of prostaglandins on sperm motility. *Prostaglandins*, **21**, 87–99

SHANNON, P., CURSON, B. and PITT, C.J. (1982). Fertility of inactivated bovine sperm. *Proceedings of the New Zealand Society of Animal Production*, **42**, 91–92

SMITH, J.F., BOYS, P.T.S., DROST, H. and WILLSON, S.G. (1975). AI of sheep with frozen semen. *Proceedings of the New Zealand Society of Animal Production*, **35**, 71–77

TERVIT, H.R., GOOLD, P.G., JAMES, R.W. and FRAZER, M.D. (1984). The insemination of sheep with fresh or frozen semen. *Proceedings of the New Zealand Society of Animal Production*, **44**, 11–13

TERVIT, H.R., JAMES, R.W., SHANNON, P. and TILLOT, M. (1982). Fertility of inactivated ram sperm. *Proceedings of the New Zealand Society of Animal Production*, **42**, 93–94

TERVIT, H.R., SMITH, J.F., GOOLD, P.G. and DROST, H. (1978). Insemination of sheep with fresh or frozen semen. *Proceedings of the New Zealand Society of Animal Production*, **38**, 97–100

TIWARI, S.B., SRIVASTAVA, A.K. and SAHNI, K.L. (1977). Some metabolic changes in ram semen stored in milk diluent. *Indian Veterinary Journal*, **54**, 111–115

VISSER, D. and SALAMON, S. (1973). Fertility of ram spermatozoa frozen in tris-based diluent. *Australian Journal of Biological Science*, **26**, 513–516

VISSER, D. and SALAMON, S. (1974). Fertility following insemination with frozen-thawed reconcentrated and unconcentrated ram semen. *Australian Journal of Biological Science*, **27**, 423–425

ZLATAREV, S.T. (1976). Optimal number of spermatozoa for artificial insemination of sheep with semen stored for 24 hours at 0–3 °C. *Proceedings of the Eighth International Congress of Animal Reproduction and Artificial Insemination*, Cracow, pp. 1104–1107

Part IV

Genetic improvement techniques

New fields are opening up in the area of genetic improvement. Genetic engineering and biotechnology, which until recently were confined to the realm of the theorist or the microbiologist's laboratory, have suddenly become techniques that are not only technically feasible with sheep but may soon be part of the sheep breeder's armoury. Several fascinating possibilities are discussed in this section including not only gene manipulation and gene transfer but the use of chromosome analysis. In particular the so-far unattainable mirage of sex determination is discussed. Whilst embryo sexing is becoming a reality, thus giving the possibility of a real boost to embryo transfer techniques, semen sexing, the major prize, is still tantalisingly elusive.

Other interesting developments in genetic improvement techniques, discussed in this section, include the important development of indirect estimation. This applies to the possibility of early assessment of characters such as prolificacy, the assessment of males for female characters and the assessment of carcass quality in live breeding stock at the time of selection.

Chapter 10

Genetic engineering, chromosome analysis and sex determination

J.W.B. King

Summary

The manipulation of genotypes is an age-old procedure but is now provided with new methods by advances in molecular biology. So far DNA transfer techniques in farm animals have used the direct microinjection of DNA into fertilised ova, but other transfer methods are feasible. What proves to be more difficult is the discovery of suitable candidate genes for transfer and their regulation to perform in the right tissues at appropriate times. Many implications can be anticipated but it is stressed that all aspects of performance need to be investigated in order to evaluate the agricultural value of transgenic animals. Greater knowledge of the chromosomes will be needed for many new techniques.

It is pointed out that in the sheep the use of fertile hybrids with wild species represents an existing method of gene transfer worthy of further exploration. The production of chimaeras from sheep–goat hybrids, which normally die *in utero*, may also prove to be a way of introducing goat genes to the sheep.

Sex determination by the separation of X-bearing and Y-bearing sperm remains a major research goal. Methods with results that are both reproducible and maintain the viability of separated sperm have yet to be described.

Introduction

The advent of biotechnology has many potential impacts on sheep breeding. The purpose of this chapter is not to attempt any comprehensive review of possibilities, but rather to enquire into the consequences of a few technologies which are either with us now or may be with us shortly, and to study the ramifications of such developments.

Genetic engineering

The term 'genetic engineering' embraces a variety of techniques, some of which have been with us for a very long period of time. The first people to employ genetic engineering were those who selectively bred their stock during and after the process of domestication and gradually evolved many of the improved breeds of domestic livestock that we use today. Although the process was a slow one, the fashioning of the genotype to produce more useful end-products has been very successful as a comparison of present breeds with their wild progenitors will show. What is more usually meant by 'genetic engineering' is the use of recombinant DNA techniques to effect more radical changes of a kind which could not be accomplished previously. These include the crossing of species barriers and the manipulation of genes in ways which are completely new.

Methods of gene transfer

Several methods have been developed for effecting gene transfer, and the important procedures that emerge are as follows:

(1) use of retrovirus vectors;
(2) use of embryonic cells;
(3) direct microinjection of DNA into fertilized ova at an early stage of development.

Of these methods, the first has the promise of providing what could be a very efficient and precise method of gene transfer. Despite some limitations on the length of DNA which can be incorporated into a retrovirus, the method has the great potential advantage of enabling the production of transgenic animals with single additional copies of the gene to be transferred.

The culture of preimplantation embryonic cells also has potential advantages in allowing the establishment of a culture which could be genetically transformed and then selected with a suitable marker, to ensure that the desired genetic change had been accomplished. This method does, however, have attendant disadvantages in requiring the reintroduction of transformed cells to developing embryos with the likelihood that resulting offspring will be chimaeric, and that only some of these chimaeras will contain the required gene in the germ line.

The method of direct microinjection has so far proved to be the most effective method of transfer for foreign DNA and one which has now been established as not only working in laboratory animals, but also in sheep and pigs. Although some transfers may occur after injection of DNA into the general cytoplasm of the egg, the direct injection into the pronucleus has proved the key to efficient gene transfer. In the mouse where there is most experience, the efficiency of microinjection into the pronucleus has reached a high level of efficiency with about 25% of transgenic individuals among surviving progeny. With farm animals, success rates have been far lower, due in the first instance to difficulties in visualizing the pronucleus, but clearly also due to other factors which are not yet fully appreciated. The method avoids the production of chimaeras, but does have difficulties in that the number of foreign genes integrated into the genome is not controlled and can vary from one to several hundred, usually arranged in tandem order.

Choice of genes for transfer

The existence of methods for DNA transfer has highlighted the difficulty of finding suitable candidate genes for that procedure. The number of major genes affecting performance is rather limited and still subject to many problems. In the majority of cases, discovering the DNA sequences corresponding to the observed genetic differences may represent a major problem. When a gene product is known, or can reasonably be looked for in a particular tissue, the prospects are quite good, but when these are unknown, the task is a daunting one, even with present methods of molecular biology.

In the absence of knowledge about segregating genes, it is not surprising that most attention has been focused on structural genes where the DNA sequences can be deduced from knowledge of protein structure. Among such genes, that for growth hormone has been a frequent candidate for transfer, even though the outcome of such changes may not be predictable. The dramatic growth shown by mice to which additional rat growth hormone genes had been transferred (Palmiter et al., 1982) led to expectations that similar changes might be effected in farm animals. Although similar transfers have been made in pigs, the outcome so far appears not to be any increase in growth rate but rather many developmental problems of an undesired kind (Hammer et al., 1985).

Regulation of gene expression

The example of the effect of added growth hormone on pig performance emphasizes the need to know more about the way in which genes function, how they are turned on and off and how expression is regulated to occur only in certain organs. In the giant mice produced by Palmiter et al. (1982), the normal processes of regulation were completely swamped with gross amounts of growth hormone being produced (up to 800 times the normal amount) and with dramatic effects on growth. In these mice reproduction was subnormal with transgenic females proving to be sterile.

Considering the complexity of this subject, good progress seems to have been made in the understanding of gene regulation. In early experiments with mice, foreign genes were integrated into the germ line but in many instances the resultant proteins could not be demonstrated. This led to the use of constructs in which strong promotors were used and fused to the structural gene. Attention is being increasingly focused on regulatory sequences and the way in which they are controlled in respect of timing and level of activity. For these studies, transgenic animals are in themselves the most informative kind of experimental material so that rapid increases in knowledge can be expected. The prospects are that promotor regions can be designed in a way that will be appropriate for expression in particular tissues at particular times. The ability to use an external signal, introduced as a dietary constituent, or by injection, would add further refinement to such a system. In addition, for genes with many different effects, such as the growth hormone gene, there is the possibility of being able to modify part of the structural gene itself to change the pattern of gene expression (Wagner, 1985).

Applications of genetic engineering

It is difficult not to share the great excitement shown by biologists about the manifold possibilities opened up by advances in molecular biology. Nevertheless, it is necessary to consider how these developments are likely to affect animal production. Ways in which the new methodology might be applied can be divided as follows:

(1) production of new products;
(2) enhanced efficiency in the production of existing products;
(3) modification of present products to enhance or reduce certain components.

New products
The use of farm animals for the production of new and novel products is an intriguing one. For certain pharmaceutical products, the use of mammals rather than microorganisms would have their advantages both in terms of ease of production and relative purity of the final product. The use of milk as the medium for production is attractive. Although the value of the product might be potentially very high, the amount of product required is likely to be small, and the total operation could be confined to the laboratories of a few pharmaceutical companies. The impact on agriculture in general would be negligible except through the development of methodology that could be applied to more widely based animal production. If the new products were new kinds of food or new fibres, then the agricultural impact would clearly be that much greater. For example, the possibility of being able to produce vicuña-like fibre from sheep might be an interesting prospect (except to present owners of vicuña!).

Improved efficiency
Of the many characteristics for attention, improved efficiency probably has the greatest general attraction as the objective to be changed. Improved efficiency of food, milk or fibre production would all be of benefit to the farmer and ultimately to the consumer. In this general area, milk production must be one of the prime topics for consideration because of the startling increases in milk yield that have been found in dairy cattle through the daily administration of exogenous growth hormone (Baumann *et al.*, 1985). The prospect of being able to build such advantages into the animal would seem to be promising and a likely field for intense activity.

Modification of present products
The third general area of application for gene transfer would be in the modification of present products. This might take the form of either increasing a component, such as for example protein in milk, or reducing a component such as fat in the carcass. The prospects for both kinds of change are probably quite good. To enhance milk proteins the transfer of additional casein genes or the introduction of more effective promoters seem obvious choices for initial trial. Relatively little activity seems to be directed so far to the prospect of reducing various components. To my mind this procedure ought to be relatively easier than enhancing production and will be greatly assisted by the use of antisense RNA to negate particular coding messages.

Consequences of genetic engineering for animal production

Those working in the field of animal breeding may feel in times of depression that theirs is a relatively neglected method of effecting change. The advent of markedly different transgenic animals could change that viewpoint overnight but there may be dangers in not recognizing the importance of assessing all aspects of performance in the new animals. The transgenic changes that are effected can be likened in most ways to major mutations and carry with them obligations to study not only the most obvious alterations in performance but also detailed studies on reproduction, longevity, etc. This need, coupled with the requirement to examine the long-term stability of transgenic animals with multiple gene copies, makes the prudent adoption of these new techniques a longer term undertaking than it might otherwise seem.

All animal breeding in future will not take place in the laboratory—large-scale farm resources will still be necessary so that animals can be evaluated and then disseminated for general farm use. These evaluations may require increased expenditure on animal breeding and will certainly raise questions about appropriate methods for financing this type of activity. Although the ultimate beneficiary will be the consumer, ways and means are necessary to recoup those making financial investment in producing the changes. One byproduct of genetic engineering may be the provision of numerous genetic markers thus facilitating a royalty system for improved animals which could be policed, and used as a possible finance mechanism. This business consideration is but one ramification of a new subject with many long-term consequences for animal breeding.

Chromosome analysis

With the coming of genetic engineering, older techniques such as the analysis of chromosomes by counting and banding techniques may seem to be outmoded. However, knowledge of the chromosomes and linkage groups of naturally occurring and transplanted genes will be necessary knowledge for many future techniques. Renewed activity in cytogenetic analysis may therefore be a necessary adjunct to genetic engineering. However, this is for the future and I want instead to discuss an existing opportunity already thrown up by chromosome studies.

Extensive research in many countries demonstrates the possibility of using some of the interspecific variation available among sheep species. From cytological analysis of wild sheep and goats, it appears that these species shared a common ancestor with the chromosome number $2n = 60$. During the course of evolution chromosome numbers have been reduced by acrocentric translocation to the number now found in domestic sheep of $2n = 54$. The divergence of species is, however, far from complete in that hybrids between many of them can be produced. Apart from one report (quoted by Gray, 1972), in which the domestic sheep had been crossed with the smaller arkhar (*O. ammon kaselini*) to produce F_1 hybrid ewes commercially, these opportunities do not appear to have been exploited or indeed explored very fully. One attractive idea (Short, 1977) would be to use the Marco Polo sheep to introduce some genes for rapid growth into the domestic species.

Even wider crosses, as for example between the goat and the sheep, should not be ruled out as a possibility. Even though the majority of workers have found the cross to be non-viable, this appears to be due to immunological problems experienced by the fetus (Hancock, McGovern and Stamp, 1968), rather than to chromosome imbalance, since there is a report of one hybrid female surviving and producing lambs (Bunch, Foote and Spiller, 1976). Even if it is not possible to find a means of overcoming the immunological problem, or using breeds of sheep and goat which avoid it, new methods of embryonic micromanipulation make it possible that more hybrids of this kind could be produced to order. The suggested method would be to introduce hybrid cells into a developing blastocyst of the sheep (or goat) thus producing chimaeras in some of which the hybrid might be expected to produce germ-line cells. Any genes introduced in this way could be exploited by selection in the normal manner with particular emphasis on those characteristics thought likely to be conferred by the introduced species.

In many ways this procedure will be similar to that used by the genetic engineers in that the effects of introduced genes will be difficult to predict and can only be properly evaluated by detailed recording and selection of desired genotypes. The possibilities of doing this then without the need for complex technology should not be overlooked, particularly in those countries where wild species of sheep occur and demonstrate their adaptation to the local environment.

Sex determination

A method for predetermining the sex of offspring has long been the ambition of many research scientists. The ability to separate X-bearing and Y-bearing sperms into separate fractions so determining the sex of offspring would indeed be a major achievement. Failing this, methods for sexing the developing embryo may also be advantageous, particularly if coupled with the use of embryo transfer methods. Ideally the methods employed should be non-destructive so that conception rates are maintained and embryo survival not adversely affected by the treatment.

Progress in sex determination is reported regularly, almost every year. The reports have in the course of time been shown to have one common feature—they do not work! The reasons for the failure are often not far to seek. For example, the difference in size between X-bearing and Y-bearing sperm is such that electrophoretic separation is unlikely to be effective. Similarly, attempts to detect the HY antigen on sperm are handicapped because of the secretion of this antigen from the somatic cells of the male parent. Future progress in the area is, however, not without hope, For example, Keeler, MacKenzie and Dresser (1983) have reported the separation of sperm into two fractions according to their fluorescence after staining with a vital dye. Preliminary evidence is presented of X and Y separation but unfortunately with present equipment the method is too slow to be of any practical consequence. For the future, more radical methods of genetic engineering may be necessary to reach the desired endpoint. For example, McClaren and Burgoyne (1983) have shown that by the combination of a sex reversal gene and a chromosome translocation it is possible to produce mice with progeny of predominantly one sex.

General perspective

With so many potential prospects in view, it may help to give some very personal assessment of priorities and of probabilities for advancement in particular areas. Such predictions are extremely hazardous but nevertheless there are constraints which will not easily be removed and will temper the rates of change achieved.

The top priority goes to study of methods of gene transfer and regulation in farm animals. Although the initial results in reducing the growth hormone to pigs have not been encouraging, the example of mice is still with us and more efforts to extend this work to sheep would seem to be justified, although possibly using alternative promoters regulated in different ways. The example, again from mice, is that growth hormone releasing factor may be a better prospect than growth hormone itself (Hammer *et al.*, 1985). Examples of useful transgenic animals are rather urgently needed in the animal production field if the momentum gained by the initial progress of the methodology is to be maintained. Long-term support will be necessary for such bold ventures as those of Ward *et al.* (1985) in attempting to alter the biochemical architecture of the rumen epithelium to remove limits to current levels of wool production.

Since there is difficulty in identifying many genes known to be of potential value, there is still probably merit in producing alternative genes from other species. Surveys of germ plasm resources such as those described in the USA, National Research Council (1983) paper show how the range of domestic genotypes might be extended. With improving techniques of hybridization, genes from them might also be secured for use in more commonplace species.

Finally, as a research topic, sex determination probably generates enough interest in its own right to avoid the necessity of awarding it any special research priority. The prospects for any practical method do not appear promising to this author, but that kind of prophesy is one which is likely to be held up to ridicule by new developments in such a rapidly developing field as biotechnology.

References

BAUMANN, D.E., EPPARD, P.J., DEGEELER, M. and LANZA, G.M. (1985). Responses of high-producing dairy cows to long-term treatment with pituitary somatotropin and recombination somatotropin. *Journal of Dairy Science*, **68**, 1352–1362

BUNCH, T.D., FOOTE, W.C. and SPILLETT, J.J. (1976). Sheep–goat hybrid karyotypes. *Theriogenology*, **6**, 379–385

GRAY, A.P. (1972). Mammalian hybrids. *Technical Communications of the Commonwealth Bureau of Animal Breeding and Genetics*, No. 10 (2nd revised edition). Farnham Royal, Slough, UK

HAMMER, R.E., BRINSTER, R.L., ROSENFELD, M.G., EVANS, R.E. and MAYO, K.E. (1985). Expression of human growth hormone-releasing factor in transgenic mice results in increased somatic growth. *Nature*, **315**, 413–416

HANCOCK, J.L., McGOVERN, P.T. and STAMP, J.T. (1968). Failure of gestation of goat and sheep hybrids in goats and sheep. In *Immunological Aspects of Pregnancy. Proceedings of the First Symposium of the Society Study Fertility, 1967. Journal of Reproduction and Fertility* **Supplement 3**, 29–36

KEELER, K.D., MacKENZIE, N.M. and DRESSER, D.R. (1983). Flow microfluorometric analysis of living spermatozoa stained with Hoechst 33342. *Journal of Reproduction and Fertility*, **68**, 205–212

McCLAREN, A. and BURGOYNE, P.S. (1983). Daughterless Sxr/Y Sxr mice. *Genetic Research*, **42**, 345–349

PALMITER, R.D., BRINSTER, R.L., HAMMER, R.E., TRUMBAUER, M.E., ROSENFELD, M.G., BIRNBERG, N.C. and EVANS, R.M. (1982). Dramatic growth of mice that develop from eggs microinjected with metallothionein-growth hormone fusion genes. *Nature (London)*, **300**, 611–615

SHORT, R.V. (1977). The introduction of new species of animals for the purpose of domestication. *The Ark*, **4**, no. 2

USA, NATIONAL RESEARCH COUNCIL (1983). *Little-known Asian animals with a promising economic future*. Washington, DC: National Academy Press

WAGNER, T.E. (1986). Biotechnology and Livestock Agriculture. Sir John Hammond Memorial Lecture at the *British Society of Animal Production Winter Meeting*, 1986

WARD, K.A., MURRAY, J.D., NANCARROW, C.D., SUTTON, R. and BOLAND, M.P. (1985). The practical aims and recent progress in the transfer of recombinant DNA to ruminants. In *Biotechnology and Recombinant DNA Technology in the Animal Production Industries. Proceedings in a Symposium held at the University of New England, Armidale, Australia 27–30 November 1984* (jointly organized by the University and CSIRO), editors R.A. Leng, J.S.F. Barker, D.B. Adams and K.J. Hutchinson. *Reviews in Rural Science Series*, **6**, 39–47

Chapter 11

Indirect selection

C.S. Haley, N.D. Cameron, J. Slee and R.B. Land

Summary

Indirect selection of traits of no intrinsic economic value, which are genetically correlated with economically valuable traits, may have an important contribution to make to improvement in overall economic efficiency. Indirect selection is potentially particularly valuable for traits which are age-limited or sex-limited in their expression, such as litter-size and milk yield. For example, in some cases use of an indirect trait in males could double the rate of response to selection for litter size. Indirect selection could make practicable improvements in traits which are difficult or expensive to measure, such as carcass composition or stress and disease resistance. Possible indirect traits require thorough research to define the optimum way in which to use them, and to enable prediction of the effects of their use on all components of overall merit. A good general strategy to fulfil these requirements is to produce selection lines for the major components of economic merit, in which potential indirect traits can be identified and evaluated.

Introduction

The merit of livestock is a combination of many biological traits, and while simple economic merit may be argued to be the breeding objective, selection is inevitably based on an index of a limited number of biological characteristics. The genetic improvement of livestock is therefore always via indirect selection and it is the aim of the geneticist to identify the index upon which the breeding objective has the highest genetic regression. The animal scientist is concerned with the integration of physical and physiological traits into an index of performance which can be used as a criterion for indirect selection.

The methodology of construction of any index is now widely accepted and the question of the definition of breeding objectives in terms of outputs minus inputs or outputs relative to inputs has been largely resolved (Smith, James and Brascamp, 1986); the two become similar when production is considered in terms of finite markets. The animal breeder is now largely concerned with the discovery of appropriate traits. This research includes both the better or more convenient measurement of traits which can already be measured, but also measures of traits in

113

circumstances where they are not normally expressed. Examples of the former would be the use of technical methods to measure carcass characteristics of the live animal or biochemical measures of efficiency. Examples of the latter would be the use of physiological characteristics to measure the merit of males for reproduction or lactation traits which are normally expressed only in the female, or the merit of individuals of both sexes for resistance to disease or environmental stress without the imposition of disease or distress.

General considerations

Selection for overall merit will generally be on an index of traits which have apparent intrinsic economic value (such as body weight and litter size) and others of no intrinsic economic value (such as hormonal or physiological characteristics) which act as indirect predictors of the desired traits. In this case, the indirect traits act to increase the regression of genetic merit on the index. The index may be the same for both males and females, but when sex-limited traits such as litter size and milk production are of importance, the index may differ between sexes. In this latter case indirect traits may allow the evaluation of one sex for desired traits expressed in the other. The use of indirect selection in males, both for traits sex-limited to females and those which are not, will be particularly valuable because of the high selection pressures which can be applied to males.

Indirect traits may also allow selection to be performed in two stages. For example, where a limited number of males can be progeny-tested, selection on an index of indirect traits prior to progeny testing could increase the overall effectiveness of selection.

The indirect traits used in the construction of an index have to be chosen with some care, particularly when the traits have no intrinsic economic value, as may be the case for physiological characteristics. The relative efficiency of indirect to direct selection for the improvement of a single trait when the two are considered as alternatives is:

$r_A h_I / h_D$ (see Hill (1985) for further discussion)

where h^2_D = heritability of the 'desired' trait

h^2_I = heritability of the 'indirect' trait

r_A = genetic correlation between the traits

Thus for any one trait, the benefits of using indirect selection increase with the heritability of the indirect trait and the genetic correlation, but decrease as the heritability of the desired trait increases. Similar conclusions apply when using desired and indirect traits in an index together or when using an indirect trait to select males for a trait sex-limited to females (Walkley and Smith, 1980). Indirect selection may be particularly useful, then, when the desired trait has a low heritability or is sex-limited in its expression. The benefits from indirect selection may be further enhanced where repeat records of the indirect trait can be taken to increase its effective heritability.

Caution must be exercised when choosing traits to act as indirect predictors of genetic merit. All traits cost money to measure, so it is important to focus on the most effective predictors of merit. More seriously, selection pressure is wasted by the use in an index of a trait uncorrelated with genetic merit.

To provide examples of possible applications of indirect selection, this chapter focuses on aspects of three major areas of sheep production: growth and lean meat production, reproduction, and resistance to stress and disease.

Two other areas of greater importance outside Britain are wool and milk production. Much work on the genetics of wool production has been conducted in Australia, and indirect traits from fibre diameter to sulphur metabolism have been studied. One example, which also sounds a cautionary note, is selection for skinfolds, aimed at increasing skin surface area and thus wool production (see Atkins, 1980). In fact increased skin surface area was to some extent compensated for by reduced production per unit area and, furthermore, shearing was made more difficult. More dramatic was the negative genetic correlation of skinfolds with reproductive rate, the flock selected for a low level of skinfolds having almost twice the reproductive rate of the high level. Little work has been performed on indirect indicators of milk production in sheep; however, by analogy with cattle, one very promising trait could be the use of the blood urea concentration after fasting (Tilakaratne et al., 1980).

Growth and lean meat production

Although lamb growth rate can be measured directly, genetic improvement in growth rate is difficult due to its low heritability (Wolf et al., 1981) and so alternative indirect selection criteria have been studied.

Croston et al. (1983) selected on ram 18 month weight, as adult weight tends to be more heritable than juvenile weight. However, the gains from indirect selection were offset by the increased sire generation interval. Selection on lamb growth rate with artificial rearing, in order to reduce maternal effects and give a better indication of a ram's breeding value, was examined by Cameron and Smith (1985a). The genetic correlation between lamb growth rate under natural and under artificial rearing was markedly less than unity, which diminished the effectiveness of reducing maternal effects. There was no advantage in these alternative selection criteria compared to selecting on lamb growth rate directly.

Lean meat production cannot be measured directly and, unlike growth rate, is reasonably heritable (Wolf et al., 1981). Several methods of estimating body composition in the live animal have been assessed and reviewed by Alliston (1983). The use of ultrasonics may have a valuable role in breeding programmes to improve lean meat production. Phenotypic correlations between body composition and ultrasonic measurements are of the order of 0.4 with a repeatability of measurements of 0.7.

New techniques for estimating body composition in the live animal are becoming available due to recent advances in computer technology. Computerized X-ray tomography on sheep has shown encouraging results, and a correlation with body composition of 0.79 has been reported (Sehested, 1984). Nuclear magnetic resonance (NMR) techniques may identify specific tissues in the live animal (Foster et al., 1984), such as intramuscular fat, which may be an important component of meat quality.

The use of physical techniques, such as ultrasonics and NMR technology, approach direct selection on the carcass trait itself. Physiological traits, by contrast, contribute through selection on the biological processes underlying the trait. Selection of broilers on the triglyceride content of both very low density lipoprotein in plasma (VLDL) and low density lipoprotein (LDL) produced significant differences in body fat content but little difference in body weight (Griffin, Whitehead and Broadbent, 1982). However, the site of lipogenesis in poultry (liver) is different from that in the sheep (adipocyte). Appreciating this, it was suggested that carcass leanness in rams may be indirectly reflected in the rate of fat metabolism on fasting, as detected by plasma VLDL concentrations. However, triglyceride concentrations before, during and after fasting showed no relationship with carcass leanness (Cameron and Smith, 1985b).

In sheep, approximately 40% of the energy intake is used in the degradation and resynthesis of existing protein. Therefore, a physiological measure of the rate of protein turnover would be a measure of the efficiency of lean tissue growth. Concentrations of 3-methylhistidine reflect the role of protein degradation in the rat but not in sheep (Harris and Milne, 1980). Similarly, acyl-CoA Δ^9 desaturase may reflect the fatty acid composition of subcutaneous fat in the pig, which may be related to the eating quality of meat (Enser and Buller, 1984).

Reproduction

Reproductive traits are prime candidates for indirect selection since merit can be measured only in the one sex, after reproductive maturity; also heritability of reproductive traits is generally low. The genes for reproduction are carried throughout life by both sexes, and possible indicators of reproductive potential have been reviewed by Land (1982). Walkley and Smith (1980) have assessed the increase in the rate of response to selection which could be achieved if favourable alleles for litter size could be detected through the physiological characteristics of males or females. Some of their results are summarized in *Table 11.1* for the case when the heritability of prolificacy is 0.1, a reasonable consensus figure for sheep.

TABLE 11.1 Predicted relative rates of improvement in prolificacy using indirect selection (litter size alone = 100). The table shows the percentage change in prolificacy relative to selection on litter size alone. The heritability of litter size is taken as 0.1 and selection on litter size gives a predicted annual rate of response of 0.056 lambs/litter/year. All selection is based on using individual and half-sib information. For further details see Walkley and Smith (1980)

Genetic correlation between prolificacy and indirect traits		0.3				0.7		
Heritability of indirect trait		0.1		0.35		0.1		0.35
Number of records on indirect trait per individual	1	5	1	5	1	5	1	5
Relative response								
Male indirect trait only	50	71	75	93	112	166	173	200
Female indirect trait only	38	57	59	68	95	136	139	163
Both sexes indirect trait only	57	82	86	105	132	191	198	218
Male indirect trait plus litter size	111	123	127	138	150	200	196	234
Female indirect trait plus litter size	109	118	120	123	138	170	173	193
Both sexes indirect trait plus litter size	116	130	129	146	168	207	214	258

For some combinations of genetic parameters, the use of indirect traits can more than double the rate of response obtained as compared with selecting on prolificacy in females alone. Selection on a single record of a physiological trait in the male alone, with a moderate heritability of 0.35 and a genetic correlation with prolificacy of 0.7 would enable the rate of genetic change to be increased by 73%. In addition the use of such a trait would facilitate selection, obviating the need to record the females themselves. This would be particularly relevant in extensive husbandry circumstances.

Several physiological traits which can be measured in males or in females or in both sexes have been suggested. A simply measured trait in males is the size of the testes. Selection for ovulation rate in sheep led to a correlated response in testis size (Hanrahan and Quirke, 1982), and selection for testis size adjusted for body weight led to a change in the number of lambs born per ewe put to the ram (Lee and Land, 1985). The heritability of testis size, at 10 weeks of age, is around 0.4 in Finn-Dorset ram lambs (Lee and Land, 1985); Purvis (1985) found similar values, with no discernible age-related trend, in Merinos. The genetic correlation of testis size at 10 weeks with ovulation rate is around 0.4 in Finnish Landrace sheep (Hanrahan and Quirke, 1982); Purvis (1985) again found similar values, with no age-related trend, in Merino sheep. These results suggest that selection for testis size could prove effective in increasing ovulation rate and thus prolificacy. One remaining problem is the exact nature of the genetic relationships when weight is taken into account. Purvis (1985) has demonstrated how some adjustments for weight may remove or even reverse the positive genetic correlation between testis size and ovulation rate.

In females, selection for ovulation rate itself should be effective in increasing litter size in less prolific sheep breeds. Estimates of the heritability of ovulation rate are generally around twice that of litter size and the two traits are highly genetically correlated; also ovulation rate can be measured several times per season. Selection in Finnish Landrace sheep increased ovulation rate by almost 50%, but litter size did not increase in this prolific breed (Hanrahan, 1982). However, selection for ovulation rate should be more effective in less prolific breeds.

Other physiological characteristics, such as the concentration of a gonado-trophin, could prove to be useful indirect traits in either or both sexes, depending on the difficulty of measurement and the increased precision gained. The low repeatability of natural luteinizing hormone (LH) level makes it unlikely to be useful. However selection for LH response to stimulation of ram lambs by gonadotrophin releasing hormone (GnRH) has been successful, although the genetic correlation of this trait with ovulation rate and litter size appears to be low. Research in France has shown that the concentration of follicle stimulating hormone (FSH) in young females is correlated with their subsequent ovulation rate (Ricordeau, Blanc and Bodin, 1984); the trait has a moderate heritability of around 0.4 in ewe-lambs (Bodin et al., 1986), but the genetic correlation with ovulation rate and the genetic parameters for males are as yet unknown.

Resistance to stress and disease

Whereas the expression of genes controlling variation in reproductive performance is limited by sex, the expression of genes affecting resistance to disease or stress is limited by the chance incidence of the particular disease or the variability of the conditions which predispose to stress.

A good example comes from the effects of severe weather upon newborn lamb mortality, particularly in Britain, Ireland, Australia, New Zealand and South America (Slee, 1986a), and upon the mortality of newly shorn adult sheep in Australia (Hutchinson, 1968). In both cases variation in weather can produce extreme variations in mortality. Data on lambs (Obst and Day, 1968) show mortality varying from 15% to 90% with short-term changes in rainfall, windspeed and environmental temperature. In situations where there is high mortality, the effect of genes determining resistance to climatic stress is exhibited (and direct genetic selection can be applied) whereas in kinder conditions such selection is not so easy. Clearly it would be morally and economically unacceptable to devise or to deliberately tolerate conditions leading to high mortality in order to permit genetic selection. On the other hand, lamb mortality is a serious problem of which cold exposure can be an important component (Slee, 1981). Moreover, the option of improving viability by genetic selection is attractive because the costs are small and non-recurrent in contrast to the alternative of providing environmental improvements involving nutrition, shelter or increased labour inputs.

The possibility of selecting indirectly for improved lamb viability by means of the component character of cold resistance has been investigated at the Animal Breeding Research Organisation (ABRO) in the United Kingdom. Evidence from breed comparisons (*Table 11.2*) has indicated that newborn lambs of different breeds, reared in the same location, differed significantly in average mortality and in resistance to hypothermia in cold weather. Further work revealed breed differences in resistance to body cooling as determined by a laboratory test involving part-immersion in a water bath. The performance of different breeds in this test corresponded fairly closely to their cold resistance in the field, suggesting that a standardized and controllable laboratory test might be used to identify animals with genetically superior characteristics for cold resistance and potentially, therefore, for survival ability. Such a technique would remove the need to rely on unsatisfactory test criteria such as high field mortality rates or on weather of variable severity in order to select for improved lamb viability.

TABLE 11.2 Viability and cold resistance in newborn lambs of eight breeds. Field data came from lambs born on the same station over a period of years. Laboratory data came from different lambs of the same breeds maintained on the same station

	Field observations				Laboratory measurements	
	Mortality within 24 h of birth		Hypothermic lambs (rectal temperature below 37°C 1 h after birth)		Cold resistance: time (min) to reduce body temperature	
	Number of lambs	Per cent	Number of lambs	Per cent	Number of lambs	Time
Hill breeds						
Welsh Mountain	181	3.3	47	8.5	21	89
Scottish Blackface	307	4.2	64	1.6	33	87
South Country Cheviot	105	9.5	22	9.1	35	98
Small lowland breeds						
Merino	104	9.6	16	68.8	21	45
Finnish Landrace	165	10.9	13	64.6	23	38
Southdown	68	13.2	17	47.1	26	51
Large lowland breeds						
Oxford	50	2.0	12	0	21	79
Border Leicester	77	9.0	17	36.4	23	80

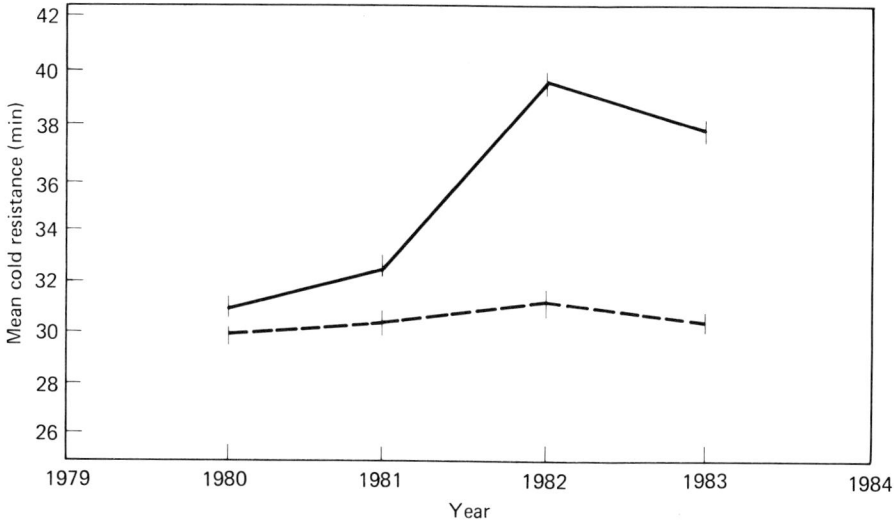

Figure 11.1 The response to 3 years of upwards and downwards selection for cold resistance, using a water bath test, in newborn lambs of the Scottish Blackface breed. ———— high selection line; ---- low selection line

An experiment in ABRO involving genetic selection for cold resistance in newborn Scottish Blackface lambs produced a clear response to selection (*Figure 11.1*) with an estimated heritability of 30% (Slee and Stott, 1986). So far, the relative viability in the field of selected high and low resistance lambs is unknown, but suitable experiments to obtain this information have been designed. Earlier studies with shorn adult sheep in climatic chambers have also revealed significant genetic variation for cold resistance with a similar heritability of about 30% (Slee, 1972, 1986b). This work is relevant to the problem of off-shears losses of adult sheep in Australia, due to exposure, which was referred to earlier.

Other factors likely to be related to lamb survival include behavioural characteristics of the ewe and lamb. There is evidence for genetic variation among breeds in both these components (Hanrahan, 1986). Slee and Springbett (1986) also showed clear breed differences in some specific behavioural characteristics—such as early udder-seeking—which are likely to be related to survival. Within breeds, heritability estimates for lamb survival as a characteristic of the ewe or of the lamb are very low (Hanrahan, 1986). However, Haughey (1983) claims a response to selection for the maternal contribution to lamb survival. Moreover, a specific lamb characteristic—the ability to stand and suck quickly—was found to be related to subsequent survival in Booroola Merinos (Owens *et al.*, 1980).

The general conclusion to be drawn is that direct selection for improved lamb viability would be difficult, but that indirect selection for components of survival such as cold resistance and, perhaps, early sucking, might well be useful in some environments.

Development and application

Animal breeding is less exact than the sophisticated statistics and computing involved might imply. Genetic correlations are not precise and are likely to vary among themselves in robustness, among populations and among environments. The choice of indices for indirect selection will therefore be a compromise between the allocation of resources to estimate parameters and the initiation of selection on the basis of limited knowledge.

The acknowledgement that in most circumstances successful animal breeding benefits the wider community rather than just the breeders themselves, and that animal breeding programmes are of a long-term nature, imply that animal breeding should have a high priority for public sector support. It is perhaps then the role of the public sector development agencies to indicate the direction and approximate magnitude of the genetic correlations between indirect traits and desired traits. The successful breeders will then tend to be those who use this information wisely.

Experimental designs for the assessment of indirect criteria of merit have recently been thoroughly addressed by Hill (1985) in the context of improving reproductive performance. His conclusions are generally applicable and some of the main points are summarized here.

Where the heritabilities of the desired traits are known, it is not necessary to estimate the heritability of an indirect trait to assess its worth and to predict the benefits of its inclusion in an index. It is only necessary to estimate the genetic covariance between the desired and indirect traits and the phenotypic variance of the indirect trait. These parameters can be estimated using a number of different designs which can be broadly classified as using an unselected pedigreed population or selecting on the indirect traits or selecting on the desired traits, and these are considered in turn.

Unselected pedigreed population

Measurements of the desired and the indirect traits in an unselected population can be used to estimate the required genetic covariances and the heritabilities of these traits. Estimation can be based upon specific relationships such as parent–offspring or half-sib, or preferably combining information from all relationships (Thompson, 1982). The approach generally needs a large population to obtain reasonable estimates. Also the study needs to be repeated as new or altered indirect traits are incorporated (even for an alteration as simple as changing the age of recording the indirect trait). The genetic covariances obtained are population estimates, the covariances realized under selection may differ somewhat from these.

Selection on the indirect traits

Observing the correlated responses for the desired traits from selection on the indirect trait for one or more generations will enable the estimation of the required genetic covariances. It also allows the calculation of the heritabilities of the indirect traits although these are not strictly required. The main advantage of this method is that it gives the realized effect of selection on the indirect trait—both beneficial and detrimental. Since heritability of the indirect trait can be estimated and the heritabilities of the desired traits are assumed replication may not be essential as the effects of random genetic drift can be approximately predicted. The main disadvantage of this method is that a new selection experiment is required for each indirect trait. As with the study of unselected pedigree populations, the effects of small changes in the indirect selection criterion, such as changing the age of measurement or adjustments for some other trait cannot be assessed without new experiments.

Selection on the desired traits

Lines produced by selection for individual desired traits for several generations can be examined for correlated changes in indirect traits to estimate the relevant genetic covariances. The effects of measuring the indirect traits in different ways can be assessed and new indirect traits can be evaluated as other data suggest them. When the indirect traits are only measured in a single generation, no estimates of their heritability will be available and so replication is required to allow detection of the effects of drift. For full effectiveness, lines selected for all the major desired traits are required in order that promising indirect traits can be evaluated over all the lines for detrimental as well as beneficial genetic correlations. The cost of a full complement of replicated lines, selected for major desired traits, would be initially high, too high perhaps for any one organization or perhaps even any country. The potential economic benefits in terms of increased selection response are, however, very large. Furthermore, the lines would provide information about the genetics of the desired traits themselves and could provide a basis for the general study of the biology of the traits.

The foregoing discussion has concerned indirect traits which are quantitative in nature and can be used in a classical selection index to enhance selection for desired traits. Some recent work has focused on marker-assisted selection, using polymorphic loci linked to quantitative trait loci to aid selection. The feasibility of such techniques has been increased by the recent discovery of abundant marker variation such as restriction fragment length polymorphisms and hypervariable DNA sites. However, the cost of the within-family studies required to detect the relevant linkages would be enormous. Smith and Simpson (1986) have recently reviewed the possibilities and problems of such approaches and concluded that the advantages are likely to be small.

Conclusions

Indirect selection may have an important contribution to make to improvement in overall economic merit, particularly where components of merits are age-limited or sex-limited in expression or are difficult or expensive to measure. Possible indirect traits require thorough research to find the most appropriate way in which to measure them and to predict the effects of their use on all components of overall merit. The optimum strategy to fulfil these requirements is to produce replicated selection lines for the major components of economic merit (the desired traits) in which the indirect traits can be evaluated. Whilst the cost of such a strategy may be too great for a single country, the potential benefits should make cooperative effort worthwhile.

References

ALLISTON, J.C. (1983). Evaluation of carcass quality in the live animal. In *Sheep Production*, edited by W. Haresign, pp. 75–95. London: Butterworths

ATKINS, K.D. (1980). Selection for skin folds and reproduction. *Proceedings of the Australian Society on Animal Production*, **13**, 174–176

BODIN, L., BIBE, B., BLANC, M. and RICORDEAU, G. (1986). Parametre génétiques de la concentration plasmatique en FSH des agnelles de race Lacaune viande. *Génétique, Sélection, Evolution*, **18**, 55–62

CAMERON, N.D. and SMITH, C. (1985a). Responses in lamb performance from selection on sire 100-day weight. *Animal Production*, **41**, 227–233

CAMERON, N.D. and SMITH, C. (1985b). Estimation of carcass leanness in young rams. *Animal Production*, **40**, 303–308

CROSTON, D., READ, J.L., JONES, D.W., STEANE, D.E. and SMITH, C. (1983). Selection on ram 18-month weight to improve lamb growth rate. *Animal Production*, **36**, 159–164

ENSER, M. and BULLER, K.J. (1984). Pig backfat consistency: regulation of unsaturated fatty acid synthesis. In *BSAP Winter Meeting*, Scarborough, 25–27 March 1985. Paper number 29

FOSTER, M.A., HUTCHINSON, J.M.S., MALLARD, J.R. and FULLER, M. (1984). Nuclear magnetic resonance pulse sequence and discrimination of high- and low-fat tissues. *Magnetic Resonance Imaging*, **2**, 187–192

GRIFFIN, H.D., WHITEHEAD, C.C. and BROADBENT, L.A. (1982). The relationship between plasma triglyceride concentrations and body fat content in male and female broilers—a basis for selection? *British Poultry Science*, **23**, 15–23

HANRAHAN, J.P. (1982). Selection for increased ovulation rate, litter size and embryo survival. *Proceedings of the Second World Congress of Genetics Applied to Livestock Production*, **5**, 294–305

HANRAHAN, J.P. (1986). Maternal effects on lamb survival. *Proceedings of Seminar on Factors Affecting Lamb Survival*. Brussels: CEC

HANRAHAN, J.P. and QUIRKE, J.F. (1982). Selection on ovulation rate in sheep aided by the use of superovulation and egg transfer. *Proceedings of the World Congress on Sheep and Cattle Breeding*, **1**, 329–335

HARRIS, C.I. and MILNE, G. (1980). The urinary excretion of N-methyl histidine in sheep: an invalid index of muscle protein breakdown. *British Journal of Nutrition*, **44**, 129–140

HAUGHEY, K.G. (1983). Selective breeding for rearing ability as an aid to improving lamb survival. *Australian Veterinary Journal*, **60**, 361–363

HILL, W.G. (1985). Detection and genetic assessment of physiological criteria of merit within breeds. In *Genetics of Reproduction in Sheep*, edited by R.B. Land and D.W. Robinson, pp. 319–331. London: Butterworths

HUTCHINSON, J.C.D. (1968). Deaths of sheep after shearing. *Australian Journal of Experimental Agriculture and Animal Husbandry*, **8**, 393–400

LAND, R.B. (1982). Indicators of reproductive potential. *Proceedings of the World Congress of Sheep and Beef Cattle Breeding*, **1**, edited by R.A. Barton and W.C. Smith, pp. 365–373

LEE, G.J. and LAND, R.B. (1985). Testis size and LH response to LH-RH as male criteria of female reproductive performance. In *Genetics of Reproduction in Sheep*, edited by R.B. Land and D.W. Robinson, pp. 333–342. London: Butterworths

OBST, J.M. and DAY, H.R. (1968). The effect of inclement weather on mortality of Merino and Corriedale lambs on Kangaroo Island. *Proceedings of the Australian Society of Animal Production*, **7**, 239–242

OWENS, J.L., BINDON, B.M., EDEY, T.N. and PIPER, L.R. (1980). Neonatal behaviour in high fecundity Booroola Merino ewes. *Review of Rural Science*, **8**, 113–116

PURVIS, I.W. (1985). Genetic relationships between male and female reproductive traits. *Unpublished PhD thesis*, University of New England

RICORDEAU, G., BLANC, M.R. and BODIN, L. (1984). Teneurs plasmatiques en FSH et LH des agneaux males et femelles issus de beliers Lacaune prolifiques et non-prolifiques. *Génétique, Sélection, Evolution*, **16**, 195–210

SEHESTED, E. (1984). Evaluation of carcass composition of live lambs based on computed tomography. In *35th Annual Meeting of the EAAP*, The Hague, 6–9 August 1984. Vol. 1. Summaries, Paper number G5.21

SLEE, J. (1972). Responses and adaptations of sheep to cold. *Animal Breeding Research Organisation Report*, pp. 39–46. Edinburgh: ABRO

SLEE, J. (1981). A review of genetic aspects of survival and resistance to cold in newborn lambs. *Livestock Production Science*, **8**, 419–429

SLEE, J. (1986a). Survival of the ovine neonate under cold stress. *World Meteorological Organization, Technical Report on Animal Health and Production at Extremes of Weather*, Chapter 2 (in press)

SLEE, J. (1986b). The genetics of adaptation to cold climates with particular reference to sheep. *Proceedings of the Third World Congress of Genetics Applied to Livestock Production*, **XI**, 462–473

SLEE, J. and SPRINGBETT, A.J. (1986). Early postnatal behaviour in lambs of ten breeds. *Applied Animal Behaviour Science*, **15**, 229–240

SLEE, J. and STOTT, A.W. (1986). Genetic selection for cold resistance in Scottish Blackface lambs. *Animal Production,* **43**, 397–404

SMITH, C., JAMES, J.W. and BRASCAMP, E.W. (1986). On the derivation of economic weights in livestock improvement. *Animal Production,* **43**, 545–551

SMITH, C. and SIMPSON, S.P. (1986). Use of genetic polymorphisms in livestock improvement. *Zeitschrift für Tierzüchtung und Züchtungsbiologie,* **103**, 205–217

THOMPSON, R. (1982). Methods of estimation of genetic parameters. *Proceedings of the Second World Congress of Genetics Applied to Livestock Production,* **5**, 95–103

TILAKARATNE, N., ALLISTON, J.C., CARR, W.R., LAND, R.B. and OSMOND, T.J. (1980). Physiological attributes as possible selection criteria for milk production. 1. Study of metabolites in Friesian calves of high or low genetic merit. *Animal Production,* **30**, 327–340

WALKLEY, J.R.W. and SMITH, C. (1980). The use of physiological traits in genetic selection for litter size in sheep. *Journal of Reproduction and Fertility,* **59**, 83–88

WOLF, B.T., SMITH, C., KING, J.W.B. and NICHOLSON, D. (1981). Genetic parameters of growth and carcass composition in crossbred lambs. *Animal Production,* **32**, 1–7

Chapter 12

Carcass evaluation in sheep breeding programmes

G. Simm

Summary

Genetic improvement of terminal sire breeds offers a permanent, cumulative and relatively cheap method of improving the carcass composition of crossbred lambs. Leaner carcasses of a given weight may be produced by using a terminal sire of higher mature weight than that currently used (either by breed substitution or within-breed selection). However, further improvement can probably be made by selecting for slower rates of maturing in fatness and for reduced fat content at maturity. *In vivo* estimates of carcass composition, together with immature live weights, may be the most appropriate selection criteria to improve overall carcass merit. Of the *in vivo* techniques available some are sufficiently mobile, robust and cheap to be used to screen large numbers of animals on farms (such as pulse-echo or transmission ultrasonic techniques). Other techniques are expensive and less mobile and will probably be used only in research or for second-stage selection following screening of animals with a cheaper technique. X-ray computed tomography (CT) seems the most immediately applicable of these 'expensive' techniques. However, in the long term, nuclear magnetic resonance (NMR) techniques should provide more comprehensive information on body composition, particularly if imaging and spectroscopy are combined. Similarly, neutron activation analysis (NAA) could provide detailed chemical analysis *in vivo*.

Introduction

Consumption of mutton and lamb in Britain has fallen from about 11 kg/head/year in the 1940s to around 7 kg/head/year in the 1980s (*Figure 12.1*). Lamb now accounts for only about 11% of total meat consumption compared to over 25% in the 1940s. Similar trends are apparent in some other countries, and one of the major reasons appears to be consumers' perception of lamb as excessively fat (Kempster, 1983). This trend in consumption is likely to be reinforced by concern over the suggested role of animal fats in coronary heart disease. Kempster, Cook and Grantley-Smith (1986) estimated that there was 44 000 to 57 000 tonnes of waste sheep fat produced in Britain in 1984, representing 20–26% of the total carcass weight produced. This waste fat production has serious economic

125

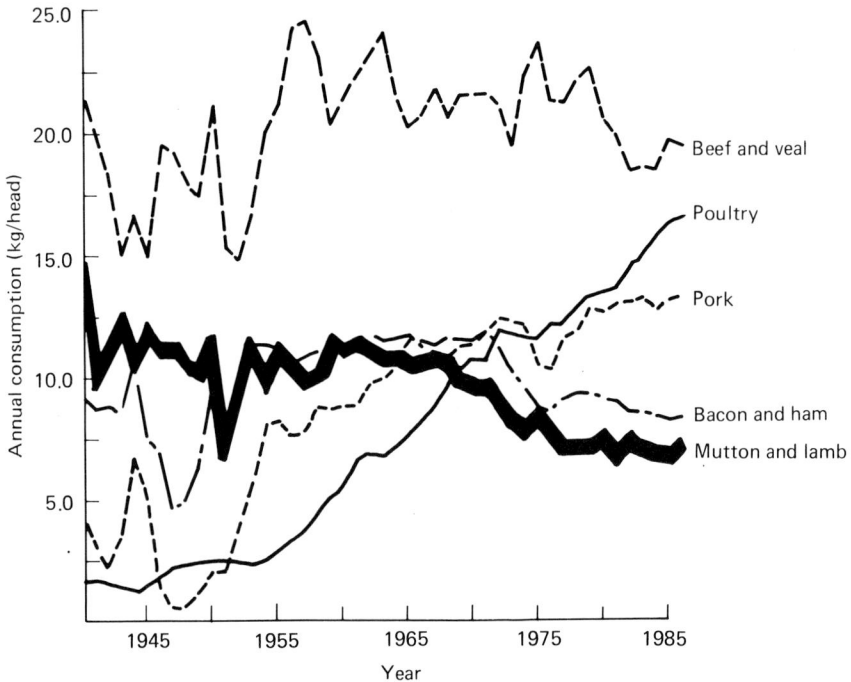

Figure 12.1 Meat consumption in Britain 1940–86 (source: Ministry of Agriculture, Fisheries and Food)

consequences at a *national* level, both in terms of the effect excessive fatness is having on lamb consumption, and in terms of the resources wasted in producing excess fat. However, until recently there has been little financial incentive for *individual* producers to change the carcass composition of their lambs. This situation is changing, and sheep breeders must adjust their breeding programmes accordingly, hence the relevance of *in vivo* estimation of carcass composition.

In this chapter the prospects for genetic improvement of carcass composition in sheep, the choice of selection criteria to incorporate *in vivo* estimates of composition, and some of the techniques available for *in vivo* assessment in breeding programmes are examined. The aim is not to give a comprehensive review of methods of *in vivo* estimation, but rather to highlight those techniques which are likely to have most impact, either in practical breeding programmes, or in animal breeding research. Comprehensive reviews of *in vivo* estimation methods are given by Bech Andersen (1982), Alliston (1983) and Lister (1984).

Prospects for genetic improvement in carcass composition

Growth and development

Most of the variation in body composition in live animals is associated with differences in the amount of fat in the body. The amounts of protein, water and ash are log-linearly related to weight of the fat-free body for animals of different genotypes subjected to a range of nutritional treatments (Black, 1983; Searle and

Griffiths, 1983). As an animal increases in live weight or degree of maturity, the weight and proportion of fat in the body and carcass generally increase, with a reduction in the proportion of lean tissue. Thus, carcasses with a higher proportion of lean could be produced without recourse to genetic improvement, simply by slaughtering lambs at lighter weights.

When breeds of different mature sizes are compared at the same live weight, the larger breeds generally grow faster, contain less fat and more protein and bone in the body and carcass than do breeds of small mature size (McClelland, Bonaiti and Taylor, 1976; Searle and Griffiths, 1976a, 1976b; Black, 1983). However, when the same breeds are compared at similar degrees of maturity, most of the differences between breeds disappear (McClelland, Bonaiti and Taylor, 1976). Breeds which differ from this general relationship include the Soay and the Texel. The Soay has less fat and more bone than expected, whilst the Texel has less fat and more lean than expected at a given degree of maturity (McClelland, Bonaiti and Taylor, 1976; Wolf, Smith and Sales, 1980). A knowledge of differences amongst breeds in mature size, and of which breeds deviate from the general relationship between composition and degree of maturity is therefore of great importance in matching breeds to production systems (Croston et al., 1983; Wolf, Smith and Sales, 1980). For example, leaner carcasses of a given weight could be produced from a given production system by switching to a breed of larger mature size, other things being equal.

Between-breed versus within-breed selection

In the United Kingdom about 30% of the genes of lambs slaughtered originate in terminal sire (Down) breeds, yet purebred ewes of these breeds represent less than 5% of the total UK breeding sheep population (Meat and Livestock Commission, 1972; J.L. Read, personal communication). Theoretically, genetic improvement should be relatively easy since: (1) selection or substitution can be concentrated on a numerically small group of animals which have a large effect on the animals used for meat production; (2) selection between or within breeds can be concentrated on growth and carcass traits, since relatively small numbers of crossbred progeny from terminal sire breeds are retained for breeding.

Breed substitution, particularly amongst the terminal sire breeds, is the most obvious route to make rapid genetic progress. However, there appear to be only small differences in rates of lean tissue deposition amongst the larger breeds, though breeds differ in the way they achieve this (Wolf, Smith and Sales, 1980; Kempster et al., 1983). For example, the Texel has a relatively high lean percentage, which is counterbalanced by a lower growth rate than that of the Suffolk. Thus, for producers already using one of the larger terminal sire breeds, within-breed selection is likely to be the most useful way of making genetic progress in lean growth rate.

Within-breed selection criteria

Although mature size is a useful selection criterion amongst breeds, it is less practicable for within-breed selection, because of the length of time needed to measure it. Earlier live weights give an indication of mature size, and simultaneous in vivo estimates of body composition may also provide information on rates of maturing in fatness and, indirectly, of composition at maturity. There is evidence of genetic variation in rates of maturing in fatness, and of fatness at maturity in laboratory animals and sheep (Roberts, 1979; Butterfield *et al.*, 1983).

Possible selection criteria to improve carcass composition by genetic means include

(1) growth rate;
(2) estimated carcass lean percentage;
(3) estimated carcass lean weight;
(4) estimated lean tissue growth rate; or
(5) economic selection indices.

In sheep, where production is usually linked to seasonal grass growth, the main objective is to maximize production in a given time interval. For this reason, and for operational reasons, selection decisions will often be made at a fixed age, rather than a fixed live weight. At a given age there may be a positive genetic correlation between growth rate and carcass fat weight or proportion, and a negative correlation between growth rate or live weight and carcass lean proportion. Hence, selection on growth rate may increase fatness, and selection on estimated lean proportion may reduce carcass weight at a given age. Neither outcome is likely to be desirable in the foreseeable future. Even with relatively precise *in vivo* estimation of carcass composition, prediction equations for lean weight tend to be dominated by live weight (Sehested, 1984). Similarly, lean tissue growth rate,

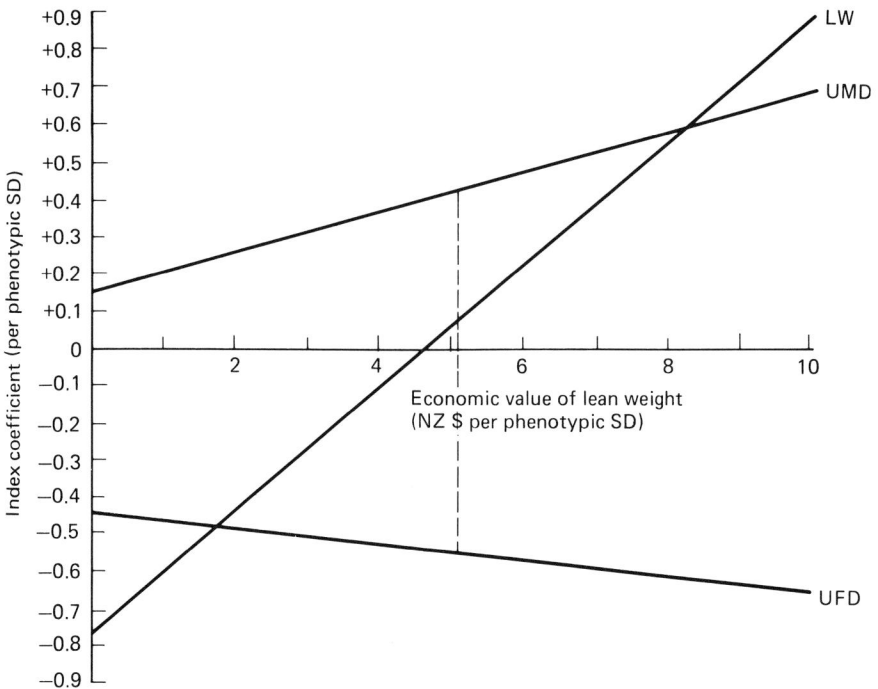

Figure 12.2 Change in index coefficients for live weight (LW), ultrasonic fat depth (UFD) and ultrasonic muscle depth (UMD) as the economic value for lean weight changes relative to that for fat weight (NZ$ − 4.62 per phenotypic SD). (After Simm, Young and Beatson, 1987)

estimated as the product of growth rate, killing-out and lean proportions, tends to be highly correlated with growth rate, especially when precision of *in vivo* estimation is low (Simm, 1983a). This is a consequence of the part–whole relationship between lean weight and live weight, and also the relatively high coefficient of variation of growth rate or live weight compared to that of killing-out and lean proportions. Economic selection indices for lean meat production theoretically give optimal weightings to live weight and *in vivo* measurements to maximize the rate of genetic change in profitability. Simm, Young and Beatson (1987) derived an economic selection index for New Zealand sheep, which included ultrasonic measurements of fat and muscle depth. *Figure 12.2* illustrates the contribution which these measurements make to the index, as the relative economic value of lean weight changes compared to that for fat weight. With reasonably accurate estimates of economic values and genetic parameters, economic indices may be very useful selection criteria in sheep breeding.

With moderately heritable traits, such as growth and carcass composition, rates of genetic change of 1.0–1.5% of the mean per annum should be achievable (Smith, 1984). With improved technology, embryo transfer could increase these rates of genetic change by 60–80% by reducing female generation intervals and increasing the selection intensity amongst females (Smith, 1986).

Requirements of *in vivo* assessment techniques

In this section the requirements for a suitable *in vivo* assessment technique for use in animal breeding research and practical improvement programmes are examined.

The precision and repeatability of the technique are obviously of prime importance. With less precise techniques, of low to moderate repeatability, there may be worthwhile gains in selection response by repeating measurements either spatially (for example, measuring fat depths on both sides of an animal) or over time (Falconer, 1981).

In commercial breeding schemes there will be a trade-off with other factors such as initial and running costs, speed of operation and mobility. In national terms, a technique of low precision may be more effective than a precise technique, if it is used more widely as a result of its lower cost or greater speed and mobility. Nevertheless, expensive, slow and immobile methods of *in vivo* assessment may still be cost-beneficial, especially if they are used in a centralized breeding scheme with effective dissemination of improved stock. Similarly, they may be useful as the second stage in a two-stage selection process, after an initial screening with a less precise, quick, mobile method.

Obviously for mass selection purposes, the technique must be non-destructive. Also, if the technique is to gain wide acceptance in the industry, it should also be non-invasive and painless. Requirements of *in vivo* methods are discussed further by King (1982).

Cost-benefit analysis is potentially useful in selecting amongst *in vivo* assessment techniques (see, for example, Kempster, 1984). Simm, Young and Beatson (1987) used a formula derived by Smith (1978) to estimate national discounted returns from a single round of selection on an index including ultrasonic measurements of fat and muscle depth. With reasonable penetration of the improved stock into the national flock, discounted returns from the index exceeded the cost of measuring potential breeding stock. However, such national cost-benefit analyses ignore the financial benefits which may accrue to individual breeders who adopt *in vivo* techniques earlier than the rest.

Techniques for *in vivo* assessment

Live weight

As mentioned previously, the composition of an animal's body changes in a more or less predictable fashion as it increases in maturity. Hence, live weight can be a valuable measurement in predicting proportions of different tissues in the carcass, particularly when comparisons are made amongst animals of similar breed and sex, from similar feeding regimes. Similarly, because of the part–whole relationship, live weight is a useful predictor of tissue weights in the carcass (*Table 12.1*). Because of this fundamental relationship between carcass composition and live weight, and because it can be measured simply and cheaply, live weight is usually included as the reference point in any comparison of *in vivo* techniques. Failure to account for variation in carcass composition due to live weight differences can lead to overestimation of the value of *in vivo* techniques, particularly if tested on a population of animals varying widely in live weight and carcass composition.

Gutfill can have an appreciable effect on the live weight of ruminants, and starved weight may be a better measurement to use in predicting carcass composition.

Ultrasound

Pulse-echo techniques
Ultrasonic pulse-echo techniques are used to map tissue boundaries, using changes in acoustic impedance (resistance to the transmission of ultrasound). The velocity of ultrasound in a biological medium depends on the temperature and the physiological condition of the medium. When a wave of ultrasound meets a boundary between two tissues, partial reflection may occur. The greater the difference between tissues in acoustic impedance, the higher the reflection. For example, more energy is reflected from a muscle: bone interface than from a muscle: fat interface. The time taken for a reflected wave to reach the source of ultrasound is proportional to the depth of the boundary, and inversely proportional to the velocity of ultrasound in that tissue. Hence the depths of tissue boundaries can be estimated from the time between emission of a pulse of ultrasound, and receipt of echoes.

In pulse-echo ultrasonic machines, pulses of ultrasound are produced by a transducer which also receives reflected waves of ultrasound and converts them to electrical energy. After modification, these electrical signals can be displayed on a screen in various ways. In so-called A-mode machines, echo amplitude (A) is displayed against time. Echoes appear as spikes on an oscilloscope screen, and the distance between successive spikes is related to the distance between tissue interfaces (*Figure 12.3*). This type of scanner has been used to estimate sheep carcass composition *in vivo* in New Zealand recently (Gooden, Beach and Purchas, 1980; Purchas and Beach, 1981; see also *Table 12.1*).

Electrical signals from echoes regulate the brightness (B) of the time-base line on a cathode ray tube in B-mode scanners. Here the distance between successive

Authors	Predictors	Trait predicted	Mean of trait predicted	SD of trait predicted	Residual SD fitting live weight alone	Residual SD fitting live weight plus in vivo measurement	Original coefficient of variation (CV%)	Residual CV fitting live weight plus in vivo measurement (%)
Purchas and Beach (1981)	A-mode ultrasonic Fat-depth 12/13 rib	Carcass fat %	24.9	4.5	3.9	2.6 to 2.9	18	10–12
Nicol and Parratt (1984)	A-mode ultrasonic Fat-depth 12/13 rib	Fat-free carcass weight (kg)	24.7	2.32	1.49	1.24	9	5
Kempster et al. (1982)	B-mode ultrasonic Fat area 12th rib	Lean %	54.9†	4.1	3.9	3.2(S)* 3.6(D)*	7	6 7
		Subcutaneous fat %	13.4†	3.4	3.2	2.2 (S) 2.8 (D)	25	16 21
Cuthbertson et al. (1984)	B-mode ultrasonic Fat area/muscle area 12th rib (and A-mode fat depth)	Lean % Flock 1	64.5	2.7	1.4	1.4/1.3 (D) 1.5/1.5 (S) 1.5 (M)*	4	2 2 2
		Flock 2	59.9	2.5	1.9	1.7/1.8 (D) 1.9/1.9 (S) 1.9 (M)	4	3 3 3
		Lean gain/day (g) Flock 1	85.0	12.0	3.8	3.7/3.9 (D) 3.8/3.9 (S) 4.0 (M)	14	4 4 5
		Flock 2	74.0	9.3	4.2	4.4/3.5 (D) 4.0/3.8 (S) 4.3 (M)	13	5 5 6
Cameron and Smith (1985)	B-mode ultrasonic Fat depth 12th rib	Lean %	55.4	3.3†	2.6†	2.1 (V)* 2.4 (D)	6	4 4
Sehested (1984)	Frequency of X-ray CT values in different ranges, from scan at 4th lumbar	Protein (kg)	2.54	0.45	0.20	0.15	18	6
		Fat (kg)	2.24	0.95	0.57	0.33	42	15
		Fat-free lean (kg)	11.49	2.04	0.84	0.60	18	5

*S = Scanogram (B-mode), D = Danscanner, V = Vetscan (both real-time B-mode), M = Merrytronics (A-mode)
†Estimated from data presented

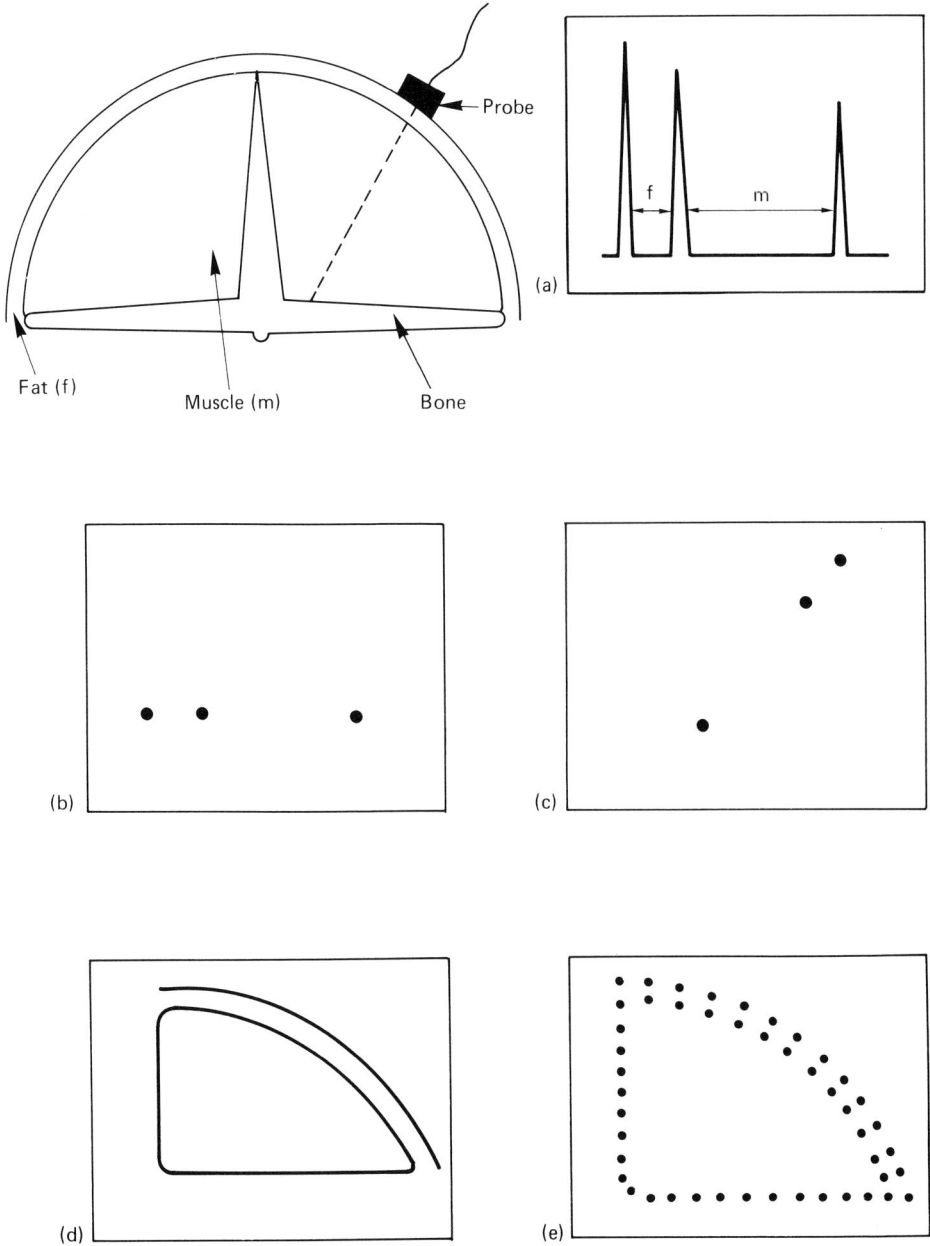

Figure 12.3 Schematic diagram illustrating how results from A-mode and B-mode pulse-echo ultrasonic scanners are displayed. (*a*) A-mode presentation; (*b*) B-mode presentation; (*c*) B-mode with direction of the timebase linked to the direction of the ultrasonic beam; (*d*) B-mode presentation built up as the probe moves across the back of the animal (e.g. Scanogram); (*e*) B-mode presentation from a multi-element transducer (e.g. Danscanner) (after Wells, 1969; Simm, 1983b)

(b)

Skin

Subcutaneous fat layer

M. longissimus lumborum

Transverse process of third
lumbar vertebra

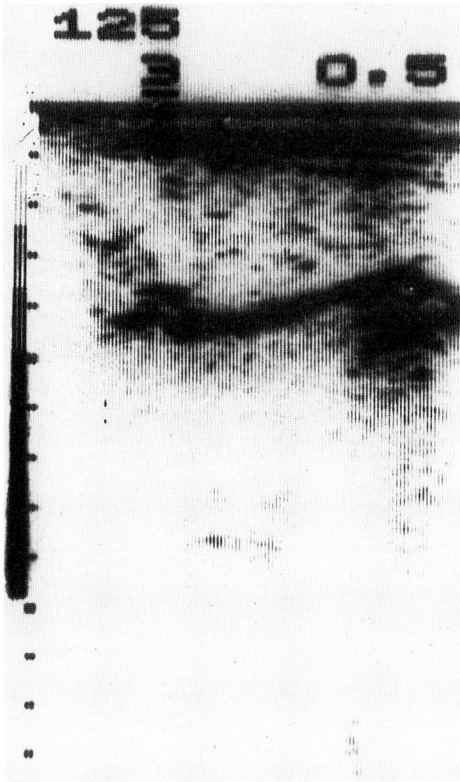

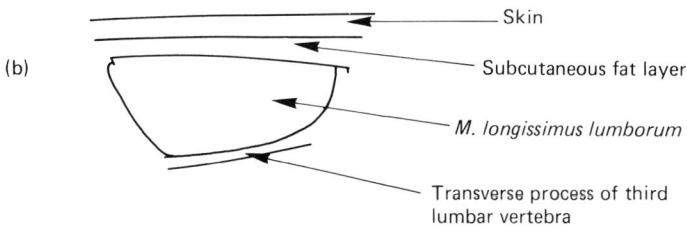

Figure 12.4 (*a*) A B-mode ultrasonic scan taken at the level of the 3rd lumbar vertebra on a live sheep;
(*b*) explanatory diagram of the ultrasonic scan

bright spots represents the distance between successive tissue interfaces (see *Figures 12.3* and *12.4*). Two-dimensional scans can be produced in a number of ways. For example, the Scanogram has a spring-loaded transducer which moves along a track across the animal. The movement of the transducer along the track is mechanically coupled to a polaroid camera aimed at the oscilloscope. The camera moves in phase with the probe, building up a two-dimensional scan photograph. In contrast, real-time scanners have a fixed array of transducers which operate in quick succession, producing an almost instantaneous two-dimensional display on the oscilloscope.

Two-dimensional scans from B-mode machines are either interpreted on the screen at the time of scanning, or the screen may be photographed for later interpretation. Generally, scans with these machines are taken in the thoracic and lumbar regions, close to the vertebra. The musculature at this location is relatively simple, and hence interpretation of scans is easier. Fat and muscle depths, or areas, are usually measured from these scans and related to corresponding carcass measurements or to dissected carcass composition. Some results are shown in *Table 12.1*.

Ultrasonic measurements of fat or muscle depth have been applied successfully in the pig industries of several countries, but the precision achieved with cattle, and particularly sheep, has been rather low. This is partly because of the dependence of pulse-echo techniques on subcutaneous fat. A lower proportion of the total carcass fat of cattle and sheep is in the subcutaneous depôt, compared to that of pigs (*Table 12.2*). Also, in sheep the absolute depths of fat and muscle are lower than in cattle and pigs.

TABLE 12.2 Typical proportions of lean and proportions of fat from different depôts in the carcasses of sheep, cattle and pigs (after Kempster *et al.*, 1986)

	Sheep	Cattle	Pigs
		Proportion of tissue in carcass*	
Lean	56.1	60.3	52.1
Subcutaneous fat	11.5	7.9	16.1
Intermuscular fat	11.0	12.9	5.0
Internal fat†	3.1	1.4	1.6
Total separable fat in carcass	25.6	22.2	22.8

*NB: Definition of carcass varies from species to species
†Kidney, knob and channel fat (KKCF) for sheep; cod/udder fat for cattle, flare fat for pigs

Speed of ultrasound transmission

In general, there is a linear relationship between the reciprocal of the speed of ultrasound and the fat content of tissue (Miles and Fursey, 1974). These authors therefore developed a technique for directly measuring the velocity of ultrasound across the hind limb of live animals. The machine consists of two transducers arranged in line at opposite sides of a caliper. One transducer acts as a transmitter and the other acts as a receiver. The distance between the two transducers is variable, and can be measured precisely. The machine incorporates an electronic timer which automatically measures the time of flight of the pulse of ultrasound across the animal, and computes the reciprocal of the speed of transmission. The technique gives a digital reading, and hence avoids the errors inherent in subjective interpretation of B-mode scans. Also, the velocity of ultrasound depends on the amount of fat in the transmission path, regardless of the depôt, and hence should give a better description of overall fatness. The technique has produced good results for cattle (Miles, Fursey and York, 1984). However, from *preliminary* studies on sheep, the technique does not appear to give higher precision than pulse-echo techniques (*see below*).

X-ray computed tomography

X-ray computed tomography (CT) is a technique developed for whole-body scanning in human medicine. The principle behind X-ray CT is that absorption of X-rays differs from one tissue to another, depending on the tissue density. This property has been used to produce images of 'sections' through the human body which clearly differentiate amongst tissues. Images are produced by transmitting X-rays through a thin 'slice' of the body, from a source rotating around the patient. A rotating detector, or a band of detectors aligned with the source, but on the opposite side of the patient, then measures the degree of attenuation of transmitted X-rays. During each rotation, the attenuation of a large number of projections is recorded. From these data, the attenuation can be computed for each point in the scan plane, which is divided into tens of thousands of elements ('pixels'). The degree of attenuation in each pixel, designated a 'CT value', is therefore dependent on the density of the tissue at the corresponding point in the patient's body. Different tissues have characteristic ranges of CT values. By assigning various colours, or shades of grey, to different ranges of CT values, images can be constructed in which tissues are differentiated by distinct shades or colours (*Figure 12.5*). Most efforts in the development of X-ray CT have been directed at improving the visual quality of images, rather than of the quantitative information on which the images are based. However, researchers at the Agricultural University of Norway recognized the potential for using this quantitative information for predicting carcass composition of animals (Skjervold *et al.*, 1981).

Sehested (1984) used live weight, together with the frequency of CT values in different ranges, from scans at one or more points on the body, to predict the carcass composition of sheep quite precisely (*see Table 12.1*). Serial 'slices' along the body can be used to produce three-dimensional information on muscle or organ shape and volume (Knopp, 1985). However, the large quantity of data produced from a single scan presents major computing and statistical problems. Further details of X-ray CT, and possible future developments are discussed by Herman (1980) and Wells (1984).

Nuclear magnetic resonance (NMR)

NMR imaging

Nuclear magnetic resonance is a consequence of the angular momentum or spin shown by atomic nuclei that have an odd number of protons or neutrons (Loeffler and Oppelt, 1981; Gadian, 1982; Foster and Hutchison, 1985). The most numerous of these in the body of man and animals is the nucleus of the hydrogen atom, which consists of a single proton. Since nuclei are charged particles, their spin is always associated with a magnetic moment. When a sample of protons is brought into a magnetic field, all spins will tend to align parallel, like compass needles in a magnetic field. In NMR imagers, this strong, uniform magnetic field is supplied by a magnet (permanent, resistive, electromagnetic or superconducting electromagnetic) with a central opening large enough to accommodate a patient (Wells, 1984). If the magnetization is tilted by an external force, such as an alternating current pulse passing through a coil surrounding the patient, the angular momentum of all spins prevents the magnetization from turning back to its equilibrium position. Instead it will 'precess', at a frequency proportional to the magnetic field strength, around an axis parallel to the magnetic field (like a spinning top in a gravitational field).

The precession of the magnetization decays in time, a process called relaxation. Two components of relaxation can be distinguished and measured. The component of the magnetization parallel to the static magnetic field approaches equilibrium because of the interaction of the nuclear spin system with the surrounding 'lattice'. The characteristic time constant is called the longitudinal or 'spin-lattice' relaxation time (T_1). The transverse component of magnetization is reduced due to

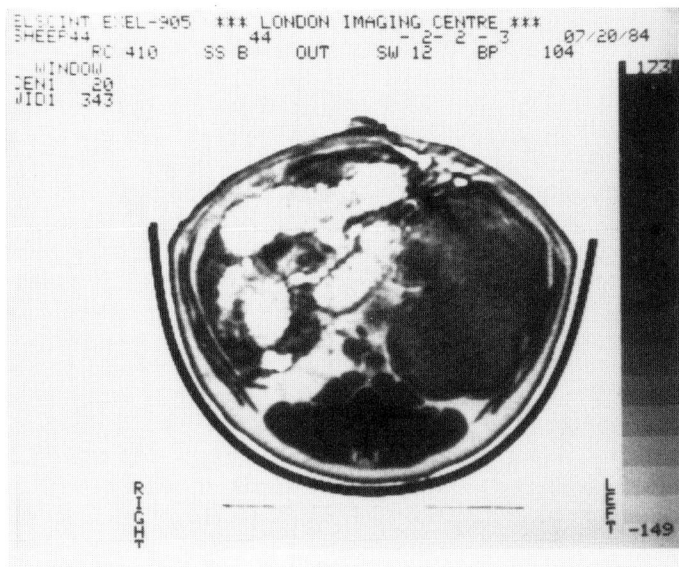

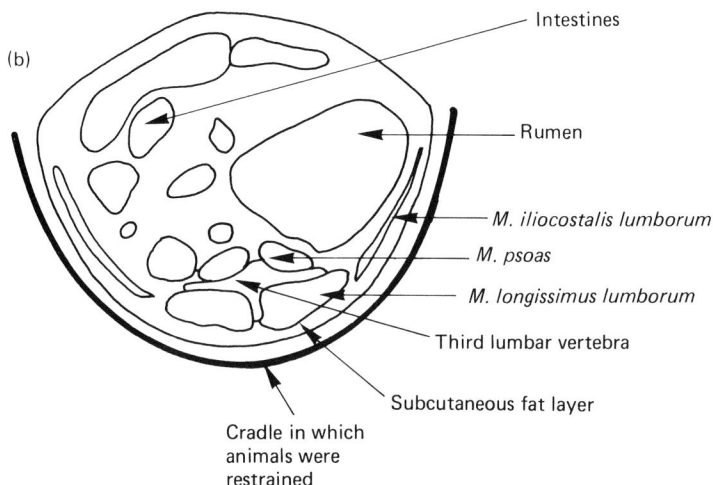

Figure 12.5 (a) An X-ray CT image of a live sheep, at the level of the 3rd lumbar vertebra;
(b) explanatory diagram of the X-ray CT image

interactions with neighbouring spins. This is called the transverse or 'spin-spin' relaxation time (T_2) (Loeffler and Oppelt, 1981). The relaxation time of protons depends on the chemical background of the tissue, such as the state of hydration and fat content. For example, muscle contains about 75% water. Water protons have a moderate longitudinal relaxation time (T_1). On the other hand, fat protons in the -CH_2- groups of triglycerides are less mobile than water protons, and hence have a shorter T_1 value (Foster et al., 1984).

The density, distribution and properties of protons can be examined by measuring the electromagnetic signals which they induce in the coil(s) surrounding the subject, in response to changes in the magnetic field characteristics. As well as displaying the basic information on proton density (q) and relaxation times (T_1 and T_2), it is possible to obtain a variety of images with proton density weighted to different extents by relaxation processes. Foster *et al.* (1984) examined a series of

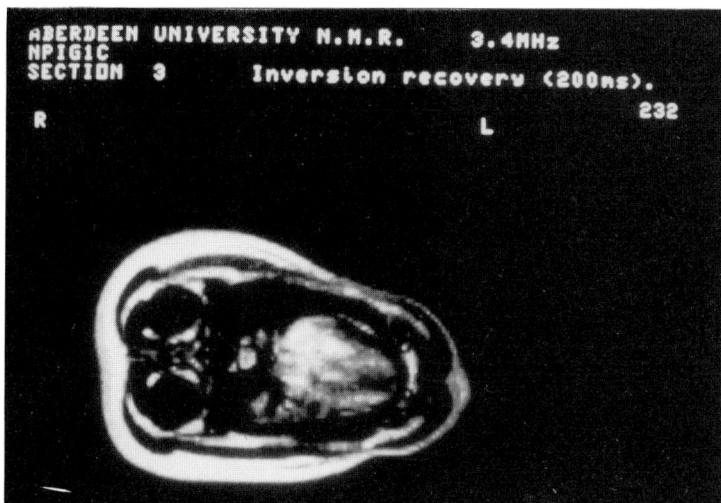

Figure 12.6 NMR scan of a live pig obtained using inversion recovery (from Fuller, Foster and Hutchison, 1984)

such NMR image types, and found that the type known as 'inversion recovery' gave the best discrimination between fat and muscle, with a contrast ratio of about 6 : 1 (*Figure 12.6*). The results of Groeneveld *et al.* (1984) also show that both X-ray CT and NMR imaging discriminate effectively between tissue types.

Fuller, Foster and Hutchison (1984) examined T_1 values for various pig tissues both *in vivo* and *in vitro*, and suggested several potential developments in NMR imaging to increase its value in animal science.

NMR spectroscopy
NMR spectroscopy has been used for many years in *in vitro* chemical analysis. Theoretically spectroscopic analysis is possible for any element with an odd number of protons or neutrons. However, very high field strengths are required, and spectroscopy operates on fairly small volumes of material. Development of superconducting magnets and horizontally polarized magnets means that spectroscopy can now be applied to laboratory animals and human limbs. In addition to ^1H, NMR spectra are most readily available for ^{31}P and ^{13}C. Spectroscopy has important applications in human medicine (Wells, 1984), but also has great potential for studying metabolic processes in live animals. At least one company (Oxford Research Systems) now offer NMR scanners which combine imaging and spectroscopy.

Neutron activation analysis (NAA)

The only naturally radioactive essential body element is ^{40}K. However, other body elements can be made radioactive by exposing them to a source of neutrons (neutron activation). The elements of interest are then measured by comparing spectra and counting rates to those obtained from neutron-activated 'phantoms' containing known amounts and distributions of the relevant elements. To date, calcium, phosphorus, sodium, potassium, chlorine, iodine, cadmium, aluminium, silver, nitrogen, carbon and oxygen have been measured by neutron activation analysis (Boddy, 1984). No single facility has been developed to measure all elements but there are facilities which measure seven elements (Preston *et al.*, 1985). The technique has been used experimentally to estimate the body composition of rats, sheep and pigs fairly accurately, in addition to estimation of human body composition (Preston, East and Robertson, 1984; Preston *et al.*, 1985).

Comparison of techniques

There are several problems in trying to compare published results on the precision of different *in vivo* assessment techniques:

(1) Few trials involve a range of techniques and this makes it difficult to get a 'reference point'.
(2) The animals used in different trials often differ widely in breed, live weight, age and carcass composition.
(3) The experience of the people using the *in vivo* assessment technique may vary from one trial to the next.
(4) The carcass trait chosen as a dependent variable in prediction equations often varies, and may be very specific to a given trial, for example, use of trimmed cuts. Complete separation of tissues is preferable, both in terms of biological interpretation, and repeatability in different trials.

Ideally, the choice of *in vivo* assessment technique should be based on direct comparison of measurements made with a range of methods on the same animals, by experienced operators. The measurements should be made on a sufficiently large group of animals to permit discrimination between techniques, and the animals should be representative of those on which the *in vivo* assessment will be made in future.

Tables 12.3 to 12.5 show some preliminary results of a recent trial conducted at the Edinburgh School of Agriculture. The objective of the study was to examine the relationships between a range of live animal measurements and carcass composition in Suffolk ram lambs, at the end of a performance test. Following live animal measurements, 50 of the rams not selected for breeding were slaughtered

TABLE 12.3 Means and standard deviations of important traits at slaughter ($n = 50$) (Simm *et al.*, unpublished observations)

Trait	Mean	SD
Age (days)	179	6
Live weight (kg)	72.8	7.6
Half carcass lean weight (kg)	9.03	0.91
Half carcass lean percentage	47.20	3.05
Half carcass fat weight (kg)*	7.25	1.22
Half carcass fat percentage*	37.64	3.57

*Subcutaneous, intermuscular fat and kidney, knob and channel fat (KKCF)

TABLE 12.4 Precision of predicting half carcass lean weight from live animal measurements (n = 50) (from Simm et al., unpublished observations)

Original SD	0.912 kg	
Independent variables*	*RSD*	R^2
Live weight, age	0.623	0.55
+ Velocity of sound, through hind limb	0.568	0.64
+Scanogram FA	0.614	0.58
+Technicare MA	0.577	0.63
+Vetscan MD†	0.576	0.63
+X-ray CT MA, FD	0.496	0.73

*LW = live weight, M = muscle, F = fat, A = area, D = depth
†Mean of measurements on 2 consecutive days

TABLE 12.5 Precision of predicting half carcass lean percentage from live animal measurements (n = 50) (from Simm et al., unpublished observations)

Original SD	3.05%	
Independent variables	*RSD*	R^2
Live weight, age	2.90	0.13
+ Velocity of sound, through hind limb	2.50	0.37
+ Scanogram FA	2.63	0.30
+ Technicare MA, FA	2.38	0.44
+ Vetscan MD, FD*	2.44	0.41
+ X-ray CT MA, FD	2.00	0.60

*Mean of measurements on 2 consecutive days

and half carcasses were dissected into lean, fat and bone. During the performance test ram lambs were given a high energy, high protein feed *ad libitum* to allow each animal to fully express its potential for lean growth. Consequently, the animals were fatter and showed a higher variation in lean proportion (*Table 12.3*) than that observed in animals of the same breed, sex and similar age under a lower plane of nutrition (Cuthbertson, Croston and Jones, 1984).

The *in vivo* predictors examined included live weight, velocity of ultrasound through the hind limb and measurements from three B-mode ultrasonic scanners, and X-ray CT.

The B-mode scanners used were: (a) the Scanogram—a purpose built animal scanner; (b) the Vetscan Mark I—a real-time scanner developed for pregnancy diagnosis in animals; and (c) the Technicare—a real-time scanner developed for diagnostic use in human medicine. With each of these ultrasonic machines scans were taken at the level of the 13th rib and the 3rd lumbar vertebra. Fat and muscle depths or areas (FD, FA, MD, MA) were recorded from polaroid photographs of scans. X-ray CT scans were also obtained at these two sites. Measurements obtained from CT images were:

(1) Fat depths over the *M. longissimus*, at 1.5 cm from the midline, and at the end of the muscle; fat depth at a vertical distance of 7.5 cm below the midline (depths were obtained on both left and right hand sides, then summed).
(2) The sum of the left and right *M. longissimus* areas.
(3) 'Carcass area' obtained by outlining the carcass soft tissues to exclude the viscera and bone.
(4) Mean CT values for the carcass area and *M. longissimus* area.

TABLE 12.6 A comparison of some characteristics of *in vivo* assessment techniques

	Ultrasound			X-ray computed tomography (CT)	Nuclear magnetic resonance (NMR)		Neutron activation analysis (NAA)
	A-mode	B-mode	Transmission		Imaging	Imaging plus spectroscopy	
Potential precision of predicting carcass composition	**	***	***	*****	******	*****	*****
Potential applications apart from animal breeding	**	***	***	****	****	*****	*****
Throughput of animals	*****	****	*****	***	**	**	*
Potential mobility	*****	*****	*****	**	**	*	*
Cost (capital and running)	*****	*****	***	**	**	*	**

Scores: * = unfavourable   ***** = favourable

For each machine, multiple regression equations were constructed to predict carcass lean weight and lean percentage, from live weight (LW), age, fat and muscle (and CT) measurements, either at the 13th rib alone, at the 3rd lumbar vertebra alone, or from the sum of measurements at the two positions. Live weight and age were retained in all equations, but other measurements were dropped sequentially if they did not significantly improve the precision of prediction ($P <$ 0.05). *Tables 12.4* and *12.5* show the precision of equations using the sum of measurements from B-mode scans at the 13th rib and 3rd lumbar vertebra. These preliminary results have not been corrected for the bias introduced by dissecting those animals not selected for breeding. Also, this analysis does not reveal the full potential of X-ray CT, since the distributions of CT values which the Norwegian workers find most useful, were not available at the time of analysis. Despite this, prediction equations including CT measurements were considerably more precise than those including ultrasonic measurements (for both lean weight and percentage). The two real-time scanners and the velocity of sound technique produced similar results to one another. Results from the Scanogram were generally the poorest.

The usefulness of live animal measurements, in addition to live weight, in this study, may be partly explained by the high variation in carcass composition, and the high absolute level of fatness.

Table 12.6 compares some of the important features of the *in vivo* assessment techniques reviewed in the preceding section. The information in *Table 12.6* is intended as a very approximate guide. Many of the techniques are developing rapidly and information on prices, scanning times etc. soon become outdated.

Conclusions

Declining consumption is a major problem in the sheep industries of the United Kingdom and many other countries. In the United Kingdom, overfatness of lambs is one of the major reasons for the decline in consumption. Genetic improvement of terminal sire breeds offers a permanent, cumulative and relatively cheap method of improving the carcass composition of crossbred lambs. Additionally, manipulation of carcass composition through genetic improvement is unlikely to receive the adverse public reaction that is often associated with hormonal or immunological treatment, etc. The carcass composition of lambs slaughtered at a given live weight may be improved by using a terminal sire of higher mature weight than that currently used. This may be achieved either by breed substitution, or by within-breed selection. However, further improvements may be possible, by selecting for slower rates of maturing in fatness, and for reduced fat content at maturity. *In vivo* estimates of body composition, together with immature live weights, may be the most appropriate selection criteria to improve carcass composition. Given reasonably accurate estimates of economic values and genetic parameters, economic selection indexes combine these measurements in an optimum way.

Of the *in vivo* techniques examined here, some are sufficiently mobile, robust and cheap to be used to screen large numbers of animals on farms (for example, ultrasonic scanners). Others are expensive and less mobile, and at present only used in research. However, such techniques may be cost-beneficial if used for second-stage selection, after a large population of animals have been screened by ultrasonics, for example. X-ray CT seems the most immediately applicable of the 'expensive' techniques. In the long term, however, NMR has the potential to provide more comprehensive information than X-ray CT, particularly if combined NMR imaging and spectroscopy become commonplace. Similarly, NAA could provide detailed chemical analysis *in vivo*. At current prices, the demand for X-ray CT and NMR imagers, and to a lesser extent ultrasonic scanners, for use in animal science is too low to warrant the adaptation by machine manufacturers of equipment specifically for use on animals. However, future development of these techniques for medical use may well have spinoffs for animal science. The greatest barriers to adoption of these techniques are the capital and running costs. Unless cheaper versions of X-ray CT or NMR images become available, then *in vivo* estimation of carcass composition in sheep breeding programmes is likely to depend on ultrasonic methods, despite their lower precision. The problem of lower precision may be reduced by testing animals under feeding regimes designed to maximize the variation in carcass composition, provided that there is no evidence of a genotype–environment interaction.

Acknowledgements

I am grateful to Drs M.A. Fuller, M.A. Foster and J.M.S. Hutchison for permission to reproduce their slide in *Figure 12.6*. I am also grateful to co-workers in the trial reported here, particularly Dr W.S. Dingwall, Professor J.W.B. King, Dr E. Avalos (ESA), Mr A. Cuthbertson, Mr M.G. Owen (MLC), Dr C.A. Miles and Mr G.A.J. Fursey (FRIB), and to the Meat and Livestock Commission who funded part of the trial. My thanks also to Drs J.D. Oldham, W.S. Dingwall, C. Smith and I.A. Wright for their helpful comments on the text.

References

ALLISTON, J.C. (1983). Evaluation of carcass quality in the live animal. In *Sheep Production*, edited by W. Haresign, pp. 75–94. London: Butterworths

BECH ANDERSEN, B. (1972). *In vivo Estimation of Body Composition in Beef*. Beretning fra Statens Husdyrbrugs Forsøg, No. 524

BLACK J.L. (1983). Growth and development of lambs. In *Sheep Production*, edited by W. Haresign, pp. 21–58. London: Butterworths

BODDY, K. (1984). Measurement of body elements and their metabolism. In *In vivo Measurement of Body Composition in Meat Animals*, edited by D. Lister, pp. 36–38. London: Elsevier

BUTTERFIELD, R.M., GRIFFITHS, D.A., THOMPSON, J.M., ZAMORA, J. and JAMES, A.M. (1983). Changes in body composition relative to weight and maturity in large and small strains of Australian Merino rams. 1. Muscle, bone and fat. *Animal Production*, **36**, 29–37

CAMERON, N.D. and SMITH, C. (1985). Estimation of carcass leanness in young rams. *Animal Production*, **40**, 303–308

CROSTON, D., GUY, D.R., JONES, D.W. and KEMPSTER, A.J. (1983). A comparison of ten sire breeds for sheep meat production. 1. Growth performance and carcass classification. *Animal Production*, **36**, 504 (abstract)

CUTHBERTSON, A., CROSTON, D. and JONES, D.W. (1984). *In vivo* estimation of lamb carcass composition and lean tissue growth rate. In *In vivo Measurement of Body Composition in Meat Animals*, edited by D. Lister, pp. 163–166. London: Elsevier

FALCONER, D.S. (1981). *Introduction to Quantitative Genetics*, 2nd edition. London: Longman

FOSTER, M.A. and HUTCHISON, J.M.S. (1985). NMR imaging—method and applications. *Journal of Biomedical Engineering*, **7**, 171–182

FOSTER, M.A., HUTCHISON, J.M.S., MALLARD, J.R. and FULLER, M. (1984). Nuclear magnetic resonance pulse sequence and discrimination of high- and low-fat tissues. *Magnetic Resonance Imaging*, **2**, 187–192

FULLER, M.F., FOSTER, M.A. and HUTCHISON, J.M.S. (1984). Nuclear magnetic resonance imaging of pigs. In *In vivo Measurement of Body Composition in Meat Animals*, edited by D. Lister, pp. 123–133. London: Elsevier

GADIAN, D.G. (1982). *Nuclear Magnetic Resonance and its Applications to Living Systems*. Oxford: Clarendon Press

GOODEN, J.M., BEACH, A.D. and PURCHAS, R.W. (1980). Measurement of subcutaneous backfat depth in live lambs with an ultrasonic probe. *New Zealand Journal of Agricultural Research*, **23**, 161–165

GROENVELD, E., KALLWEIT, E., HENNING, M. and PFAU, A. (1984). Evaluation of body composition of live animals by X-ray and nuclear magnetic resonance computed tomography. In *In vivo Measurement of Body Composition in Meat Animals*, edited by D. Lister, pp. 84–88. London: Elsevier

HERMAN, G.T. (1980). *Image Reconstructions from Projections. The Fundamentals of Computerized Tomography*. New York: Academic Press

KEMPSTER, A.J. (1983). Carcass quality and its measurement in sheep. In *Sheep Production*, edited by W. Haresign, pp. 59–74. London: Butterworths

KEMPSTER, A.J. (1984). Cost-benefit analyses of *in vivo* estimates of body composition in meat animals. In *In vivo Measurements of Body Composition in Meat Animals*, edited by D. Lister, pp. 191–203. London: Elsevier

KEMPSTER, A.J., ARNALL, D., ALLISTON, J.C. and BARKER, J.D. (1982). An evaluation of two ultrasonic machines (Scanogram and Danscanner) for predicting the body composition of live sheep. *Animal Production*, **34**, 249–255

KEMPSTER, A.J., COOK, G.L. and GRANTLEY-SMITH, M. (1986). National estimates of the body composition of British cattle, sheep and pigs with special reference to trends in fatness. *Meat Science* (in press)

KEMPSTER, A.J., CROSTON, D., GUY, D.R. and JONES, D.W. (1983). A comparison of ten sire breeds for sheep meat production. 2. Tissue growth and distribution. *Animal Production*, **36**, 504 (abstract)

KING, J.W.B. (1982). Potential use of in vivo techniques for breeding purposes. In *In vivo Estimation of Body Composition in Beef*, edited by B. Bech Anderson. Beretning fra Statens Husdyrbrugs Forsøg, No. **524**, 86–93

KNOPP, T.C. (1985). Quantitative analysis of computed tomographic images. *MSc thesis*, University of Otago, New Zealand

LISTER, D. (1984). *In vivo Measurement of Body Composition in Meat Animals*. London: Elsevier

LOEFFLER, W. and OPPELT, A. (1981). Physical principles of NMR tomography. *European Journal of Radiology*, **1**, 338–344

McCLELLAND, T.H., BONATI, B. and TAYLOR, St C.S. (1976). Breed differences in body composition of equally mature sheep. *Animal Production*, **23**, 281–293

MEAT AND LIVESTOCK COMMISSION (1972). *Sheep Improvement. Report of a Scientific Study Group*. Bletchley: Meat and Livestock Commission

MILES, C.A. and FURSEY, G.A.J. (1974). A note on the velocity of ultrasound in living tissue. *Animal Production*, **18**, 93–96

MILES, C.A., FURSEY, G.A.J. and YORK, R.W.R. (1984). New equipment for measuring the speed of ultrasound and its application in the estimation of body composition of farm livestock. In *In vivo Measurement of Body Composition in Meat Animals*, edited by D. Lister, pp. 93–105. London: Elsevier

NICOL, A.M. and PARRATT, A.C. (1984). Methods of ranking two-tooth rams for fat-free carcass growth rate. *Proceedings of the New Zealand Society of Animal Production*, **44**, 253–256

PRESTON, T., EAST, B.W. and ROBERTSON, I. (1984). Body composition measurements of rats, sheep, pigs and humans by neutron activation analysis. In *In vivo Measurement of Body Composition in Meat Animals*, edited by D. Lister, pp. 181–184. London: Elsevier

PRESTON, T., FULLER, M.F., EAST, B.W. and BRUCE, I. (1985). Preliminary experiments to assess the suitability of whole-body neutron activation for body composition analysis in 70 kg pigs. *Proceedings of the Nutrition Society*, **44**, 109A (abstract)

PURCHAS, R.W. and BEACH, A.D. (1981). Between-operator repeatability of fat depth measurements made on live sheep and lambs with an ultrasonic probe. *New Zealand Journal of Experimental Agriculture*, **9**, 213–220

ROBERTS, R.C. (1979). Side effects of selection for growth in laboratory animals. *Livestock Production Science*, **6**, 93–104

SEARLE, T.W. and GRIFFITHS, D.A. (1976a). The body composition of growing sheep during milk feeding, and the effect on composition of weaning at various body weights. *Journal of Agricultural Science, Cambridge*, **86**, 483–493

SEARLE, T.W. and GRIFFITHS, D.A. (1976b). Differences in body composition between three breeds of sheep. *Proceedings of the Australian Society of Animal Production*, **11**, 57–60

SEARLE, T.W. and GRIFFITHS, D.A. (1983). The composition of the fat-free empty body of growing sheep and the effect of weight stasis, weight loss and compensatory growth. *Journal of Agricultural Science, Cambridge*, **98**, 241–245

SEHESTED, E. (1984). Computerised tomography of sheep. In *In vivo Measurement of Body Composition in Meat Animals*, edited by D. Lister, pp. 67–74. London: Elsevier

SIMM, G. (1983a). Selection of beef cattle for efficiency of lean growth. *PhD thesis*, University of Edinburgh

SIMM, G. (1983b). The use of ultrasound to predict the carcass composition of live cattle—a review. *Animal Breeding Abstracts*, **51**, 853–875

SIMM, G., YOUNG, M.J. and BEATSON, P.R. (1987). An economic selection index for lean meat production in New Zealand sheep. *Animal Production* (in press)

SKJERVOLD, H., GRONSETH, K., VANGEN, O. and EVENSEN, A. (1981). *In vivo* estimation of body composition by computerised tomography. *Zeitschrift für Tierzüchtung und Züchtungsbiologie*, **98**, 77–79

SMITH, C. (1978). The effect of inflation and form of investment on the estimated value of genetic improvement in farm livestock. *Animal Production*, **26**, 101–110

SMITH, C. (1984). Rates of genetic change in farm livestock. *Research and Development in Agriculture*, **1**, 79–85

SMITH, C. (1986). Use of embryo transfer in genetic improvement of sheep. *Animal Production*, **42**, 81–88

WELLS, P.N.T. (1969). *Physical Principles of Ultrasonic Diagnosis*. London: Academic Press

WELLS, P.N.T. (1984). Introduction to imaging technology. In *In vivo Measurement of Body Composition in Meat Animals*, edited by D. Lister, pp. 25–32. London: Elsevier

WOLF, B.T., SMITH, C. and SALES, D.I. (1980). Growth and carcass composition in the crossbred progeny of six terminal sire breeds of sheep. *Animal Production*, **31**, 307–313

Part V

Breed development

The end result of genetic improvement research is to influence the development of new genotypes into widespread use in the sheep industry. The work of pioneers like Robert Bakewell in the 18th and early 19th century culminated in a great movement in the 19th and early 20th century towards the organization of local genotypes into the formal structure of breed societies. It is only relatively recently that new breeds, based on combining several established breeds and based on more objective recording, have been created.

Some of these developments are discussed in the present section by breeders who have been themselves closely associated with such schemes.

Another development, in some ways similar to the original establishment of breed societies, is the establishment of group breeding schemes. These are mainly developed within the parent breed societies as a more effective arrangement for the application of objective methods in sheep improvement.

Attention is also given in this section to the problem of sheep recording and of the development of simplified recording capable of being used to screen a much wider spectrum of the current sheep population.

Chapter 13

Formation of new breeds

H.R. Fell

Summary

The historical and geographical background of the British sheep industry is discussed, which explains the existence of a multiplicity of breeds and crossbreeds. From this analysis, it is clear that this presents two concentrated points where the animal breeder can work to greatest effect: the sire of the halfbred ewe, and the terminal sire.

Economic pressures are forcing change, and if sheep are to maintain their position in British agriculture, these two critically important breed groups must be improved.

In the first category, the Colbred, the ABRO Damline, and the Cambridge have come forward to challenge the current favourites, the Border Leicester and the Bluefaced Leicester. Their performance is well documented. Amongst the terminal sire group, the Suffolk is dominant. Sadly, most flocks are selected on visual assessment of good looks rather than economic factors such as speed of growth and carcass quality. The Texel was imported from France and Holland and is not a new breed. The only really new breed is the Meatlinc, bred by the author, and a full description is given of the work leading to the creation of this promising terminal sire.

Introduction

There are some 40 or so recognized pure breeds of sheep in the British Isles and they are supplemented by innumerable crossbreds and regional types. It may be wondered why, amongst a total population of some 15 million ewes, it should be necessary or desirable to have such a wide range; and why the British situation should be so different from other major sheep industries of the world such as Australia, New Zealand and South America where there is narrow specialization into very few breed types. Bearing this in mind, it may be a cause of some surprise that prolonged and expensive effort is being put into the creation of yet more new breeds.

British breeds developed over the centuries on a purely regional basis. Each breed type evolved to suit the particular climatic conditions of its own locality. Poor communications in the hill areas prevented any crossbreeding and we still see this very clearly today with breeds such as the Herdwick, the Welsh Mountain and the Derbyshire Gritstone as examples. The lowlands, however, saw a different pattern under the influence of market forces. Much of the prosperity of Britain in the Middle Ages was built on wool and although regional variation was still very evident, the market place, rather than climate, coupled with local idiosyncrasy, began to determine what selection took place.

The great agricultural revolution of the 18th century marked a turning point. England's population was increasing and urban industry was pulling people into the towns. Manorial subsistence farming had to change to supply the food needs of industry, and mutton began to overtake wool as a priority. The need produced the men and in retrospect we can see what a truly remarkable influence men such as Townsend, Coke and, in livestock breeding, Robert Bakewell of Dishley, had. The 100 years from 1750 saw a transformation of agriculture which has only been paralleled in the last 40 years. Bakewell and many others left a legacy which is still seen today, worldwide, in the influence of the Leicester and Border Leicester breeds. The purpose of these breeds was to act as improvers and their use foreshadowed an essential part of the modern British sheep industry, the stratification of breeds.

Stratification

Britain, although small in area, has considerable geographical and climatic variation—from the mountains of Scotland, Wales, and northern England where the climate is severe and food supply very limited; to the uplands where climate and soil quality are still limiting factors, and the only crop is grass; to the lowlands where conditions favour, and indeed demand, intensive production.

Not surprisingly, breeds have developed to meet these three broad bands of differing conditions. However, a system has been developed, perhaps more by accident than by design, which is unique and which interrelates these three areas and breed types to make a composite whole. This is called 'stratification', and it is of the greatest importance to the British sheep industry. It is one of the main reasons for the present high level of productivity and it provides the opportunity for further advances in that productivity. Briefly the structure is as follows.

The hills

The breeds are kept pure and the dominating influence in selection is that of hardiness needed to survive and produce on heather and poor quality pastures. The older ewes who cannot cope with these conditions are 'drafted' to areas where the conditions are somewhat kinder and where they can still produce perhaps another two crops.

The uplands

Here the draft hill ewes are not bred pure but are crossed with rams of the so-called 'longwool' breeds, such as the Border Leicester and the Bluefaced Leicester. These are breeds selected for their ability to transmit prolificacy to their female offspring—and it is these 'halfbred' ewe lambs which form the basis of lowland sheep production, and are a prime target area for genetic improvement which can affect the whole lowland sheep industry.

The lowlands

This land is fertile and productive growing both grass and a wide range of crops. It is also expensive land where the high level of costs make it essential that sheep productivity, if sheep are to have any place at all, must be at a high level. The prolific halfbred ewes, brought in regularly from the upland farms, are the basis of nearly half the British sheep industry. But she is not a carcass sheep, she has been selected for maternal qualities such as hardiness, milk and prolificacy. At this stage, she is crossed with rams from the so-called 'Down' group who are selected primarily for their meat qualities—thus putting onto the last stage of all, the butchers' lamb, the high meat qualities that the market demands. Again it should be noted that this category of terminal sire breeds provides a prime target area for genetic improvement.

This structure of the British sheep industry is highly fortuitous, providing clearly defined areas where genetic improvement can be concentrated, but yet where that improvement can be distributed effectively throughout the national flock.

It should not be imagined that this stratification was designed, nor that it happened suddenly, nor that the pattern is completely uniform. The truth is that it evolved slowly in response to market pressures and opportunities and only became really clearly defined in the post-1945 period. Nevertheless its great importance to the geneticist is that it concentrates his efforts.

Economic pressures since 1945

After long periods of depression, the Second World War provided the stimulus for a major resurgence of British agriculture, which led to statutory recognition in the historic Agriculture Act of 1947. The scene was set for the second great agricultural revolution. Suffice it to say that the input of research and development, of education, of technology has transformed the industry. Economic conditions have also changed and the stark truth is that traditional low input systems of husbandry could not survive. The sheep sector could not stand aside from these pressures, and this was, and is still, especially true in the high cost lowlands.

Sheep were traditionally recognized as efficient utilizers of forage, whether in the form of heather at the one extreme or grass leys and roots at the other. But they were not regarded as being capable of intensive production in the same way as the pig and the dairy cow. This remained as a mental barrier to improvement, and for many years sheep lagged behind other species in their adaption to the changing circumstances. Development came eventually on two fronts: with husbandry, in particular with intensive high density stocking systems allied to inwintering; and with breeding, aimed at producing improved sire breeds.

The need for new breeds

Given the structure of the British sheep industry, the most effective improvement overall can come by influencing two strategic male lines.

The longwool breeds

The main breeds in this category are the Border Leicester and the Bluefaced Leicester, the sires of the lowland halfbred ewe. The objective is to concentrate on the maternal characters of prolificacy and milk yield, without worrying too much about carcass characteristics. These two established breeds are recognized as being particularly successful so the question, 'why new breeds?', is especially pertinent.

The economic competitiveness of lowland sheep depends almost entirely upon output. That output is a combination of stocking density per hectare (which is largely a matter of husbandry) and individual ewe prolificacy (which is mostly a matter of breeding).

Any examination of performance records (MLC) shows, even with the best managed flocks, that individual productivity per ewe is stuck at a maximum of 1.75 measured in terms of lambs weaned per ewe put to the ram. Furthermore, the average performance of lowland flocks is less than 1.5. The overhead costs of maintaining the ewe over the 12 months has to be borne by this output of lambs plus a minor contribution from wool. Clearly even small increases in prolificacy will have considerable effects on the marginal profitability. If the national average could be brought nearer to 1.75 and that of the best managed flocks to 2, the effect on profitability would be dramatic.

It must be emphasized that there need to be changes in management to cope with increased ewe productivity—but it seems that improved management cannot push performance higher than current achievement without genetic change. This has been the stimulus to new breed development in this category over the past 40 years.

The Down breeds

Here the choice of breed is much wider, but it is dominated by the Suffolk. The objective is to produce a terminal sire, where all the progeny, male and female, are destined for slaughter. It should be remembered that this terminal sire will, in the main, be crossed with halfbred ewes with little claim to carcass quality. The terminal sire must transmit highly developed carcass qualities to his offspring to compensate for the quite different genetic characteristics of the halfbred ewe.

The animal breeder is therefore aiming to develop different characteristics in this category from those in the longwool breeds; it is the market place for the finished product that he must satisfy primarily. In this respect fat must be at a minimum, and flesh concentrated on the highest value cuts, at the same time improving the growth rate, which is important for the producer.

The need for new breeds in this category is not because a plateau of performance has been reached but that selection in the pedigree flocks has been based upon false criteria, on the superficial judgement of the show ring, rather than on objective assessment of performance. The result has been that the Down breeds as a whole have failed to give the housewife what she really wants—and the consequences of that failure are now to be seen in the falling per capita consumption of lamb.

New breed development

The longwool category

Four new breed types have emerged: the Colbred, the Cadzow improver, the ABRO Damline and the Cambridge. Of these, the Cadzow improver has disappeared due to financial difficulties. All have a common objective—pushing prolificacy significantly over 2 with accompanying advantages in milk yield and early maturity. All have introduced into the mix an infusion of genetic material from one of the world's highly prolific breeds, mainly the Finnish Landrace and the Romanov and to a lesser extent the East Friesland milk sheep. Both these breeds are capable of litter production of between 3 and 5 but, in their pure form, have significant disadvantages of conformation, size and temperament.

The Colbred

This was the first on the scene, indeed the first work of constructive breed creation that had been carried out for over a century. Oscar Colburn of Northleach saw the need and combined material from four breeds: the Border Leicester, the Clun Forest, the Dorset Horn and the East Friesland milk sheep. As a breeder, Colburn was undoubtedly successful, for the Colbred ram produced halfbred ewes of considerable merit, of higher prolificacy and earlier maturity. Unfortunately, the breed was acquired by Thornbers who sought to repeat in the sheep world their undoubted success with poultry. The venture was not successful, possibly because of the attempt to go too far too fast, without much regard for stockmanship. When Thornbers pulled out the breed was left in the hands of a few enthusiasts. It has taken a long time for the Colbred to get over this setback but it may now be on the way to recovery (*Table 13.1*).

TABLE 13.1 The Colbred

*Prolificacy**	Ewe lambs		110%	
	Shearling ewes		165%	
	Mature ewes		182%	
*Weight gain***	Adjusted 56 day weight in kg			
		Males		*Females*
	Singles	25.0		22.7
	Twins	22.2		20.8
	Triplets	19.8		17.7
Miscellaneous				
	Bodyweight of mature ewe		70 kg	
	Bodyweight of mature ram		100 kg	
	Wool clip		3.85 kg	

*Lambing percentage is calculated on the basis of ewes put to the tup compared with lambs sold or weaned
**Weights are based on the averages of 4 years (1978–81) MLC weighings of more than 2000 lambs

The ABRO Damline

As its name suggests this is the creation of the Animal Breeding Research Organization (part of the ARC, now AFRC). The composition is similar to the Colbred but includes Finnish Landrace as well as the Border Leicester, Dorset Horn and East Friesland milk sheep. This work was well within the terms of reference of ABRO, which includes extending the frontiers of breeding in a way that is often not possible for the individual breeder who has to earn his living in the short term.

The performance of the Damline is naturally well documented (see *Table 13.2*) and there is little doubt that the Damline had the ability to influence the production of halfbred ewes.

TABLE 13.2 The ABRO Damline: performance of 4 and 5 year olds on the commercial farm (Source: Smith *et al.*, 1979)

	Damline cross	Border Leicester cross
Number of ewes	168	205
Per cent barren or culled	7	7
Number born/ewe lambing	2.35	2.11
Number surviving/ewe lambing	2.02	1.90

However, despite its qualities this objective was not achieved and the ABRO flock has been dispersed to three private breeders since 1983. The lesson to be learned here is the reverse of that learned with the Colbred. A research organization is very well equipped to do the basic work of creation and evaluation, but not to carry out the essential next stage, that of commercialization.

The Cambridge

This is probably the most prolific of all British breeds. John Owen and Alun Davies both then at Cambridge University in the early 1960s also saw a gap in the market. Unlike many others at the time, however, they did not believe that it was necessary to go outside the British breeds to get highly prolific stock. They recognized that, within our own breeds, there were outstanding individuals whose performance was consistently above average. Was it not possible to collect together a flock of these individuals to form the basis of a new breed?

This is exactly what John Owen and Alun Davies did. They advertized in the agricultural press for ewes that were proven triplet producers, irrespective of size, shape or colour. In fact most of them came from the Clun, Llanwenog and Lleyn breeds. The next stage was to use Finnish Landrace rams, the only imported blood, as a once-and-for-all cross to produce the F_1 generation. This was followed by 20 years of rigorous culling and selection based on performance records, until now the Cambridge is an established breed.

Like the Damline, the performance of the Cambridge is well documented (*Table 13.3*). Unlike the Damline, Owen (now at Bangor) and Davies (at Liverpool) saw the dangers of restricting the breed to the academic environment and set about creating a breed society with an annual sale.

The question is whether the Cambridge will succeed where others have failed. The outcome is not yet certain for the conservative resistance, to say nothing of low

TABLE 13.3 The Cambridge: ewe performance—1968–81

Year of birth	2 Years old		3 Years old	
	Number of ewes	Litter size	Number of ewes	Litter size
1968–69	122	2.20	89	2.38
1970–71	257	2.27	225	2.62
1972–73	296	2.27	261	2.58
1974–75	354	2.51	276	2.85
1976–77	285	2.78	279	2.97
1978–79	537	2.50	372	2.85
1980–81	348	2.65	300	2.90

husbandry standards, amongst many sheep producers is a considerable barrier. The potential is certainly there to produce halfbreds out of the hill breeds that can consistently produce 200% lambs sold. Economic pressures in agriculture generally may be the potent factor in pushing up shepherding skills and management to the point where significant numbers of triplets can be reared.

The future

It almost goes without saying that breeds in this category are a vital link in the structure of our sheep industry. The question is, which breed can serve the industry best. There is tremendous potential for improvement by selection based on recording within breeds. Professor Owen has shown us what can be done by creating the Cambridge from amongst a motley collection of individuals. How much easier it ought to be within a pure breed which is already established in public favour. Sadly there are the barriers of pedigree tradition, based largely on small flocks, where the selection is based on individual preference for some visual characteristic. Performance recording is not part of this tradition and it is not surprising that there has been no progress. Fashion has its place in the world of *haute couture*, but this is a luxury that cannot be afforded in sheep improvement. Perhaps the Cambridge and the Colbred will provide the competitive stimulus to push the lowland halfbred up into the 200% lambing league, where sheep can then compete with the 10 tonne/ha wheat yield.

The Down breed category

If the Down breeds have indeed failed to produce the lamb the housewife wants, what has been done to produce something better? Currently there are really only two contenders, the Texel and the Meatlinc. Both have the same objective, that is, a genetically potent terminal sire which will transmit high carcass quality to its progeny out of commercial halfbred ewes.

The Texel

The Texel is of course not a new breed, only a relatively new import to the United Kingdom from France and Holland where it has been established for a long time. However, it warrants a brief mention because of its deserved reputation for producing very lean, well-fleshed carcasses although with the penalty of slow growth. After an intensively promoted start in the United Kingdom its popularity has declined somewhat—largely due to lack of disciplined control in its production, leading to considerable variability.

The Meatlinc

The only truly new breed in the terminal sire category, this is the creation of the author on his farm at Worlaby in South Humberside. Now a pure breed, the Meatlinc originated from a crossing programme using a base flock of British ewes onto which were put imported stock from three French breeds. The base flock of ewes selected as the starting point was a mixture—the common features were: size, since high adult weight is positively associated with fast liveweight gain; and conformation, that is, carrying plenty of flesh down the hind leg and across the loin, and doing this under plain, commercial conditions of grass feeding. Breed was of no importance, and no emphasis was placed on the colour of the head. This female base flock was mated to individual rams of different breeds, chosen not because of the reputation of their respective breed, but because as individuals they had the necessary qualities. Over the course of a number of years, individuals from the Suffolk, Dorset Down, Île de France, Berrichon du Cher, and Charollais breeds were used. In addition there was a short flirtation with the Texel, but it was quickly discarded because it slowed down growth rate and visibly shortened the sheep.

The main ingredients were size and growth rate plus high carcass quality and the aim was to find the best combination.

To do this the flock was divided into five different female families, each containing around 60 ewes, to give a total of over 300. This size is important so as to avoid running out of genetic manoeuvring room. This is the problem with very many purebred flocks—they are not big enough to select from within and the breeder is forced to go outside to buy rams with all the risks that entails. It was a fundamental principle to have a closed flock, selecting from within, once the best ingredients had been found.

The five families then were set up and they were recorded under the MLC individual ewe recording scheme. Recording is the means of identifying those animals which are superior and on which to base future breeding. The first priority was to sort out and eliminate the poor performers. This brought about a dramatic and rapid change. This is the first—and easiest—stage in the improvement of any population. The second stage—that of selecting steadily better and better individuals from within the population and spreading their influence—takes much longer. The better one gets, the slower the progress.

One of the ways by which it is possible to speed up change is by going for a rapid generation turnover—by always using ram lambs and by keeping the average age of the ewe population low. It is really essential to do this, or a lifetime will not be long enough to get anywhere. As it was, from the start in 1964, it was 12 years before a saleable product in the form of shearling rams was produced.

For many practical reasons the weight of selection pressure was put onto the male side. It was not possible to select the females too hard without running out of numbers. The procedure for doing this involved an assessment each August of that year's crop of lambs. Ram and ewe lambs from each family are penned separately, and it is one of the most interesting days of the year to have a good look at the different families, as distinct from the total population. Then it is possible to sort out the high performers and the good lookers.

Selection procedure

First of all the five pens of ram lambs are examined. Using the MLC records, all those individuals whose performance is not good enough are discarded. It needed considerable self-discipline, especially in the early days, to disregard the pleas of the shepherds to keep certain rams that they particularly liked the look of. This leaves the better performers. It is only at this stage that physical appearance and meat conformation is examined and a few more are rejected. Then out of this remaining population, all of which will be kept on for eventual sale as shearling rams (although in the intervening time, they are subject to at least three more selection checks), the few outstanding individuals who will be used for stud breeding that season are selected. Generally, the number is reduced to three, based both on performance and on looks.

The same basic procedure is followed with the ewe lambs and it is on the female side that uniformity as well as quality is sought. At the same time as looking at the ewe lambs, a close look is taken at last year's crop, the gimmers, to see how they have grown out since selection a year previously. Inevitably, some will have disappointed and must go. The same is true of the older ewes, and where rapid improvement is sought then the flock must be kept young. That is not to say that the outstanding 'matrons' should not be kept, subject to very close scrutiny indeed.

The breeding flock is thus set up for the following year. Steps must be taken to avoid getting too close in-breeding. There are five families, numbered 1 to 5. The ram lambs finally chosen for stud use from family number 1 are used on family number 2, those from number 2 are used on family number 3, and so on. The females always remain in their original family; the males are always crossed onto the next family line. The other precaution is that once any individual ewe has produced a ram lamb that is eventually selected for stud use, that ewe is never used as a stud mother again. She is kept in the flock, of course, because she is obviously an outstanding individual and she will breed both ewe lambs that can be kept as well as ram lambs that can be sold.

The objectives of the Meatlinc are:

(1) It must be a breed which, when crossed with commercial halfbred ewes, is prepotent, it must stamp its quality on its progeny.
(2) It must be quick growing, and therefore must be big.
(3) The size must, however, be combined with quality. If selection is on growth rate figures alone, there is a real danger that big, rawboned animals will result.
(4) It must be heavily fleshed, especially deeply down the hind leg.
(5) This flesh must be meat and not fat.
(6) It must produce 'finished' lambs when used with commercial ewes at a wide range of slaughter weight.
(7) The killing-out percentage must be high.

All the recording, performance figures and the visual judgements must always have these objectives clearly in mind.

Growth rate is easy to measure, but in a meat animal growth that is affected by the milking capacity of the mother should be distinguished from that growth which is clearly the genetic potential of the animal itself when fed on normal forage. Weight at birth is thus recorded, then at 56 days (which is mostly a measure of the influence of the ewe's milk yield), and finally at 110 days when a true overall picture is gained. There would be something to be said for a further weighing, a year later, at the shearling stage. But this would slow down the generation turnover.

The evaluation of carcass quality internally is much more difficult. Selected animals from each family are slaughtered to cross check on progress. But this is not positive selection. The Ultrasonic Backfat Scanner is now in use, and if the same progress with sheep can be made that has already happened with pigs, much can be achieved.

Finally, all the selected ram lambs are used on commercial ewes as a vital part of the Meatlinc selection programme. There are a thousand ordinary commercial ewes on the farm, of the same general type that the customers are likely to have, and these ewes are all crossed with that year's crop of ram lambs. This gives an annual judgement in the commercial lamb crop, year by year, of what has been done—and this judgement is backed up by an abattoir assessment.

Although it is too early to be dogmatic, early indications are that the market is giving a favourable response to the Meatlinc. Its value depends on the soundness of the breeding policy, based as it is on market quality, without the distraction of the minutiae of breed points. However, attention has been paid to developing a breed that is uniform, recognizable and repeatable.

The future

As 1947 marked a watershed in agricultural history, further dramatic developments are now taking place in agriculture. The crisis of surplus production leading to severe economic pressure, increasing emphasis on environmental protection, and a demand for 'healthy' food are all potent factors for change. *Fat* lamb is not wanted; but lean, tasty, well-presented lamb joints sell at a premium. The housewife and the controllers of the EEC budget act in an unholy alliance to put severe pressures on the British sheep industry. Yet there is great potential to make use of this most adaptable of farm animals, all of whose qualities have the outstanding ability to convert cheap forage into meat.

There will be improvements in sheep husbandry. More and more, lowland sheep will be inwintered so enabling high density stocking rates of the order of 20 ewes per hectare to become commonplace. The demand will grow for the right sheep to make use of these husbandry improvements. It is inconceivable that sheep breeds and crosses no better than they were 40 years ago can continue in use. No other branch of agriculture has stood still in this way. Even the breeding of beef cattle, as deeply traditional as sheep breeding only 20 years ago, has made great strides.

So the stimulus will be there: from the lowland sheep farmer demanding from his upland colleagues better and more highly productive halfbreds; and from the housewife, determined that her husband shall not die of heart disease and, faced with the alternative of buying cheap poultry meat, demanding lean, heavily fleshed joints ready prepared for her microwave oven. The breeder and the husbandman face a bleak future if they do not respond to these stimuli.

Reference

SMITH, C., KING, J.W.B., NICHOLSON, D., WOLF, B.T. and BAMPTON, P.R. (1979). Performance of crossbred sheep from a synthetic dam line. *Animal Production*, **29**, 1–9

Chapter 14

Group breeding schemes and simplified recording in sheep improvement

J.B. Owen

Summary

Group breeding schemes have been developed to enable several small-scale breeders to combine into an effective breeding operation, involving the screening of large numbers of females for use in a nucleus flock. Developments associated with the University College of North Wales are described and some detail given of the operation of the Lleyn group breeding scheme.

The scheme operates on a total group membership of approximately 1200 ewes, associated with an open nucleus of 150 breeding females. Emphasis in selection is placed on prolificacy and milk yield and the results indicate a level of performance of two lambs weaned per mature ewe over a 5 year period at a mean 56 day weight of 17.3 kg for twins. Computer simulation studies indicate that the structure of the scheme, involving the use of 7 month old rams only as the nucleus sires, is in the main appropriate for the type of breed and flock structure involved.

The use of simplified recording, particularly for screening extensively managed flocks, is an important aspect; it is concluded from the work of Fadel (1986, personal communication) that systems that involve no individual identification and only one visit, shortly post-weaning, is a most efficient way of ranking ewes, on the basis of milk yield, in dairy ewes of the Awassi breed. Similar applications are also noted for Welsh Mountain sheep in temperate conditions (Phillips, 1986).

Introduction

Most countries have well-adapted native sheep that can form the basis of improvement schemes, relevant to the efficient use of resources in that environment. Group breeding is a major advance in the organization of breeding schemes to promote within-breed improvement (Smith, 1976). It can also be an efficient means by which native sheep stocks may be screened for superior females, and a vehicle for the efficient incorporation of foreign genetic material (rams, semen or embryos) into adapted sheep breeds (Bichard, 1971).

157

Group breeding schemes

Group breeding schemes commonly involve the cooperative action of 10–15 breeders who may individually only possess relatively small flocks. The structure of these schemes can vary but the most common form involves the initial screening of the female population in the members' flocks to establish a nucleus. A two-way flow of genetic material between the member flocks and the nucleus is usually maintained. This is achieved mainly by the transfer of breeding rams from the nucleus flock to the members and by continuous systematic screening of females in the member flocks to augment the nucleus (James, 1977).

TABLE 14.1 The characteristics of some group breeding schemes under operation in the southern hemisphere in the early 1980s (source: C.A. Ellis, private communication)

Number of groups	18
Countries of operation	Australia (4) New Zealand (14)
Year of formation	1967–82; 1960s (5), 1970s (12), 1980s (1)
Number of members	3–25, average 12
Breed of sheep	Merino (4), Romney (9), Coopworth (5)
Nucleus size	150–1800; average 1408
Breeding females in member flocks per scheme	4500–425 000; average 58 579
Nucleus females as per cent of all females	0.5–5.6%; average 2.7%
Proportion of members' females regularly screened for nucleus	7–100% average 48%
Proportion of nucleus females contributed from member flocks (openness of nucleus)	0–80%; average 28%
Selection criteria	Wool production (100%), weaned lamb production, (100%), easy care (28%)

Table 14.1, based on data by Ellis (private communication), gives some of the details of group breeding schemes operating in the southern hemisphere in the early 1980s.

The nucleus flock may be located on the farm of one of the members or on a separate farm, possibly associated with a university or other research institution.

The main functions of the group activity in the running of the breeding scheme include:

(1) The establishment and review of breeding objectives and their incorporation in routine recording procedures; in this function they may solicit the advice and assistance of scientists and recording organizations.
(2) To establish a nucleus flock and to organize and operate agreed procedures for the flock's management, particularly for the selection of breeding stock and to deploy the selected breeding stock within agreed mating procedures.
(3) To operate a flow of selected genetic material from the member flocks to the nucleus, from the nucleus to the member flocks and from the member flocks to the breed at large.
(4) To provide a focus for communication within the group, for application of new knowledge, for trading purposes and for public relations.

The University College of North Wales, Bangor, is involved with three major group breeding schemes: the Camda scheme for the improvement of Welsh

Mountain sheep (established in 1976); the Cambridge scheme for the development of the Cambridge (established in 1969); and the Haulfryn scheme for the improvement of the Lleyn sheep breed (established in 1978).

The Lleyn sheep scheme

This scheme is described in some detail because it illustrates some important features of a highly developed and successful scheme that has had a major impact on the development of a breed where average flock sizes are relatively small.

The twelve members of the group own between them 1200 breeding ewes in flocks varying in size from 20 to 250. The flocks had been recording with the Meat and Livestock Commission for at least 2 years before the institution of the scheme with the screening of the flocks and the establishment of a nucleus flock in 1978. The nucleus consists of 100 mature breeding ewes, including twelve 3 year old ewes, one from each member, on loan annually to the nucleus from the members. In addition there are 30–40 replacement ewe lambs, breeding at 1 year old.

The nucleus is in the ownership of the cooperative group of breeder members, formally organized as a cooperative under the auspices of the Welsh Agricultural Organization Society, and was established by the donation of the foundation ewes to the group by the individual members. In addition to the initial donation of ewes, each member delivers his top ranking 3 year old ewe (as judged by the MLC index) to the nucleus in early December each year. This ewe, in lamb to a mating in the member's own flock, remains on loan in the nucleus until she has dropped her lambs and these have been weaned. The ewe is then returned and the lambs become the property of the cooperative group. Each member in return is entitled to receive a yearling breeding ram from the nucleus for his own use.

The nucleus is located on a university farm and managed by a university staff member who works closely with the group. All the costs of running the flock, including the financing of the initial establishment loan provided by the university, are reimbursed by the group from the income generated by sale of surplus stock. Some of the initial experimental costs were grant-aided by the Agricultural and Food Research Organization.

The group is organized as a committee, with its chairman and officers, who meet at 2 monthly intervals to transact business, including all necessary decision-making and the selection of stock. The group is augmented at its meetings by advisory members from various organizations.

Replacements for the nucleus are selected on the basis of an index which is mainly weighted on prolificacy and ewe milking ability. Twelve ram lambs are selected annually for use as stock sires at 7 months of age; these are subsequently wintered and made available free of charge to the members the following summer. Ewes are maintained in the twelve ram families and all ewes culled are replaced by ewe lambs, which are expected to lamb at 1 year old.

Table 14.2 summarizes the flock performance in recent years. Although the data do not confirm any marked trend over the short timescale involved, they illustrate a high level of performance achieved, in relation to ewe body size, particularly in relation to lamb survival and ewe rearing ability.

It is intended, as soon as a satisfactory level of conception is obtained from the method of storing frozen semen, that a sample of the semen of all stock rams will be retained to enable controlled comparisons to be made in estimating the rate of genetic response in the scheme.

TABLE 14.2 Production performance (56 day lamb) of ewes in the Lleyn sheep nucleus 1981–85 (excluding ewes on loan from members)

		Year of performance					
		1981	1982	1983	1984	1985	Mean 1981–85
	Mating weight (kg)	35	37	36	35	32	35.0
	Number mated	35	30	38	42	46	
	Number lambed	18	21	23	18	34	
	Per cent fertile ewes	51	70	61	43	61	57.2
1 year olds:	Lambs weaned/ewe	1.39	1.14	1.30	1.17	1.06	1.21
	Mean weaning weight (kg)						
	Singles	21.1	17.9	18.3	17.2	16.4	18.2
	Twins	15.3	12.0	14.9	13.9	15.1	14.2
	Mating weight (kg)	49	50	41	47	46	46.6
	Number mated	27	29	25	29	31	
	Number lambed	23	27	22	25	26	
	Per cent fertile ewes	85	90	88	86	84	86.6
2 year olds:	Lambs weaned/ewe	1.48	1.89	1.73	1.88	1.77	1.75
	Mean weaning weight (kg)						
	Singles	19.8	22.1	20.7	22.2	17.6	20.5
	Twins	18.6	17.1	14.9	15.8	17.0	16.7
	Triplets	—	12.6	—	10.4	—	11.5
	Mating weight (kg)	56	58	52	55	53	54.8
	Number mated	70	70	75	73	52	
	Number lambed	63	65	65	71	50	
	Per cent fertile ewes	90	93	87	97	96	92.6
3 year and older:	Lambs weaned/ewe	1.89	1.95	2.06	2.06	1.94	1.98
	Mean weaning weight (kg)						
	Singles	20.3	19.0	19.9	23.0	19.5	20.3
	Twins	17.7	17.9	16.5	16.7	17.8	17.3
	Triplets	16.1	14.0	15.4	13.4	16.8	15.1

The operation of the scheme over the last 8 year period has confirmed the organizational feasibility of operating a cooperative group breeding scheme with close financial control and the achievement of perceived benefits to the breeders involved.

Computer simulation exercises carried out by C.A. Ellis (private communication), based on the structure operated in this scheme, have been carried out to examine the optimum structure on a basis similar to that of Jackson and Turner (1972). These have confirmed the appropriateness of several features of the scheme, including the use of twelve sires and of 7 month old sires for rapid generation turnover. The simulation also suggests that the annual rate of genetic gain would increase with an increase in the size of the nucleus flock and that benefits would be gained if members minimized generation intervals in their own flocks.

The case for maintaining an 'open' nucleus is not very strong on the basis of genetic gain but there are several operational advantages, particularly with a small nucleus flock, including the opportunity to minimize inbreeding.

Preliminary assessment of the value of multiple ovulation and ova transfer, in the context of a group breeding scheme, suggest that whilst significant increases in annual genetic gain may result, it is less clear whether these increases would be cost effective.

Simplified recording

A major obstacle to the efficient improvement of sheep breeds is the difficulty of making sophisticated measurements on any but a tiny fraction of the population represented by any breed. Even in breeds of dairy cattle, which are intensively managed and handled, a similar problem exists (Maarof, 1980; Mosi, 1984). The activities of many group breeding schemes, which rely on the intensive screening of base populations, are hindered by the difficulty of intensive recording, particularly for traits such as milk production. The problem is worst in situations of extensive husbandry under difficult range conditions. Examples are the mountain conditions in temperate areas, where grazing is on unenclosed common areas and under semiarid range conditions, especially where nomadic flock husbandry is practised. However, sheep managed under such conditions are valuable sources of breeding stock because of their adaptation to those conditions and their success in surviving, reproducing and yielding useful products.

Recently Fadel (1986), in relation to the improvement of Awassi dairy sheep in Syria, has made a significant contribution to the solution of this problem.

Following a study of over 1700 Syrian Awassi ewes in the Arab Centre for the Study of Arid Zones and Dry Lands (ACSAD) at their ElKraim experimental station in mid-Syria, Fadel has demonstrated the possibility that ewes can be selected for their milk production ability on the basis of one daily record (that is, am and pm weighings) taken on one day in one flock, shortly after the lambs are weaned at approximately 2 months old. She reported a high correlation (Spearman's rank correlation coefficient) between the ranking of all the ewes in the group on their performance on the one recording day and that based on their fully recorded total milk yield (from fortnightly recording during the lactation). The correlation value of approximately 0.9 was not significantly improved by correcting for the stage of lactation in ewes where the range of lambing date in the flock was of the order of 8–10 weeks.

If these findings are applicable to other flocks in this and other areas it suggests that extensively managed, nomadic range flocks can be efficiently screened for ewes of high milk production without individual identification of sheep. Flocks would be visited by arrangement on one day shortly after weaning and milk production recorded on that day for each ewe, which would be given temporary identification. At the end of the day high performance ewes would be identified by ranking on the milk yield record (possibly corrected for any known major age effects).

Phillips (1986) has adopted a similar procedure in the Welsh Mountain sheep ram selection programme carried out at the University College of North Wales, Bangor. The programme involves a two-stage sire performance testing procedure whereby single born ram lambs are presented for testing at weaning. The lambs in any one flock are all weighed on the same day and selection based on the ranking of lambs based on body weight. Where possible, lambs born to 2 year old ewes, and those born early or late in the season, are separately identified by a paint mark so that selection does not unfairly discriminate against younger lambs and those with immature dams. Selected ram lambs then participate in a central performance test on winter grazing, ending at approximately 13 months of age.

References

BICHARD, M. (1971). Dissemination of genetic improvement through a livestock industry. *Animal Production*, **13**, 401–411

JACKSON, N. and TURNER, H.N. (1972). Optimal structure for a cooperative nucleus breeding scheme. *Proceedings of the Australian Society of Animal Production*, **9**, 55–64

JAMES, J.W. (1977). Open nucleus breeding systems. *Animal Production*, **24**, 287–305

MAAROF, N.N. (1980). The use of milk records in dairy sire evaluation with special reference to simplified recording practice. *PhD thesis*, University of Aberdeen

MOSI, R.O. (1984). The use of milk records in cow evaluation and dairy cattle improvement in Kenya. *PhD thesis*, University of Wales

PHILLIPS, C.J.C. (1986). Performance testing of Welsh Mountain rams. *Handbook of the Welsh Mountain Sheep Society*, Aberystwyth

SMITH, C. (1976). *Group Breeding Schemes*. ARC Animal Breeding Research Organization

Chapter 15

Improving fecundity in subtropical fat-tailed sheep

M.B. Aboul-Ela and A.M. Aboul-Naga

Summary

Subtropical fat-tailed sheep possess the ability to breed more than once per year. Most of these breeds are non-prolific. The few prolific subtropical breeds combine good prolificacy (>1.6 lambs per ewe lambed) with high rebreeding ability.

Fecundity of fat-tailed sheep can be improved either through increasing the number of lambings per year utilizing their rebreeding ability, or by increasing their prolificacy. Producing a lamb crop every 8 months resulted in increases of 33–49% in the number of lambs born per ewe per year. Prolificacy of fat-tailed sheep can be effectively improved through crossing with prolific breeds, preferably using the first cross with subtropical prolific breeds or using a lower contribution of temperate prolific breeds.

Crossing with subtropical prolific breeds (Chios and D'man) gave variable increases in the number of lambs born per ewe per year ranging from 11 to 45%. Crosses with temperate prolific breeds (Finn and Romanov) gave detectable increases in the number of lambs born per ewe per year. It ranged in the different trials reported, under accelerated lambing system, from 56 to 90% in the half prolific half local crosses and reached 25–30% in the quarter prolific three-quarter local crosses.

The possibilities of improving fecundity through flushing or use of hormonal treatments are also discussed.

Introduction

Fat-failed sheep are raised in many countries throughout the world but they are concentrated mainly in the subtropical region where they are raised mostly under unfavourable management and environmental conditions, including less availability and seasonal fluctuations in feed resources, heat stress and diseases which affect their productivity.

Improving fecundity, defined as the number of lambs born per unit time, in subtropical fat-tailed breeds, has two dimensions; increasing the number of lambings per year, and the number of lambs per lambing.

Developing fecundity may be achieved in different ways which are either (a) managerial (improving nutritional status, early breeding, accelerated lambing, and use of hormonal treatments), or (b) genetic (selection and crossbreeding). In this chapter some of these aspects and the possibility of their application in subtropical fat-tailed sheep are reviewed.

Fecundity of subtropical fat-tailed sheep

Age at first lambing and possibility of early breeding

Most of the subtropical fat-tailed breeds have their first lambing at the age of 17–22 months. In some of them, the prolific ones, ewe lambs give their first crop at an earlier age of 12–24 months (*Table 15.1*).

TABLE 15.1 Reproductive performance of subtropical fat-tailed sheep breeds

Breed	Country	Age at first lambing (month)	Lambing interval (month)	Litter size Yearling	Litter size Adult ewes	Reference
			Non-prolific breeds			
Ossimi	Egypt	15–22			1.17	
Rahmani	Egypt	15–22			1.21	Aboul-Naga and
Barki	Egypt	18–25			1.07	Aboul-Ela (1986)
TimHadit	Morocco	21–23	10–12		1.02–1.07	
Serdi	Morocco	20–23	10–12		1.00–1.07	
Beni-Guil	Morocco	21–23	10–12		1.04	Lahlou-Kassi (1985)
Beni-Hsen	Morocco	21–23	10–12		1.04–1.20	
Nejdi	Saudi Arabia	13–18			1.10–1.38	
Nejdi	Kuwait	22			1.01	
Harri	Saudi Arabia	17–18			1.06	
Kurdi	Iraq	24			1.04	
Hamadani	Iraq				1.10	Ghanem (1980)
Awassi	Syria	24	12		1.05	
Barbari	Libya				1.00–1.10	
Barbari	Tunisia	24			1.20	
Barbari	Algeria	14–15			1.00	
White karaman	Turkey				1.05	
Red karaman	Turkey				1.00–1.05	
Daglie	Turkey				1.00–1.02	
Kellakui	Iran				1.11	Osman (1985)
Bakhtiari	Iran				1.20	
Kizili	Iran				1.25	
Baluchi	Iran				1.16	Sefidbakht, Mostafari
Karkul	Iran		7–8			and Farid (1977)
Bibrik	Pakistan				1.00–1.05	Osman (1985)
			Prolific breeds			
Chios	Greece	>13	6–12		2.3	Mason (1980)
Chios	Cyprus				1.69	Constantinou (1985)
D'man	Morocco	12–14	6–8	1.67–2.00	1.98–2.67	Lahlou-Kassi (1985)
D'man	Morocco	11–17	6–8	1.85 –	2.40	
Javanese	Indonesia	6–12	6–9		1.6	Mason (1980)
Horro	Ethiopia				1.6	Galal *et al.* (1986)
Hu-Yang	China	13–16	6–8		>2.0	Mason (1980)

As in temperate sheep, subtropical ewe lambs fed on a high nutritional level grow faster and reach puberty at an earlier age (Younis, 1977; Younis *et al.*, 1978; Aboul-Naga, Ashmawy and Shalaby, 1984). Early mating can result, however, in low conception and twinning rates and higher lamb losses. In Awassi ewes first bred at an early age of 10–12 months, Younis *et al.* (1978) reported a conception rate of only 27%. Aboul-Naga, Ashmawy and Shalaby (1984) reported lower conception rate and higher lamb mortality from birth to weaning for Rahmani ewe lambs mated at 10 months compared to those mated at 18 months of age (58 versus 83% and 33 versus 24%, respectively). It was concluded from that study that early breeding should not be practised for early weaned subtropical lambs produced in an accelerated lambing production system. Early breeding, however, may be recommended in the case of good nutritional and management conditions which would allow ewe lambs to reach heavier body weight at an early age.

Oestrous activity, rebreeding ability and accelerated lambing

In a recent review, Aboul-Naga (1985) concluded that subtropical breeds, of which the fat-tailed constitute a major part, have a prolonged breeding season, they do not exhibit a clear anoestrous period, and they show some seasonal variation in oestrous activity with a drop in the spring months. Seasonal variation in ovarian activity of subtropical breeds seems to be less than in oestrous activity (Noorshadi, Bennett and Bunch, 1975; Sefidbakht, Mostafavi and Farid, 1977; Lahlou-Kassi and Marie, 1985; Aboul-Naga and Aboul-Ela, 1986). Among the subtropical sheep, some breeds, particularly the prolific ones, show consistent and rather regular oestrous activity throughout the year (Lahlou-Kassi (1985) for the D'man; and Lysandrides (1981) for the Chios).

There is good evidence that changes in daylength, although of small magnitude in the subtropical region, are a major factor causing changes in oestrous activity in subtropical sheep (Aboul-Naga *et al.* (1987) in Rahmani; and Mousa (1986) in Ossimi and Awassi fat-tailed breeds). Response to changes in daylength seem to be larger in the relatively more seasonal Awassi than the Ossimi sheep (Mousa, 1986).

It is a common practice in most subtropical countries to breed sheep more than once a year. In his review, Aboul-Naga (1985) concluded that the number of lambings per ewe per year generally ranged from 1.03 to 1.85 for subtropical sheep with an average of 1.3 lambings per ewe per year.

Several trials were carried out to utilize the rebreeding ability of subtropical fat-tailed sheep for increasing annual lamb production through producing a lamb crop every 8 months. The results of some of these are summarized in *Table 15.2*. The increase in number of lambs born per ewe per year in these trials ranged from 33 to 49%. In the Egyptian Ministry of Agriculture trial, reported by Aboul-Naga and Aboul-Ela (1986), an accelerated lamb production system was based on mating in September, May and January for a mating season of 35 days. September mating was characterized by a significant increase in conception and twinning rates over those in May, the common mating season in the country. In the Lebanese and Iraqi trials, better reproductive performance was also reported for autumn mating over spring and summer matings.

Body weight at weaning was not significantly affected by producing more than one crop per year in Barki (Sharafeldin, Ragab and Ramadan, 1968), Rahmani, Ossimi and Barki (Aboul-Naga, Afifi and El-Shobokshy, 1980) and Awassi (Ampy and Rottensten, 1968). Meanwhile, lamb losses in all trials were reported to

TABLE 15.2 Reproductive performance of fat-tailed breeds when bred either once per year (C/YR) or three times every two years (3C/2YR)

Breed	Production system	Number of records	Conception rate	Lambs born/ewe joined	Lambs weaned/ewe joined	Lambs born/ewe lambed	Weaned/ewe lambed (kg)
		Egyptian trial (Aboul-Naga and Aboul-Ela, 1986)					
Ossimi	C/YR	4269	0.83	0.95	0.83	1.14	22.8
	3C/2YR	3222	0.73	0.88	0.71	1.22	18.8
Rahmani	C/YR	3567	0.86	1.06	0.93	1.23	24.6
	3C/2YR	2905	0.77	1.01	0.80	1.33	22.7
Barki	C/YR	937	0.88	0.92	0.83	1.05	20.6
	3C/2YR	1023	0.71	0.76	0.66	1.07	20.4
		Lebanese trial (Ampy and Rottensten, 1968)					
Awassi	CR/YR	113		1.06	1.04		
	3C/2YR	147		1.41	1.21		
		Iraqi trial (Karam and Abu-al-Ma'Ali, 1975)					
Awassi	CR/YR	100		0.62	1.11		
	3C/YR	100		0.61	1.02		

increase in the accelerated lambing system, which indicates the importance of providing better management for the lambs produced under such systems, to minimize losses.

The increase in production costs by applying accelerated lambing was estimated to be 20% of the increase in cash return due to the improvement in lamb production (Aboul-Naga and Aboul-Ela, 1986). Accelerated lambing seems to be an effective and simple management method of increasing lamb production from subtropical fat-tailed sheep. However, before accelerated lambing is applied to a particular breed in the region, different aspects of such a production system should be investigated under the prevailing conditions.

Prolificacy

While prolificacy in most subtropical fat-tailed breeds is low (*see Table 15.1*), there are few prolific breeds with a number of more than 1.6 lambs born per ewe per lambing. Out of these, the Chios and the D'man (as a subtropical thin-tailed breed) are well known and their performance has been reported by many authors. Information on other prolific subtropical fat-tailed breeds (Hu-Yang, East Javanese and Horro) is rather scanty. More attention should be given to thoroughly studying the reproductive performance of these breeds and the possibility of utilizing them in programmes for increasing fecundity in subtropical sheep.

It is of interest to note that the subtropical prolific breeds have a lower age at first lambing in addition to their higher litter size. They also have a very long breeding season and can be bred more than once per year (*see Table 15.1*). Furthermore, some figures show that these breeds have a high conception rate. This may suggest a common basic mechanism in these breeds that is controlling prolificacy and other components of fecundity including precocity, seasonality and postpartum ovarian activity; this is an interesting finding to be regionally and internationally investigated.

Improving fecundity in subtropical fat-tailed sheep

Management

Lamb production from subtropical sheep could be increased through early breeding and accelerated lambing as discussed before. Other managerial means include the improvement of nutritional conditions, particularly prior to or during the mating season (flushing) and the use of hormones and hormone-related compounds.

Flushing

Only a few trials have been reported on the use of flushing to improve fecundity in subtropical fat-tailed sheep. Younis (1977) indicated that 30% increases in twinning appears to be the maximum response expected from any flushing treatment in Awassi sheep under practical farm conditions. In Turkey, the twinning rate of Akkaraman sheep was increased from 25 to 50% when the feeding level was increased from 593 to 994 starch units for 4 weeks prior to the mating season (Askin, Kaymakci and Isik, 1983). In Egypt, no effect on conception, twinning or lambing rate was observed when levels of 100, 125 or 150% protein in the diet were given in the form of either plant protein, fish meal or urea (El-Shobokshy *et al.*, 1982). Variation in the response to flushing in different sheep breeds may be related to various factors including: the status of the ewes prior to flushing, length of flushing period, magnitude of increase in the level of nutrients, composition of the diet and the level of both energy and protein (Smith, 1985).

Hormonal treatments

Use of hormones to control fertility in subtropical fat-tailed sheep has been restricted to few trials involving small number of ewes. Vaginal sponges have been tried successfully for oestrous synchronization in Ossimi and Rahmani ewes in Egypt (Aboul-Naga and Abdel-Rahman, 1981) and in Turkish breeds (Askin, Kaymakci and Isik, 1983). Commercial application of oestrous synchronization in subtropical sheep, however, has been very limited, probably due to the tendency of these breeds to have continuous oestrous activity throughout the year, and the relative costs of the materials involved.

The use of pregnant mare serum gonadotrophin (PMSG) at a dose of 400–500 IU per ewe resulted in an increase of 25% in litter size in Kivircik sheep (Askin, Kaymakci and Isik, 1983) and 42% in Ossimi and Rahmani sheep (Aboul-Naga and Abdel-Rahman, 1981). In the latter trial, the higher litter size recorded was mainly due to the increases in the percentage of quadruplets. The use of PMSG at doses higher than 500 IU (authors' unpublished data) resulted in a marked drop in both conception and twinning rates.

To our knowledge, there is no report on the use of immunization against gonadal steroids for increasing twinning in subtropical fat-tailed sheep. Preliminary trials on the immunization against androstenedione in Rahmani sheep (authors' unpublished data) indicate only small increases of less than 10% in litter size resulting from immunization.

Genetic improvement

Within-breed selection

Selection for prolificacy in subtropical fat-tailed breeds has not received much attention. Improvement of prolificacy in some flocks, however, was practised indirectly with the selection for increase in other quantitative traits, for example, body size and milk production as in the case of Awassi sheep (Epstein, 1977). Twinning was claimed to have increased from 5% in unimproved Awassi flocks in the 1930s to about 30% in the highly improved flocks raised during the 1970s in Israel. In a state flock of D'man sheep, prolificacy was reported to have increased from 150 to 216% in yearling ewes (<2 years old) and from 176 to 231% in >2 year old ewes as a result of selection over the period from 1972 to 1980 (Ouarzazate Regional Office of Agricultural Development, Morocco, 1981).

In 1974, a trial was started for selection for litter size in Rahmani ewes either on their own performance (having produced one twin at least in the first 2 years of production) or on their dam performance (born as twins or triplets) versus a control group. The preliminary results (authors' unpublished data) showed some improvement in the litter size of the second generation of the second selected group (124.3, 128.4 versus 110.8 for the control).

Some evidence has been found of correlation between regular cycling activity of Rahmani ewes and their general reproductive performance (Aboul-Naga and Aboul-Ela, 1986). Such a relationship has also been postulated in the D'man prolific breed (Lahlou-Kassi and Marie, 1985). A trial has been established this year to select within a state Rahmani flock for those ewes having regular oestrous activity in the spring (May–June), and to study its effect on their ovarian activity and general reproductive performance.

Crossbreeding for improving fecundity

With the expected slow progress in the prolificacy of subtropical sheep by selection within breeds, crossing with prolific breeds has been tried in several subtropical countries. Various factors determine the choice of the prolific breed used and the degree of crossing. These include:

(1) seasonality in the crossbred ewes produced—they should be comparable with the local breeds in their ability to produce more than one crop per year;
(2) adaptability of the crossbred ewes to the prevailing harsh conditions, climatic, nutritional and disease; and
(3) appraisal characteristics of the ewes and lambs which may affect their preference in the market.

Size and shape of the fat-tail are decisive marketing characteristics for consumers in most of the Arab countries. Some prolific breeds may have other shortcomings which may limit their potential for use in crossbreeding, for example, the coloured wool of Romanov and the small mature size of the D'man.

The crossbreeding trials involving subtropical fat-tailed sheep fall into two categories: first is the crossbreeding with subtropical prolific breeds (Chios and D'man), and second is crossbreeding with temperate prolific breeds (Finn and Romanov).

Crossbreeding the Chios with Awassi sheep in Lebanon (Ampy and Rottensten, 1968) resulted in a substantial increase of 45% in the number of lambs born per ewe joined (*Table 15.3*). Much smaller increases (11–15%), however, have been recorded when the Chios was crossed with the Awassi in Cyprus (Constantinou, 1985; Mavrogenis, 1985) or with the Kivircik in Turkey (Askin, Kaymakci and Isik, 1983).

TABLE 15.3 Reproductive performance of subtropical sheep breeds and their crosses with subtropical prolific breeds

Breed group	Number of records	Lambs born/ewe joined	Lambs born/ewe lambed	Number of lambings/ year	Number of lambs born/year
D'man trial, Morocco (Lahlou-Kassi, personal communication)					
D'man (D)	400	1.80		1.7	3.06
TimHadit (T)	600	1.05		1.2	1.26
D × T	50	1.30		1.4	1.82
Chios trial, Lebanon (Ampy and Rottensten, 1968)					
Chios (C)	69	1.96		1.12	2.19
Awassi (A)	1034	1.02		1.12	1.14
C × A	304	1.48		1.12	1.66
Chios trial, Turkey (Askin, Kaymakci and Isik, 1983)					
Kivircik		0.96	1.00		
C × K		1.00	1.11		
Chios trial, Cyprus (Constantinou, 1985)					
Chios (C)	1371	1.53	1.96		
Awassi (A)	9.8	0.96	1.11		
C × A and A × C	274	1.22	1.34		
Chios trial, Cyprus (Mavrogenis, 1985)					
Chios (C)	1415	1.69			
Awassi (A)	632	1.07			
C × A	69	1.17			
A × C	90	1.28			

In Morocco, crossbreeding the local TimHadit with the D'man resulted in an increase of 24% in the number of lambs born per ewe joined. Such an advantage was augmented further by the increase in the number of lambings per year resulting in an overall increase of 44% in the number of lambs born per ewe per year (*see Table 15.3*).

Trials for crossbreeding fat-tailed breeds with temperate prolific breeds (Finn and Romanov) have been comprehensively reviewed by Aboul-Naga (1985). In Egypt, under the accelerated lambing system, crossbreds between Finn (F) and each of the Ossimi and Rahmani (Aboul-Naga, 1985) had a higher conception rate, more frequent lambing per ewe per year and a higher litter size which resulted in overall increases of 31 and 25% in the number of lambs born per ewe per year in quarter Finn, threequarter Ossimi, and quarter Finn, three-quarter Rahmani, respectively, than the corresponding local breeds. More recent figures (Aboul-Naga and Mansour, unpublished data) indicate that the number of lambs born per year was increased by 0.16 and 0.14 in the quarter Finn, three-quarter Ossimi, and quarter Finn, three-quarter Rahmani crossbreds, respectively. The quarter Finn, three-quarter local seems to be a suitable degree of crossing under the prevailing conditions in Egypt. Validation of the crossbreeding programme with the breeders is being carried out in two Delta provinces. More recently, Romanov sheep were imported from France and were crossed with Rahmani sheep to produce quarter Romanov, three-quarter Rahmani crossbred ewes as a suitable degree of crossing. Only the first-cross ewes produced were mated, for two seasons. In their first mating season (January 1985) their litter size averaged 1.16 versus 1.04 for Rahmani (authors' unpublished data). Detectable improvement in the crossbred ewes was obtained in their second mating season (September 1985) where their litter size averaged 1.77 versus 1.39 for the local Rahmani.

In the Israeli trial (Goot *et al.*, 1980), Finn/Awassi and Romanov/Awassi crossbreds had an advantage over the Awassi in terms of earlier age at first lambing and larger litter size. This was augmented further under the accelerated lambing system, where Finn/Awassi and Romanov/Awassi crossbreds produced 69 and 90% more lambs per ewe per year than the Awassi. Despite the small number of Romanov crosses reported in that study, the results indicate that crosses with the Romanov were even more advantageous than with the Finn, and part of this was due to the higher ability to produce more than one crop per year.

Conclusions

(1) Fecundity of subtropical fat-tailed sheep can be improved in two ways: (a) utilize their rebreeding potential to produce more than one lamb crop per year; (b) increase their prolificacy through crossbreeding, using either a low contribution of temperate prolific breeds or preferably using subtropical prolific breeds.
(2) The prevailing environmental, managerial and economic conditions may influence the way in which improved fecundity in subtropical sheep is achieved. Under optimal conditions accelerated lambing could feasibly be combined with increased prolificacy. Under harsh conditions, as in some of the arid subtropical areas, increased prolificacy may be undesirable. Any proposed system for improving fecundity in subtropical fat-tailed sheep should be validated and economically evaluated under the field conditions of the breeders before being implemented on a large national scale.
(3) Attention should be given, through national or international networks, to investigate the unique potentiality of subtropical prolific breeds in combining good prolificacy with high rebreeding ability.

References

ABOUL-NAGA, A.M. (1985). Crossbreeding for fecundity in subtropical sheep. In *Genetics of Reproduction in Sheep*, edited by R.B. Land and D.W. Robinson, pp. 55–62. London: Butterworths

ABOUL-NAGA, A.M. and ABDEL-RAHMAN, H. (1981). Hormonal control of fertility in Ossimi and Rahmani ewes. *Manoufia Journal of Agricultural Research*, **16**, 136–143

ABOUL-NAGA, A.M. and ABOUL-ELA, M.B. (1986). Performance of subtropical Egyptian sheep breeds, European breeds and their crosses. I. Egyptian sheep breeds. *World Review of Animal Production*

ABOUL-NAGA, A.M., ABOUL-ELA, M.B., EL-NAKHLA, S.M. and MEHREZ, A.Z. (1987). Oestrous and ovarian activity of subtropical fat-tailed Rahmani sheep and their response to light treatment. *Journal of Agricultural Science, Cambridge* (in press)

ABOUL-NAGA, A.M., AFIFI, E.A. and EL-SHOBOKSHY, A.S. (1980). Early weaning of Rahmani, Ossimi and Barki local lambs. *Egyptian Journal of Animal Production*, **20**, 137–146

ABOUL-NAGA, A.M., ASHMAWY, G. and SHALABY, T.H. (1987). Early breeding of subtropical ewe lambs. *International Goat and Sheep Research* (in press)

AMPY, F.R. and ROTTENSTEN, K.V. (1968). Fertility in the Awassi sheep. 1. Seasonal influence on fertility. *Tropical Agriculture, Trinidad*, **45**, 191–197

ASKIN, Y., KAYMAKCI, M. and ISIK, N. (1983). Studies in Turkey on increasing reproductive performance in sheep and goats. *Proceedings of International Symposium on Production of Sheep and Goats in the Mediterranean Area*, pp. 91–100. Ankara: Semih Ofset Matbaacilik

CONSTANTINOU, A. (1985). Ruminant livestock genetic resources in Cyprus. In *Animal Genetic Resources Information*, no. 4, pp. 1–8. FAO/UNEP Publication

EL-SHOBOKSHY, A.S., ABOUL-NAGA, A.M., SWEDAN, F.Z., SAAD, N.R. and MARAI, I.F.M. (1982). Increased productivity of subtropical ewes. In *Animal Production in the Tropics*, edited by M. Yousef, pp. 212–220. New York: Praeger Publishers

EPSTEIN, H. (1977). The Awassi sheep in Israel. *World Review of Animal Production*, **13**, 19–26

GALAL, E.S.E., GOJJAM, Y., TIYU, U. and WOLDGABRIEL, K. (1986). Horro sheep: development, reproductive efficiency and productivity in Ethiopia. *International Goat and Sheep Research*

GHANEM, Y.S. (1980). *Arabian Sheep Breeds*. Publication of the Arab Organization for Education, Culture and Sciences (in Arabic)

GOOT, H., FOOTE, W.C., EYAL, E. and FOLMAN, Y. (1980). *Crossbreeding to Increase Meat Production from the Native Awassi Sheep*. Bet Dogan, Israel: Division of Scientific Publication, The Volcani Centre, Special Publication, no. 175.

KARAM, H.A. and ABU-AL-MA'ALI, H.N. (1975). A preliminary study on three lambings vs. two in two years in Awassi sheep in Iraq. *UNEP/FAO Technical Report*, no. 4.

LAHLOU-KASSI, A. (1985). Review of the Moroccan sheep breeds. In *Animal Genetic Resources in Africa: High Potential and Endangered Livestock*, pp. 103–110. Nairobi: OAU/STRC/IBAR Publication

LAHLOU-KASSI, A. and MARIE, M. (1985). Sexual and ovarian function of the D'man ewe. In *Genetics of Reproduction in Sheep*, edited by R.B. Land and D.W. Robinson, pp. 245–260. London: Butterworths

LYSANDRIDES, P. (1981). The Chios sheep in Cyprus. *World Animal Review*, **39**, 12–16

MASON, I.L. (1980). *Prolific Tropical Sheep*. FAO Animal Production and Health paper, no. 17

MAVROGENIS, A.P. (1985). The fecundity of the Chios sheep. In *Genetics of Reproduction in Sheep*, edited by R.B. Land and D.W. Robinson, pp. 63–68. London: Butterworths

MOUSA, T.M. (1986). Studies in reproduction activity of fat-tailed ewes under upper Egypt conditions. *MSc thesis*, Assuit University, Egypt

NOORSHADI, S., BENNETT, J.A. and BUNCH, T.D. (1975). Estrus and ovulation in fat-tailed sheep. *Journal of Animal Science*, **41**, 282 (abstract)

OSMAN, A.H. (1985). Near East: sheep breeding and improvement. *World Animal Review*, **54**, 2–15

OUARZAZATE REGIONAL OFFICE OF AGRICULTURAL DEVELOPMENT, MOROCCO (1981). *D'man Sheep in Ouarzazate region*

SEFIDBAKHT, N., MOSTAFAVI, M.S. and FARID, A. (1977). Effect of season of lambing on postpartum ovulation, conception and follicular development of four fat-tailed Iranian breeds of sheep. *Journal of Animal Science*, **45**, 305–310

SHARAFELDIN, M.A., RAGAB, M.T. and RAMADAN, I.A. (1968). Effect of rebreeding on lamb production. *Journal of Agricultural Science, Cambridge*, **71**, 351–354

SMITH, J.F. (1985). Protein, energy and ovulation rate. In *Genetics of Reproduction in Sheep*, edited by R.B. Land and D.W. Robinson, pp. 349–359. London: Butterworths

YOUNIS, A.A. (1977). Increasing ewe fertility in Arab countries. *World Review of Animal Production*, **13**, 31–36

YOUNIS, A.A., EL-GABOORY, I.A., EL-TAWIL, E.A. and EL-SHOBOKSHY, A.S. (1978). Age at puberty and the possibility of early breeding in Awassi ewes. *Journal of Agricultural Science, Cambridge*, **90**, 255–260

Part VI

Feeding systems

Improving the capability of the sheep must go hand in hand with adequate feeding systems. Again this can vary at the one extreme from the intensive feeding of housed sheep, entirely dependent on the shepherd, to the supplementation of flocks mainly dependent on traditional range grazing.

In the present section new developments in the intensive feeding of ewes and lambs, applicable to various parts of the world are discussed. In particular, diet form and protein nutrition are important aspects, not only in relation to growth and efficiency but also to the quality of the final carcass.

Supplementation of range pasture and the supply of forage when range pasture is not available, are major problems in the utilization of semiarid ranges. New work shows how the well-tried traditional systems can be adapted and improved to cope with the changing conditions of sheep husbandry.

Chapter 16

Nutrition of housed sheep

J.J. Robinson

Summary

The use of sheep as convenient small ruminants for fundamental studies on digestive physiology and substrate utilization together with direct observations on their nutrient requirements for maintenance, pregnancy and lactation, has provided us with a comprehensive framework for the formulation of optimal feeding strategies for the breeding ewe. What this chapter demonstrates is that rarely do these feeding strategies involve meeting the exact nutrient requirements of the ewe at each stage of her reproductive cycle. Instead, for practical, economic and indeed sound physiological reasons they involve periods in which nutrient intakes exceed requirements and others where they are below requirements, the overall aim being to achieve balance in the composition of the body over the breeding cycle as a whole.

Failure to achieve the correct body condition at mating leads to suboptimal ovulation rates. Equally damaging to overall reproductive performance are extremes of nutrition in the first month of pregnancy in that they are detrimental to embryo survival. During the second and third months of pregnancy, a period when the nutrient requirements for conceptus growth are minimal, there is now evidence that a mild degree of undernutrition is beneficial to placental growth and lamb birthweight in ewes that are in good body condition at mating. Recent advances in our understanding of the interrelationships between energy and protein and their effects on fetal growth and milk production have allowed greater flexibility in diet formulation and have provided us with a means of efficiently using body fat reserves at a time when energy requirements invariably exceed intake. Due to the fact that for most sheep production systems, housing is intermittent rather than continuous, few attempts have been made to develop an overall feeding strategy for continuously housed ewes. This chapter contains a description of the diets, feeding regimes and levels of reproductive performance achieved by a continuously housed flock of highly prolific ewes maintained on a 7 month lambing interval.

Introduction

The use of sheep as convenient small ruminants for fundamental studies on digestive physiology and substrate utilization, together with direct observations on their response to alterations in nutrient supply in a range of production systems, has provided us with a comprehensive catalogue of their nutrient requirements (see, for example, INRA, 1978; Agricultural Research Council (ARC), 1980, 1984; National Research Council (NRC), 1985). At the same time attempts have been made to incorporate observations on the influence of the numerous factors that affect nutrient requirements into computer models (Black, 1984; Neal, France and Treacher, 1986) for use in formulating optimal feeding strategies at the individual farm level. It is not the purpose in this chapter to use the approach of computer modelling of nutrient requirements to illustrate how the technique can be used to arrive at an optimal feeding strategy for housed sheep; such an exercise would be premature. Rather it is to compare some of the currently recommended nutrient allowances, discuss the nutritional principles that are important in the feeding of the pregnant and lactating ewe, give sample diets and feeding regimes for different physiological states, and draw attention to nutritional problems that tend to be more prevalent in housed sheep.

Recommended nutrient allowances

Energy

Estimates of the metabolizable energy (ME) requirements of housed sheep for maintenance were computed from observations on fasting metabolism and presented by the Agricultural Research Council (1980). Similarly, estimates for pregnancy and lactation were calculated from the net accretion of energy in the

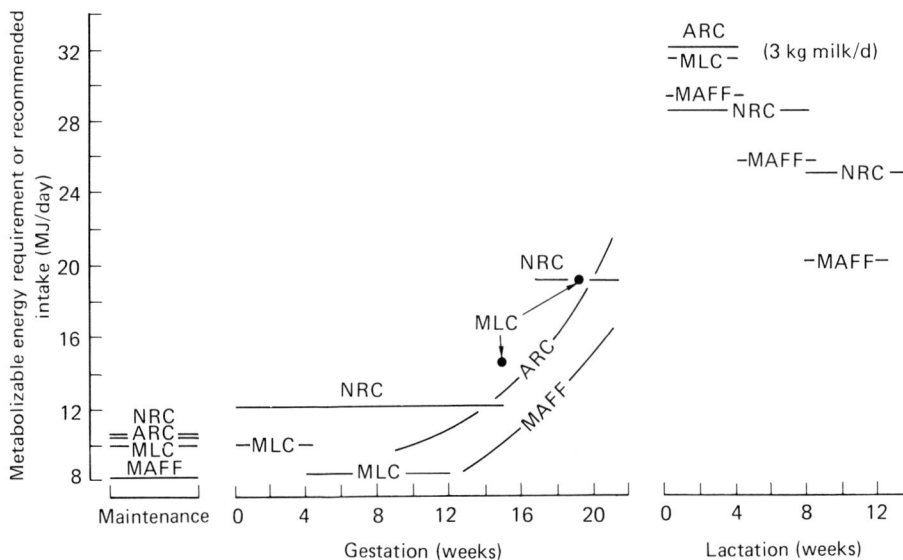

Figure 16.1 Estimates of daily metabolizable energy requirements (ARC, 1980) or recommended allowances (MLC, 1981; MAFF, DAFS, DANI, 1984 and NRC, 1985) for a 75 kg ewe with twin lambs. For this and *Figure 16.2* requirements for ewes of different weights and litter sizes are readily calculated from the data presented by ARC. Recommended allowances which cover a range of weights and litter sizes are given in the original MAFF, NRC and MLC publications

products of conception and milk respectively. In *Figure 16.1*, these estimates are plotted together with the recommended allowances for metabolizable energy given by the Meat and Livestock Commission (MLC, 1981), the Ministry of Agriculture Fisheries and Food (MAFF, 1984), and the National Research Council (NRC, 1985). The most noteworthy feature of the information given in *Figure 16.1* is the extent to which the recommended pattern of energy intake differs from the actual pattern of requirement. In some instances the pattern of intake is dictated by practical considerations. An example of this is the stepwise increase in recommended allowances for late pregnancy (NRC, 1985) compared with the steady rise in actual requirements. In other instances the pattern is governed by a combination of practical and economic factors. Here an example is the utilization of body fat reserves during early lactation when the energy requirements of prolific ewes exceed those supplied by the voluntary consumption of all but the highest quality diets. These reserves are normally replenished towards the end of lactation when milk yield declines and/or in the period leading up to rebreeding. In yet other instances there are beneficial effects of not adhering exactly to theoretical requirements. For example, due to the small demand for energy that the products of conception place on the ewe during the second and third months of pregnancy it is usual to recommend restrictions of feed intake to maintenance or indeed to levels that result in losses of up to 5% of live weight for ewes that were in good body condition (score 3–3.5) at mating (MLC, 1981). It was largely on economic grounds that this feeding strategy was first recommended although it had been observed (Robinson, 1977), that high-plane feeding in mid-pregnancy could cause dramatic reductions in birthweight. More recently Russel *et al.* (1981) have clarified the interacting effects of ewe weight at mating and plane of nutrition in mid-pregnancy on lamb birthweight with low-plane feeding of light ewes and high-plant feeding of heavy ewes both having detrimental effects on size at birth (*Table 16.1*).

TABLE 16.1 The effect in light and heavy primiparous Scottish Blackface ewes of feeding to maintain (high) or lose 5–6 kg maternal body weight (low) between days 30 and 98 of pregnancy (source: Russel *et al.*, 1981)

Mating weight (kg)	Feeding levels from 36 to 98 days of pregnancy	Lamb birthweight (low feeding level as per cent of high)
43	High versus low	86.7
55	High versus low	117.0

The detrimental effect of the low plane of nutrition during the second and third months of pregnancy on lamb birthweight in light ewes (low body condition) is understandable in terms of the extensive observations which demonstrate that severe restrictions in feed intake during this period, corresponding as they do with the time when the placenta is growing, can have a damaging effect on the growth of this organ (Robinson, 1983a). While the latter observations can be used to explain the mode of action on fetal growth of severe undernutrition in mid-pregnancy, they do not accommodate the data of Russel *et al.* (1981) in which a less severe degree of undernutrition in mid-pregnancy increased the birthweight of lambs from ewes that were in good body condition at mating (*see Table 16.1*). In this context the findings of Faichney (1981) are pertinent in that they show that a mild degree of undernutrition in mid-pregnancy can stimulate placental growth and as a consequence enhance birthweight (*Table 16.2*).

TABLE 16.2 The effects of plane of nutrition [maintenance (M) versus 0.6 M] in mid-pregnancy on placental and fetal weights on day 135 of pregnancy (from Faichney, 1981)

Plane of nutrition		Weights (kg)	
50–100 days of gestation	100–135 days of gestation	Placenta	Fetus
M	M	0.321	3.3
0.6	M	0.463	3.7

The preceding examples draw attention to the fact that the formulation of practical feeding regimes for the breeding ewe is not solely about meeting the exact energy requirements of the ewe at all times, rather the aim is to produce balance in the composition of the body over the breeding cycle as a whole.

Protein

In contrast to the ability of the ewe, particularly during early lactation, to draw on body fat reserves, when the intake of dietary energy fails to meet her needs, there is little scope for sustaining production by drawing on body protein. In comparative slaughter studies Cowan *et al.* (1979) showed that lactating ewes could lose up to 7 kg of body fat during a 4 week period in early lactation when energy intake was below requirements. For ewes on a low protein intake the maximum daily loss of protein was 26 g (Cowan *et al.*, 1980). With this principle in mind it is of interest to

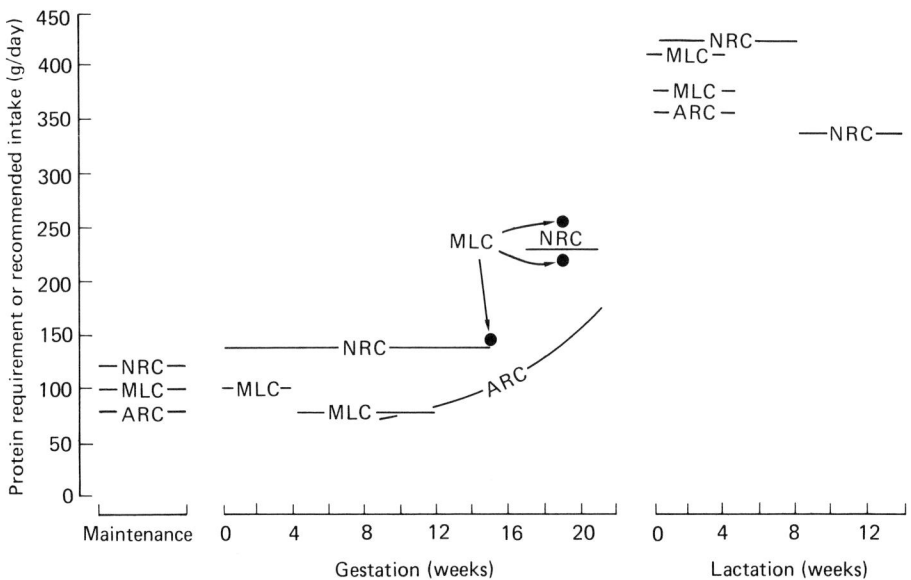

Figure 16.2 Estimates of the daily protein requirements (rumen degradable protein and undegraded dietary protein, ARC, 1980) or recommended dietary allowances (MLC, 1981; NRC, 1985) for a 75 kg ewe with twin lambs. The higher of the two protein allowances given by MLC for late pregnancy and early lactation refer to dietary protein supplement of high degradability and the lower to one of low degradability

compare estimates of protein needs with recommended dietary allowances. This is done in *Figure 16.2* using a similar format to that already used for energy.

A cursory glance at the data presented in *Figure 16.2* tends to confirm the principle that it is important to meet the protein needs of the ewe at all times in that recommended dietary allowances appear always to be higher than the ARC (1980) estimates of requirement. This, however, is an oversimplified view and is not the sole reason for the difference between estimated requirements and dietary allowances. First, in the light of more recent observations, in particular those for the nitrogen excretion of sheep maintained on intragastric infusions of nitrogen-free nutrients (Ørskov, MacLeod and Grubb, 1980), the ARC increased its estimates of maintenance protein requirements (ARC, 1984). Secondly in neither publication did it take into consideration the protein needs for udder development and colostrum production. Compared with the ARC (1980), the use of the ARC (1984) estimate of protein needs for tissue maintenance, together with estimates of the net rates of accretion of protein in the udder during late pregnancy, was shown by Robinson (1983b) to increase the total daily protein needs (that is, the rumen-degradable protein (RDP) + undegraded dietary protein, UDP) of a 75 kg ewe with twin lambs from 100 to 119 g in week 15 of gestation and from 175 to 205 g just prior to parturition. Finally, the ARC estimates, based as they are on the important principle of distinguishing between the needs of the rumen microorganisms for rumen-degradable protein and of the host animal for additional undegraded dietary protein when RDP fails to meet those requirements, represents the minimal protein needs of the animal. In practice, dietary allowances for late pregnancy and early lactation are invariably higher than the sum of RDP and UDP. The reason for this is that apart from nitrogen sources such as urea, which is completely degraded in the rumen, or heat-treated blood meal, which is completely undegraded (Gonzalez *et al.*, 1979), most other protein sources provide a combination of both RDP and UDP. Thus, when used to supply additional UDP the majority of protein sources inevitably provide an excess of RDP and consequently a higher dietary allowance of protein than the theoretical minimum given by ARC.

Minerals and vitamins

In addition to the detailed estimates for the requirements of minerals and vitamins given by the ARC (1980) there are numerous other recent publications (see for example Church, 1984; NRC, 1985) which deal with requirements, or in the case of the mineral elements, with approaches which ensure that mineral deficiencies are prevented at least cost to the farmer (Suttle, 1983). Similarly, conflicts between the various dietary allowances for calcium and phosphorus which stem from disagreement between data sources on absorbability for calcium and faecal endogenous loss for phosphorus have been dealt with by Scott (1986). In view of all this detailed and up-to-date information there is little point in dealing further with mineral and vitamin needs other than to point out that despite the high rate of accretion of calcium in the fetus during late pregnancy (0.25 g/kg birthweight; Robinson, 1983b) protein undernutrition caused a much greater reduction in lamb birthweight than a reduction in the intake of dietary calcium to less than 50% of the net accretion rate of this element in the fetus (see Robinson, 1983a).

Nutritional principles—their application in formulating feeding strategies

Attention has already been drawn to the fact that practical feeding regimes for the pregnant and lactating ewe have to take into consideration the extent to which body fat reserves can be used for fetal growth and milk production. It was with this in mind that the concept of acceptable targets for body condition at different stages of the reproductive cycle was developed (Russel, Gunn and Doney, 1968) and has proved so valuable in formulating feeding strategies (Robinson, 1986).

Assessment of body condition

This involves equating certain physical characteristics that are readily identifiable in ewes of different degrees of fatness with a particular score. For the fine-tailed sheep breeds of Britain and northern Europe this assessment is made in the lumbar region, immediately behind the last rib, and at mating the aim is to achieve a condition score of 3.0 to 3.5. At this body condition the spinous processes of the vertebrae are only detectable as small elevations, they are smooth and rounded and individual bones can only be felt with pressure. The transverse processes are smooth and well covered and firm pressure is required to feel over the ends. This corresponds to about 30% fat in the fleece-free empty body. For the fat-tailed breeds which inhabit the countries that border on the Mediterranean a modified form of subjective assessment of body condition has been devised (Hossamo, Owen and Farid, 1986).

Change in body condition during pregnancy and lactation

Following on from a body condition score of 3.0 to 3.5, which ensures maximum ovulation rate, there is now ample evidence to indicate that the optimal feeding strategy for the first month of pregnancy is one of maintenance in that it minimizes embryo mortality. For ewes with a body condition score of 3.5 at mating it is desirable to allow them to steadily lose up to 5% of their body weight, or approximately 0.5 of a condition score during the second and third months of pregnancy. As pointed out earlier (p.177), this mild degree of undernutrition enhances placental growth and in so doing establishes the basis for maximum fetal growth in the fourth and fifth months of pregnancy, this being the period when the fetus achieves over 80% of its growth. During these final 2 months of pregnancy there is a limit to the extent to which body fat reserves can be utilized as excessive mobilization of depot fats leads to pregnancy toxaemia. In contrast, as already alluded to, early lactation is a period in which body fat can be safely used to meet some of the high energy demands of lactation. During this period a loss in body condition score of 1.0 (equivalent to 5 kg of fat for a 70 kg ewe at mating) is quite acceptable. The replacement of this body fat prior to the next breeding cycle is important in achieving maximum ovulation rate and optimal reproductive performance subsequently.

Mid-pregnancy
For ewes in good body condition at mating, the period from 30 to 90 days of pregnancy is a time when they can obtain adequate intakes of energy from the *ad libitum* feeding of low quality roughages that have energy contents of 7–8 MJ of

metabolizable energy/kg dry matter. It is very important, however, that the roughage contains enough nitrogen to ensure that the maximum synthesis of microbial protein in the rumen is achieved. If this is not the case voluntary intake will decline and the ewes will be exposed not only to an unacceptable energy deficit but also to a specific protein deficiency.

A useful 'rule of thumb' for the minimum amount of crude protein that is needed in roughage feeds in order to meet the requirements of the rumen microbes for nitrogen is 10 g/MJ of metabolizable energy. This is based on the reasonable assumption that for most basal feeds (excluding the protein supplements) 80% of their protein is degraded to ammonia in the rumen, and this in turn allows maximal synthesis of microbial protein (around 8 g/MJ of metabolizable energy; *Figure 16.3*). If, for example, ewes are being maintained on a roughage containing 7.0 MJ of metabolizable energy/kg dry matter then it should contain about 70 g/kg of total protein, and 80% of this value (56 g) of rumen-degradable protein (RDP), in order to meet the nitrogen needs of the rumen microbes. In practice, the roughage might

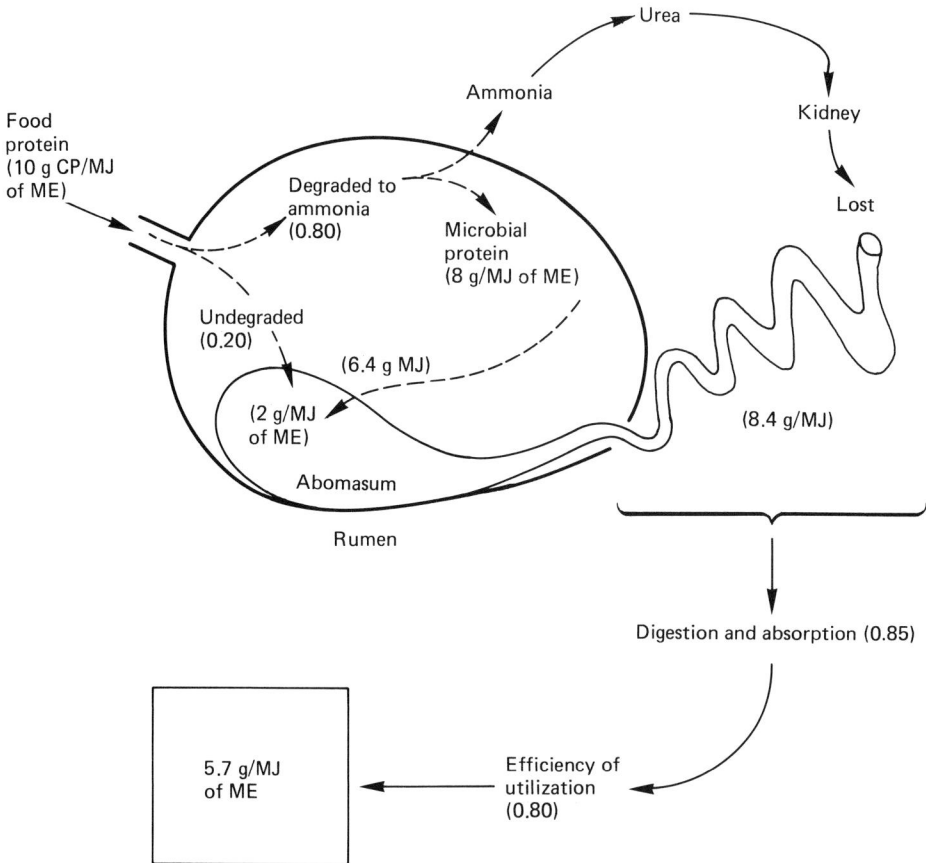

Figure 16.3 Factors used for basal feeds in estimating protein degradation in the rumen, the amounts of microbial and undegraded dietary protein reaching the abomasum, their subsequent digestion, absorption and utilization together with an estimate of net protein synthesis (5.7 g/MJ of metabolizable energy, ME) for a basal diet containing the minimum of crude protein (10 g/MJ ME) for maximal microbial protein synthesis

only contain 5.5% crude protein—that is 55 g/kg dry matter which would give a RDP content of 55 × 0.8 or 44 g/kg dry matter instead of the required 56 g. For this situation the deficit of 12 g (equivalent to 12 ÷ 6.25 or 1.92 g of nitrogen) could be corrected by adding urea. With urea containing 45% nitrogen, and an efficiency of conversion of urea nitrogen to microbial nitrogen of 80%, the amount of urea required would be 1.92 ÷ (0.45 × 0.8) = 5.3 g/kg dry matter or 0.53%. This can be easily incorporated on to the roughage by spraying with a 50 : 50 w/v solution of urea. In view of the requirement of the rumen bacteria for sulphur the benefit of a non-protein–nitrogen (NPN) supplement such as urea is often only obtained if sulphur is also added, the appropriate rate of inclusion being 1 part of sulphur for every 14 parts of nitrogen from the NPN source. Suitable sources of sulphur are sodium and ammonium sulphate. Care should be taken to avoid an excess of sulphur as it can reduce the availability of copper and may lead to copper deficiency (neonatal ataxia or swayback) in newborn lambs.

While the protein requirements for conceptus growth in mid-pregnancy are small, this should not be taken to imply that a protein deficiency which is sufficient to cause a reduction in the weights of the individual components of the conceptus (placenta, fetus) at 90 days of gestation is unlikely to occur. Indeed the numerous observations on the reduction in conceptus size as a result of very low planes of nutrition in mid-pregnancy (Robinson, 1984) could well arise from a specific protein deficiency, and some support for this view comes from the observations of Lippert, Milne and Russel (1983). The reason for suspecting a specific protein effect arises from the fact that when ewes are given basal roughages as their only source of feed and these roughages contain amounts of protein that are only adequate to meet the RDP requirements of the rumen microbes, then the net production of protein is 5.7 g/MJ of metabolizable energy (see *Figure 16.3*). This is a value which, at maintenance energy intake, is virtually the same as the sum of the net protein requirements for tissue maintenance and wool production (Robinson, 1983a). It is thus clear that under these conditions an energy deficit is inevitably accompanied by a protein deficit. To prevent this protein deficit in ewes that are in negative energy balance it is essential to supplement the roughage with a source of undegradable dietary protein such as fish meal or heat-treated soya bean meal.

Late pregnancy
The rapid growth of the fetus after 90 days of gestation and the corresponding increase in energy requirements (see *Figure 16.1*) impose a progressive limitation on the use of poor quality roughages as the sole feed. This is particularly true for ewes carrying twins and triplets. When roughage is of poor quality it is tempting to feed larger quantities of concentrates but this is by no means ideal. Large quantities of concentrates, particularly if they are given in a single daily feed, can lead to a rapid fall in the pH of the rumen contents, a reduction in the numbers of bacteria that digest fibre, and a decline in overall food intake; this in turn can trigger off pregnancy toxaemia. This sequence of detrimental effects is further accentuated when the cereal portion of the concentrates is finely milled.

An alternative approach to excessive concentrate usage is to use the information that a modest energy deficit in late pregnancy is quite acceptable provided it is not accompanied by a deficit in protein intake (Robinson, 1983a). The manipulation of this interrelationship between energy and protein provides the farmer and his feed compounder with an alternative to the high levels of concentrate feeding that can be detrimental to optimum rumen function. Taking, in the first instance, the

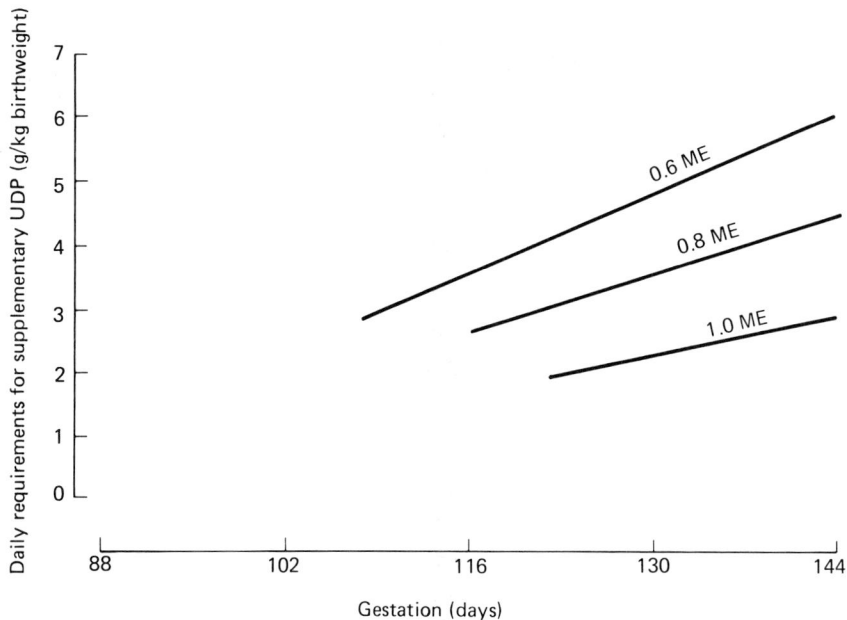

Figure 16.4 The daily amounts of supplementary undegraded dietary protein (UDP) required by ewes receiving in their diet either all their energy needs for pregnancy (1.0 metabolizable energy, ME), 80% of their needs (0.8 ME) or 60% of their needs (0.6 ME)

situation in which ewes in late pregnancy are meeting their entire energy needs from their diet, then the protein produced from microbial synthesis (equivalent to about 8 g/MJ of metabolizable energy; see *Figure 16.3*) will supply enough amino acids to meet the requirements for fetal growth up to 3 weeks before lambing. Thereafter an additional supply of undegraded dietary protein (UDP) is required. If we now take the situation in which their diet fails to supply enough energy to meet the demands for fetal growth, then the requirements for a supplement of undegraded dietary protein is not only increased but occurs earlier in gestation. This principle is illustrated in *Figure 16.4*. It is this interrelationship between energy and protein, albeit operating over a limited range of energy deficits because of the dangers of excessive fat mobilization and pregnancy toxaemia, that makes high-protein 'balancers' for on-farm mixing with whole cereal grains so attractive; they add flexibility to diet formulation in that their inclusion rate can be varied depending on energy intake and the specific needs of the flock.

Lactation
The ewe suckling two lambs, each growing at 300 g/day is as productive as the dairy cow yielding 30 kg/day of milk. To prevent loss of her body tissue, daily intakes of over 30 MJ of metabolizable energy (thrice maintenance) in the case of a 75 kg ewe with twins, are required (see *Figure 16.1*). In practice these are seldom achieved. Since three-quarters of the 30% by weight of fat in the fleece-free empty body at mating should still be present at lambing (that is, if the preceding principles of nutrition during pregnancy have been adhered to), up to half of this can be used to meet the deficit between metabolizable energy intake and the energy needs for milk

Protein escaping
from the rumen
undegraded (%)

Approximate growth rate for each of twin lambs (g/day)

Milk yield (kg/day)

Blood meal   100

                              69
         Fish meal

                              55
Linseed meal

Soyabean meal
                              42
                              60
Meat and bone
meal

                              40

Groundnut meal

                              0

Urea

Daily intake of crude protein (N × 6.25)

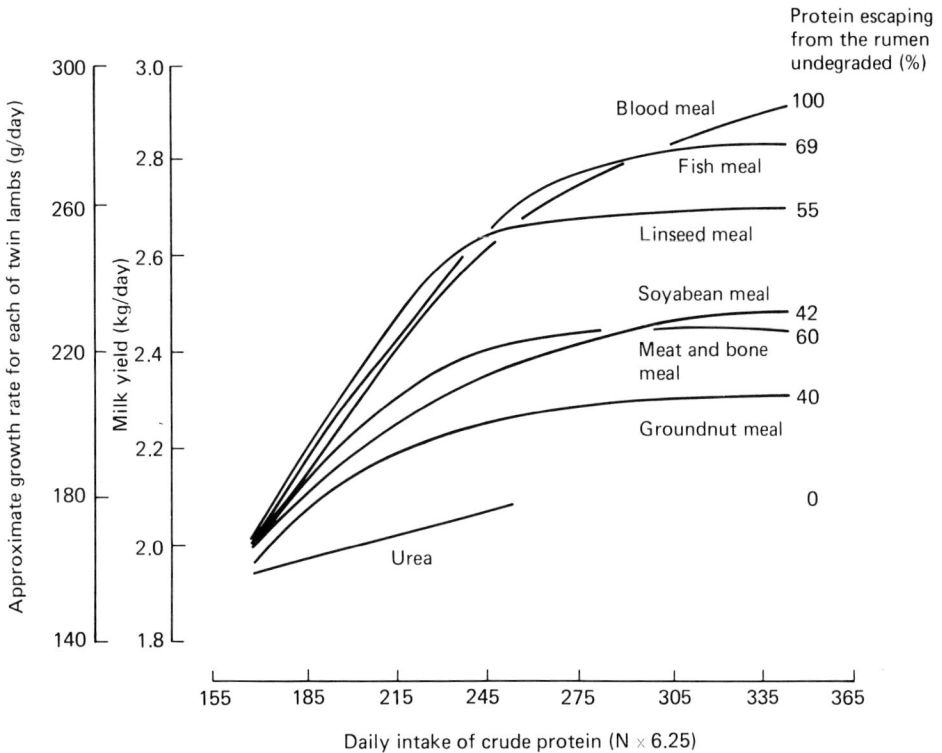

Figure 16.5 Milk yield responses of ewes of 65 kg live weight when given a basal diet of hay and barley alone (H/B), the basal diet supplemented with urea or with six protein sources. Daily intake of metabolizable energy = 18.3 MJ

production. This fat can only be used efficiently for milk synthesis if the ewe is absorbing adequate amounts of amino acids from her diet (Cowan *et al.*, 1981). It is for this reason that the ranking of the response curves for milk yield in *Figure 16.5* is in broad agreement with the extent to which each of the protein sources escapes from the rumen undegraded.

Diets and feeding regimes for continuously housed sheep

Diets

Although the practice of keeping sheep indoors is gaining in popularity, as far as the breeding ewe is concerned housing for part of the reproductive cycle is more common. Under these conditions the period of housing usually corresponds with low availability of natural grazings and high nutrient requirements by the ewe. Examples of continuously housed flocks are very few. Ainsworth *et al.* (1985) give a general description of the indoor intensive husbandry of fecund sheep maintained on an 8 month lambing interval in Canada. The diet that is used contains 50% alfalfa silage, 30% hay, 15% corn silage and 5% concentrates; 5 weeks before parturition the proportion of concentrate is increased to 18% and 3 weeks later to 30%.

At the Rowett Research Institute a continuously housed flock of highly prolific Finnish Landrace × Dorset Horn ewes have been maintained on artificial photoperiods and bred at 7 month intervals (Robinson, Fraser and McHattie, 1975). Complete diets have been used throughout. These have varied in their constituents from 40% barley straw with the remainder made up of cereal grains, high-protein balancer, molasses, minerals and vitamins to a similar diet in which the barley straw was replaced with a high-quality dried grass; the straw-based diet was used from weaning, through mating and during the entire pregnancy, and the dried-grass diet during lactation. The straw-based diet contains per kg: 8 MJ of metabolizable energy, 130 g crude protein, 8 g calcium (Ca), 7.2 g phosphorus (P), 1.7 g magnesium (Mg), 12 g sodium (Na) and 13 g potassium (K). Vitamins A, D_3 and E are included at 10×10^6, 2×10^6 and 10.5×10^3 IUs respectively per tonne. More recently a single complete diet containing a 50 : 50 mixture of milled hay and concentrates has been fed throughout pregnancy and lactation with levels of intake and protein inclusion rates being adjusted for each physiological state in accordance with the nutritional principles discussed earlier.

Feeding regimes

Using the types of diet described in the preceding section, and the adoption of feeding regimes during late pregnancy and early lactation that were some 15–20% above (high) or below (low) those given in *Table 16.3*, Robinson and Ørskov (1975)

TABLE 16.3 Recommended daily intakes of metabolizable energy (ME) for Finnish Landrace × Dorset Horn ewes kept indoors and maintained on a 7 month lambing interval

Stage of the breeding cycle	Litter size			
	1	*2*	*3*	*4*
Weaning to end of first month of gestation			15.0	
30–90 Days of gestation			9.4	
90 Days of gestation to parturition	12.4	14.5	15.8	16.6
Lactation (1 month)	18.8	24.5	26.2	—

obtained the long-term patterns of weight change shown in *Figure 16.6* for continuously housed ewes subjected to a 7 month breeding interval. The observations illustrated in *Figure 16.6* demonstrate two important points: first, that it is possible to devise feeding regimes which maintain the mating weights of prolific ewes over a succession of breeding cycles, and by so doing sustain an exceptionally high level of lamb production; second, as the data in *Figure 16.6* for ewes on the low plane of feeding demonstrate, inadequate levels of feeding lead to a steady decline in mating weight and an eventual failure to conceive.

Nutritional problems specific to housed sheep

In over 15 years experience with continuously housed breeding ewes at the Rowett Research Institute it is abundantly clear that their nutritional problems are far fewer than those for ewes kept outdoors. As with all systems of sheep production increases in feed intake, in particular those involving readily fermentable feeds, must be gradual to avoid acidosis. With regard to specific dietary ingredients it is

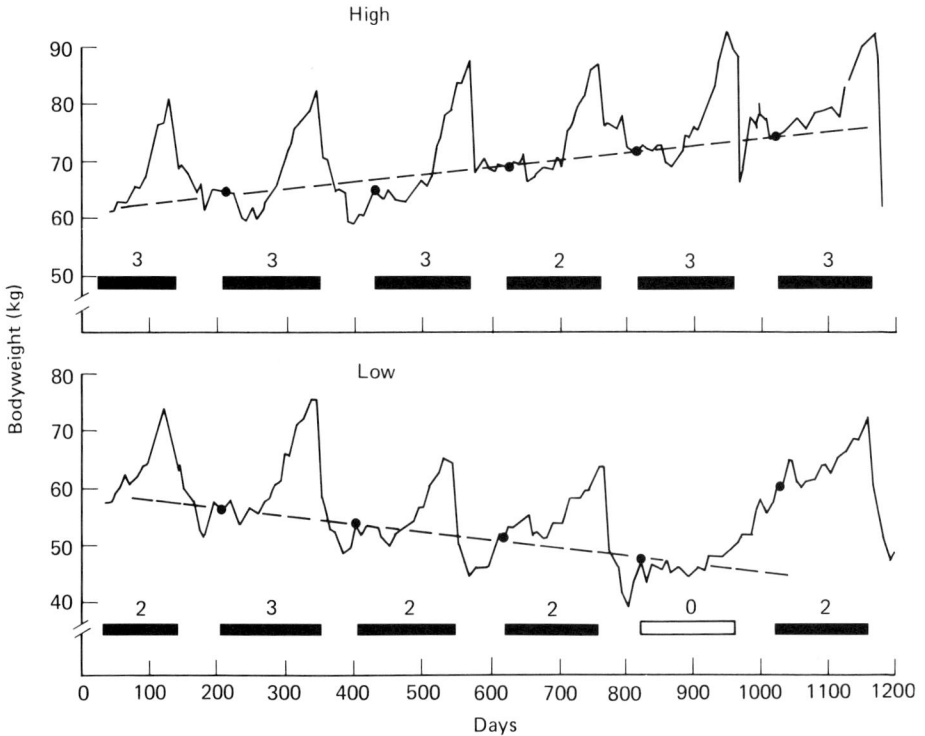

Figure 16.6 The effect of continuously providing energy intakes in late pregnancy and lactation some 15–20% above and below those suggested in *Table 16.3* on the long-term changes in bodyweight of Finnish Landrace × Dorset Horn ewes maintained on a reduced breeding interval of 7 months. Shaded areas and the figures corresponding to them refer to pregnancy and numbers of lambs respectively. ●, Mating weights

essential to monitor the copper content of all diets in order to prevent copper poisoning, suitable copper concentrations in the overall diet being 10–12 ppm.

To those familiar with formulating dietary supplements for sheep at pasture, in particular lactating ewes, the dietary magnesium concentration of 0.17% given earlier for housed sheep will appear very low. This low concentration is deliberate, not solely because the availability of magnesium is much higher on dry feed than grazed herbage thus removing the risk of hypomagnesaemia indoors, but rather to avoid urolithiasis in male lambs and breeding rams that have access to the ewe's feed.

Conclusions

In recent years enormous progress has been made in understanding the principles of food digestion and utilization by sheep and in defining their nutrient requirements for maintenance, pregnancy and lactation. Many of the observations have been made on sheep kept indoors and are therefore directly applicable to housed sheep. What the present paper has attempted to do is to demonstrate the importance of moulding together the individual research findings so as to achieve optimal feeding strategies for each stage of the reproductive cycle. Rarely do these strategies involve meeting the exact nutrient requirements of the ewe at any particular stage of her reproductive cycle. Instead, for practical, economic and indeed sound physiological reasons they involve periods in which nutrient intakes exceed requirements and others where they are below requirements; the overall aim is one of achieving balance in the composition of the body over the breeding cycle as a whole.

References

AGRICULTURAL RESEARCH COUNCIL (ARC) (1980). *The Nutrient Requirements of Ruminant Livestock*. Slough: Commonwealth Agricultural Bureaux

AGRICULTURAL RESEARCH COUNCIL (ARC) (1984). *The Nutrient Requirements of Ruminant Livestock*, **Supplement no. 1**. Slough: Commonwealth Agricultural Bureaux

AINSWORTH, L., FISER, P.S., HEANEY, D.P., LANGFORD, G.A. and SHRESTHA, J.N.B. (1985). Intensive husbandry of fecund sheep. In *Genetics of Reproduction in Sheep*, edited by R.B. Land and D.W. Robinson, pp. 399–409. London: Butterworths

BLACK, J.L. (1984). The integration of data for prediction of feed intake, nutrient requirements and animal performance. In *Herbivore Nutrition in the Subtropics and Tropics*, edited by F.M.C. Gilchrist and R.I. Mackie, pp. 648–671. Craighall, South Africa: The Science Press

CHURCH, D.C. (1984). *Livestock Feeds and Feeding*, 2nd edition. Corvallis, Oregon: O and B Books, Inc.

COWAN, R.T., ROBINSON, J.J., GREENHALGH, J.F.D. and McHATTIE, I. (1979). Body composition changes in lactating ewes estimated by serial slaughter and deuterium dilution. *Animal Production*, **9**, 81–90

COWAN, R.T., ROBINSON, J.J., McDONALD, I. and SMART, R.I. (1980). Effects of body fatness at lambing and diet in lactation on body tissue loss, feed intake and milk yield of ewes in early lactation. *Journal of Agricultural Science, Cambridge*, **95**, 497–514

COWAN, R.T., ROBINSON, J.J., McHATTIE, I. and PENNIE, K. (1981). Effects of protein concentration in the diet on milk yield, change in body composition and the efficiency of utilization of body tissue for milk production in ewes. *Animal Production*, **33**, 111–120

FAICHNEY, G.J. (1981). Amino acid utilization by the fetal lamb. *Proceedings of the Nutrition Society of Australia*, **6**, 48–53

GONZALEZ, J.S., ROBINSON, J.J., McHATTIE, I. and MEHREZ, A.Z. (1979). The use of lactating ewes in evaluating protein sources for ruminants. *Proceedings of the Nutrition Society*, **38**, 145

HOSSAMO, H.E., OWEN, J.B. and FARID, M.F.A. (1986). Body condition score and production in fat-tailed Awassi sheep under range conditions. *Research and Development in Agriculture*, **3**, 99–104

INRA (1978). *Alimentation des Ruminants*. Versailles: INRA Publications

LIPPERT, M., MILNE, J.A. and RUSSEL, A.J.F. (1983). Effect of mid-pregnancy supplementation on hill-ewe performance. *Animal Production*, **36**, 543

MEAT AND LIVESTOCK COMMISSION (MLC) (1981). *Feeding the Ewe*, revised edition. Bletchley: Meat and Livestock Commission

MINISTRY OF AGRICULTURE FISHERIES AND FOOD (MAFF), DEPARTMENT OF AGRICULTURE AND FISHERIES FOR SCOTLAND (DAFS), DEPARTMENT OF AGRICULTURE FOR NORTHERN IRELAND (DANI) (1984). *Energy Allowances and Feeding Systems for Ruminants*, Reference Book 433. London: HMSO

NATIONAL RESEARCH COUNCIL (NRC) (1985). *Nutrient Requirements of Domestic Animals: Nutrient Requirements of Sheep*, 6th revised edition. Washington, DC: National Academy Press

NEAL, H.D.St.C., FRANCE, J. and TREACHER, T.T. (1986). Using goal programming in formulating rations for pregnant ewes. *Animal Production*, **42**, 97–104

ØRSKOV, E.R., MacLEOD, N.A. and GRUBB, D.A. (1980). New concepts of N metabolism in ruminants. In *Proceedings of the Third European Association for Animal Production*, edited by H.J. Oslage and K. Rohr, pp. 451–457. EAAP Publication, no. 27

ROBINSON, J.J. (1977). The influence of maternal nutrition on ovine foetal growth. *Proceedings of the Nutrition Society*, **36**, 9–16

ROBINSON, J.J. (1983a). Nutrition of the pregnant ewe. In *Sheep Production*, edited by W. Haresign, pp. 111–131. London: Butterworths

ROBINSON, J.J. (1983b). Nutrient requirement of the breeding ewe. In *Recent Advances in Animal Nutrition*, edited by W. Haresign, pp. 143–161. London: Butterworths

ROBINSON, J.J. (1984). Ewe nutrition. In *Livestock Feeds and Feeding*, 2nd edition, edited by D.C. Church, pp. 318–338. Corvallis, Oregon: O and B Books, Inc.

ROBINSON, J.J. (1986). Feeding regimes for ewes in late pregnancy and early lactation. In *Science and Quality Lamb Production*, edited by J. Hardcastle, pp. 8–9. London: Agricultural and Food Research Council

ROBINSON, J.J., FRASER, C. and McHATTIE, I. (1975). The use of progestagens and photoperiodism in improving the reproductive rate of the ewe. *Annales de Biologie Animale, Biochimie Biophysique*, **15**, 345–352

ROBINSON, J.J. and ØRSKOV, E.R. (1975). An integrated approach to improving the biological efficiency of sheep meat production. *World Review of Animal Production*, **11**, 63–76

RUSSEL, A.J.F., FOOT, J.Z., WHITE, I.R. and DAVIES, G.J. (1981). The effect on weight at mating and of nutrition during mid-pregnancy on the birth weight of lambs from primiparous ewes. *Journal of Agricultural Science, Cambridge*, **97**, 723–729

RUSSEL, A.J.F., GUNN, R.G. and DONEY, J.M. (1968). Components of weight loss in pregnant hill ewes during winter. *Animal Production*, **10**, 43–51

SCOTT, D. (1986). Formulation of feeding strategies for sheep. In *Feedingstuffs Evaluation, Modern Aspects—Problems—Future Trends*, edited by R.M. Livingstone, pp. 76–92. Aberdeen: Rowett Research Institute

SUTTLE, N.F. (1983). Meeting the mineral requirements of sheep. In *Sheep Production*, edited by W. Haresign, pp. 167–183. London: Butterworths

Chapter 17

Early weaning and fattening of lambs

E.R. Ørskov

Summary

The recent knowledge on fattening of lambs is discussed with emphasis on feed processing, protein requirement and manipulation of body composition.

Processing of concentrate diets should be limited only to that required to ensure maximal digestion. Feeding of whole cereals is usually preferable to feeding of processed grain. Excessive processing of grain leads to problems of rumenitis, acidosis and poor carcass quality.

The need for protein in addition to that derived from microbial protein, in the fattening of lambs, depends to a large extent on previous nutrition. Store lambs, having experienced a period of low level or undernutrition, have a greater requirement for undegraded dietary protein (UDP) than well-fed animals at the same live weight.

Recent observations have shown that it is possible to utilize stored body fat as a source of energy from which to grow or maintain weight, provided a source of undegraded protein is given. This has practical applications when lambs are overfat at the time of marketing and for extending time of slaughter into the dry season in areas of fluctuating nutrient supply.

Introduction

During the past decades some progress has been made in our understanding of early weaning and fattening of lambs. Systems of early weaning have been developed which have particular importance for intensive systems of frequent breeding or in systems where ewes are milked for the sale of milk for processing. Aspects of artificial rearing are not discussed in this chapter but a few comments are made about management of early weaning. Special attention is given to aspects of protein requirement of lambs and processing of cereals, and to the manipulation of body composition in lambs.

Management of early weaning systems

In practice lambs should be at least 4 weeks old before they are weaned from the dam or from an artificial rearing system. While lambs do not normally eat dry food before they are 2–3 weeks of age, it is useful for the lambs to have access to creep feed as well as to have the possibility of nibbling feeds with the ewes. The creep feed should ideally consist of the same type of feed as they are going to be given during the fattening period. If the lambs are to be housed and fed on concentrates, they must have clean feed available with no possibilities for soiling with their feet or with faeces, they must have access to clean water and be bedded with clean straw or kept on wire floors or slats. Soiled straw bedding can lead to problems of coccidiosis.

Protein requirement

Early weaned lambs
Andrews and Ørskov (1970a, 1970b) published a series of papers dealing with protein requirement of early weaned lambs. It was concluded from this work that male animals required more protein than females due to a higher protein content in the carcasses at similar weight. It was also shown that the protein-to-energy ratio required in the feed decreased with increasing stage of maturity. Again this was to be expected, since the protein : energy ratio in the carcasses altered as the lambs increased in weight.

It was initially noticed that the higher the feeding level, the greater the response to protein. However, the latter conclusion must be questioned in light of the more recent observations. It is now known, and indeed illustrated in earlier work

TABLE 17.1 Degradability of several protein supplements at different outflow rates; degradation rates determined with roughage-fed animals (from Ørskov, 1982)

Protein source	Degradability (%) at outflow rate (k)		
	0.02	0.05	0.08
White fishmeal (unknown origin)	64.3	49.6	41.5
White fishmeal (unknown origin)	72.6	58.5	52.4
Fishmeal stale at processing	61.7	51.6	47.7
Fishmeal freshly processed	22.7	22.0	21.5
Processed and preserved fish press cake	18.0	13.9	11.3
Meat and bone meal	52.1	45.4	41.2
Cottonseed meal	80.6	69.6	62.7
Linseed meal	78.1	58.9	46.0
Soyabean meal	80.8	62.5	50.4
Groundnut meal	87.4	74.1	64.3
Sunflower meal	85.9	76.9	70.4
Guar meal	82.5	66.7	56.3
Sugar beet pulp	63.5	50.3	45.3
Ground peas	89.4	80.0	74.4
Beans (*Vicia faba*)	82.9	66.6	56.2
Spring rape	86.5	78.1	72.2
Autumn rape	88.7	78.6	71.5
Turnip rape	90.2	79.9	73.0

(Ørskov and Fraser, 1973), that the amount of protein degraded differed with different levels of feeding due to differences in rumen retention time. This has now also been accepted by the Agricultural Research Council (ARC) (1984) in so far that different outflow or different retention times of protein supplement are used for different levels of production (*Table 17.1*; *see also* Eliman and Ørskov, 1985).

The conclusion reached by many authors including Andrews and Ørskov (1970a) that protein requirement increases with increasing level of feeding to the animals may well be due to differences in effective degradability. It is in fact possible that the concentration of protein in the diet should be greater at low level feeding due to increased degradability at low level feeding.

There is, however, no doubt that protein requirement depends on the protein content of the tissues laid down, and therefore the results of differences between sexes in protein need and differences due to stage of maturity are to be expected. In the work of Andrews and Ørskov (1970a), the optimal concentration in barley based diets varied from 20% at 15 kg live weight to about 12% at 40 kg live weight. Changing concentrations during fattening of early weaned lambs is, however, not practicable, because a constant crude protein concentration of 15% throughout the fattening period could not be distinguished from a decreasing concentration from 20 to 12%. Due to differences in mature weight of different breeds and crosses, it is not possible to relate protein requirement to live weight without defining the breed cross or sex involved.

Late weaned lambs

While estimating the protein requirement for early weaned lambs may be difficult, the estimation of that of the late weaned lambs is even more difficult. The reason for this is that both the age of the lamb and its previous nutrition will influence its protein requirement, as well as its sex and breed.

During a period of energy undernutrition, the microbial protein produced is not sufficient to meet the nitrogen required for tissue maintenance. This is illustrated in *Table 17.2* indicating that when energy intake is below the amount required for energy balance, microbial protein alone becomes more and more inadequate to meet the need.

As a result of energy undernutrition, the lambs will lose protein tissue. It must be understood that the longer the period of energy undernutrition, and the more severe the undernutrition, the more depleted for protein the animal will be, thus

TABLE 17.2 The effect of energy intake on adequacy of microbial protein for tissue maintenance (see text for details)

Energy input kJ/kg$W^{0.75}$	225	350	450
Microbial nitrogen produced MgN/kg$W^{0.75}$	299	465	598
Microbial amino acids MgN/kg$W^{0.75}$	239	372	478
Absorbed amino acids MgN/kg$W^{0.75}$	203	316	406
Net supply of amino acid MgN/kg$W^{0.75}$	162	253	324

Tissue maintenance 300–400 mg net amino acid N/kg$W^{0.75}$, energy maintenance 450 kJ/kg$W^{0.75}$

TABLE 17.3 Effect of urea and fish meal on food conversion ration (FCR), live weight gain (LWG) and dry-matter intake (DMI) by lambs from 40 to 55 kg live weight, having experienced a period of undernutrition (Ørskov and Grubb, 1979)

	Level of urea (g/kg diet)	*0*	*12*	*SE*
	Level of fish meal (g/kg diet)	*80*	*0*	*of means*
	Period			
LWG (g/day)	0–14 days	452	276	22
	14 days end	328	350	11
FCR	0–14 days	3.89	5.68	0.29
	14 days end	5.86	4.84	0.59
DMI (kg/day)	0–14 days	1.73	1.52	0.06
	14 days end	1.78	1.67	0.09

the greater will be the repletion when the energy restriction is lifted. This is illustrated in an experiment (*Table 17.3*) in which store lambs having received a low level of nutrition were given diets with and without a supplement of fish meal. During the first 2 weeks, the animals were replenishing protein stores and responded positively to fish meal; this was not observed with early weaned lambs of similar live weight.

Another aspect is that of compensatory growth, which occurs where animals have been prevented from reaching their target weight for a long period. Lambs will grow fast when the restriction is lifted and it is possible for them to achieve heavy carcass weight without excessive fatness. Animals which have been kept at energy maintenance during autumn and winter months or dry periods, will reach higher carcass weight without being overfat than lambs fed intensively from birth. Such lambs have also a higher protein requirement than intensively fed lambs at similar live weight.

In practice, this means that previously undernourished lambs or lambs which have been kept at maintenance energy for a long period will normally respond to supplements of rumen undegraded proteins, such as fish meal.

Manipulation of body composition

In recent experiments using intragastric nutrition with lambs, steers and dairy cows it was demonstrated (Ørskov *et al.*, 1983; Hovell *et al.*, 1983) that all species when given only their required protein with no external sources of energy, attained protein balance. In other words, protein given to fasting lambs was not oxidized but used efficiently as a source of amino acids.

In order to test the practical implications of these observations, fat store lambs were given straw diets at different levels with or without 75 g/day of fish meal. The results are summarized in *Table 17.4* in which the increment in fat and protein deposition for the different treatments have been calculated from comparative slaughter.

It can be seen that it was possible for a loss in body fat to occur and yet to achieve increases in protein deposition and in empty body weight. This observation is quite contrary to common belief that protein deposition would always be associated with fat deposition.

TABLE 17.4 Effect of fishmeal supplements to straw diets for lambs on body gain and composition (from Fattet *et al.*, 1984)

Treatment	Intake kJ/kgW$^{0.75}$	Live weight gain (kg)	Empty body gain (kg)	Body fat (kg)	Body protein (kg)
Low straw	285	−4.32	−5.05	−3.53	−4.87
Low straw + fishmeal	307	0.29	0.64	−1.53	0.48
High straw	466	0.08	−0.08	−1.40	−0.14
High straw + fishmeal	488	6.22	4.18	−0.90	0.89

Practical implications

Overfat lambs
If lambs reaching marked weight are likely to be rejected due to overfatness, or indeed have been rejected in the live auction market from failure to achieve grading standards, then farmers are faced with problems of decreasing body fat without excessive loss of carcass weight. If the lambs are undernourished with energy and receiving microbial protein only they will lose body protein and thus carcass weight. The results in *Table 17.4* show that it is possible to decrease body fat and yet achieve an increase in body weight by using a level of energy undernutrition, but allowing a small supplement of undegraded protein. In other words, the excessive fat deposited can be considered as a source of energy for growth provided attention is given to supply of protein.

Extending slaughter time in areas with fluctuating supply of nutrients
At the end of rainy seasons many cattle or sheep may be fat and prices may be depressed. The experiments referred to suggest that it is possible to allow the sheep or cattle a period of energy undernutrition without weight loss and even some overall weight gain, provided a protein supplement is given which is not degraded in the rumen.

Rumen development and type of feeds

Lambs weaned early, or before they are about 12 weeks old, have a rumen volume which has not reached mature proportions. As a consequence, slowly fermenting feeds like fibrous roughages are generally inadequate to sustain any gain and sometimes inadequate to achieve maintenance energy intake. As a result, most systems for the production of early weaned lambs are based on concentrate feed which ferments rapidly and thus compensates for the low rumen volume. It is, however, important to achieve a rate of fermentation which is not too rapid and all too easily leads to problems of acidosis and rumen parakeratosis. Another problem with rapidly fermenting feed relates to the high proportion of propionic acid that is produced. Normally propionic acid is metabolized in the liver and only a little appears in the peripheral circulation. In lambs, however, problems can arise when the propionic acid proportion is so high that it exceeds the capacity of the liver to metabolize it.

As a result, both propionic acid and its intermediary of metabolism, methylmalonic acid, appear in the peripheral blood (Duncan, Ørskov and Garton, 1972). Here it interferes with normal fat synthesis so that a high proportion of odd-numbered fatty acids are formed, and also a high proportion of branched chain fatty acids which result in soft fat, unsuitable and undesirable for the meat trade.

Processing of cereals

Attempt to reduce the proportion of propionic acid in the rumen fermentation led to the discovery that the simplest and cheapest solution was to feed the grain whole and unprocessed. This had the effect of reducing the rate of fermentation in the rumen; it also increased rumination time and time spent eating, thus increasing saliva secretion and rumen pH. In addition, all the previous problems of soft subcutaneous fat was solved and, as can be seen in *Table 17.5*, food conversion and digestibility were not changed. Thus cereals are now generally fed to sheep and goats in the whole unprocessed form.

TABLE 17.5 The effect of processing cereals on rumen pH, proportion of acetic and propionic acids and on food utilization in lambs (from Ørskov, Fraser and Gordon, 1974)

Cereal	Form	Rumen pH	Molar proportion of Acetic acid	Propionic acid	Liveweight gain (g/day)	Digestibility of organic matter (g/kg)	Food conversion (kg/dry matter/kg gain)
Barley	Whole, loose	6.4	52.5	30.1	340	81.1	2.75
Barley	Ground, pelleted	5.4	45.0	45.3	347	77.2	2.79
Maize	Whole, loose	6.1	47.2	38.7	345	84.3	2.52
Maize	Ground, pelleted	5.2	41.3	43.2	346	82.1	2.62
Oats	Whole, loose	6.7	65.0	18.6	241	69.9	3.07
Oats	Ground, pelleted	6.1	53.2	37.5	238	67.5	3.33
Wheat	Whole, loose	5.9	52.3	32.3	303	82.7	2.97
Wheat	Ground, pelleted	5.0	34.2	42.6	323	86.6	2.56
SE of mean		0.14	2.4	3.2	15	1.2	0.11

Processing and roughage utilization

Another advantage of not processing grain for lambs is apparent if fibrous roughages are used as well. The low rumen pH created by processed grain (rolled, ground, pelleted etc.) reduces the rate of fibre digestion and decreases both intake and digestibility of the roughage, in comparison with the feeding of whole grain. This was clearly shown in experiments reported by Ørskov and Fraser (1975).

Processing and the need for roughage

In some areas of the world, for example, some Middle East countries, roughages are sometimes more expensive than cereals and are often fed to lambs because it is felt that they need a source of fibre. In fact, ruminants need structure rather than fibre *per se* in the diet. When whole grain diets are used, no roughages are required since the structure is retained in the grain. On the other hand, if highly processed cereals are used, a source of roughage is then required. Thus feeding of unprocessed whole grain has the additional advantage of reducing the requirement for other structural feeds in the diet.

References

AGRICULTURAL RESEARCH COUNCIL (ARC) (1984). *The Nutrient Requirements of Ruminant Livestock*, **Supplement no. 1**. Slough: Commonwealth Agricultural Bureaux

ANDREWS, R.P. and ØRSKOV, E.R. (1970a). The nutrition of the early weaned lamb. I. The influence of protein concentrations and feeding level on rate of gain in body weight. *Journal of Agricultural Science, Cambridge*, **75**, 11–18

ANDREWS, R.P. and ØRSKOV, E.R. (1970b). The nutrition of the early weaned lamb. II. The effect of dietary protein concentrations, feeding level and sex on body composition at two live weights. *Journal of Agricultural Science, Cambridge*, **75**, 19–26

DUNCAN, W.R.H., ØRSKOV, E.R. and GARTON, G.A. (1972). Fatty acid composition of triglycerides of lambs fed on barley-based diets. *Proceedings of the Nutrition Society*, **31**, 19A–20A

ELIMAN, M.E. and ØRSKOV, E.R. (1985). Factors affecting the fractional outflow of protein supplements from the rumen. 3. Effects of frequency of feeding, intake of water induced by the addition of sodium chloride and the particle size of protein supplements. *Animal Production*, **40**, 309–313

FATTET, I., HOVELL, F.D.DeB., ØRSKOV, E.R., KYLE, K.J. and SMART, R.I. (1984). Undernutrition in sheep. The effect of supplementation with protein on protetin accretion. *British Journal of Nutrition*, **52**, 561–574

HOVELL, F.D.DeB., ØRSKOV, E.R., MacLEOD, N.A. and McDONALD, I. (1983). The effects of changes in the amount of energy infused as volatile fatty acids on the nitrogen retention and creatinine excretion of lambs wholly nourished by intragastric infusion. *British Journal of Nutrition*, **50**, 331–343

ØRSKOV, E.R. (1982). *Protein Nutrition in Ruminants*. London, New York: Academic Press

ØRSKOV, E.R. and FRASER, C. (1973). The effect of level of feeding and protein concentration on disappearance of protein in different segments of the gut in sheep. *Proceedings of the Nutrition Society*, **32**, 69A

ØRSKOV, E.R. and FRASER, C. (1975). The effect of processing of barley-based supplements on rumen pH, rate of digestion and voluntary intake in sheep. *British Journal of Nutrition*, **34**, 493–500

ØRSKOV, E.R., FRASER, C. and GORDON, J.G. (1974). Effect of processing of cereals on rumen fermentation, digestibility, rumination time and firmness of subcutaneous fat. *British Journal of Nutrition*, **32**, 59–69

ØRSKOV, E.R. and GRUBB, D.A. (1979). Growth of store lambs on cereal based diets with protein or urea. *Animal Production*, **29**, 371–377

ØRSKOV, E.R., MacLEOD, N.A., FAHMY, S.T.M., ISTASSE, L. and HOVELL, F.D.DeB. (1983). Investigation of nitrogen balance in dairy cows and steers nourished by intragastric infusion. Effect of submaintenance energy input with or without protein. *British Journal of Nutrition*, **50**, 99–107

Chapter 18

Improving feed resources: a key step towards expanding ruminant production in semiarid north Africa and west Asia

E.F. Thomson, P.S. Cocks and F. Bahhady

Summary

A rapidly expanding population and growing per capita income in north Africa and west Asia are creating a demand for basic foodstuffs which is increasingly met from imports. Food and feed self-sufficiency is therefore falling since domestic production is expanding too slowly. Evidence that the region can produce more feedstuffs and thereby increase small ruminant production is presented.

Extra feed could come from: higher crop yields per unit area and per plant, improved utilization of feed grains and cereal crop residues, and new feed resources such as forage and pasture crops. By integrating all these resources, carrying capacity and meat and milk production could be doubled.

Introduction

While the extra demand for wheat will probably be met by improved yields, recent growth rates of domestic meat and milk production will need to nearly double for self-sufficiency to be reached by the year 2000 (Sarma and Yeung, 1985). Unless production of feedstuffs increases considerably, self-sufficiency in livestock products will only be possible through increasing imports of feedstuffs, especially coarse grains; the gap between domestic production and demand for coarse grains could reach 36 million tonnes by the end of the century (Khaldi, 1984). The non-oil-producing countries of north Africa and west Asia are, and will be, unable to finance these imports.

Meeting the extra demand for feedstuffs must therefore receive top priority. This can be achieved by using more efficient livestock, increasing productivity of rangelands, improving traditional feed resources, and introducing new crops. In this chapter research in north-west Syria on increasing productivity and utilization of barley, cereal residues, and new forage and pasture legumes is discussed. The results are considered applicable to those areas of west Asia and north Africa receiving 200–350 mm annual rainfall where small ruminants predominate.

197

Improving traditional feed resources

Barley grain

After wheat, barley is the most important cereal crop in the region (Food and Agriculture Organization of the United Nations (FAO), 1984). Much of it is grown between the old cultivated zone and the steppe where rainfall decreases from 350 to 200 mm. Indeed barley, which provides grain, straw, and stubbles for small ruminants, is currently the best crop at the driest margins of cultivation. It does not use land that would otherwise be sown to crops for direct human consumption.

Average grain yields of barley are below 1000 kg/ha with large annual variations. These low yields are the result of poor agronomic practices, such as continuous

TABLE 18.1 Yields (kg/ha) of grain, straw and hay in barley–barley, barley–fallow, and barley–vetch rotations (from Breda, 1982–83, rainfall 285 mm)

Rotation*	Barley yield		Vetch hay yields
	Grain	Straw	
Bo/Bo	450	920	—
Bo/F	810	1220	—
Bnp/F	2080	2980	—
Bnp/Bnp	1440	2240	—
Bo/Vo	700	1120	1460
Bo/Vp	1300	1540	2170
Bnp/Vo	1800	2470	2270
Bnp/Vp	2060	3010	3000

*B = barley, V = vetch, o = no fertilization, n = 20 kg N/ha, P = 60 kg P_2O_5/ha, except Bnp/Vp where P = 30 kg P_2O_5/ha

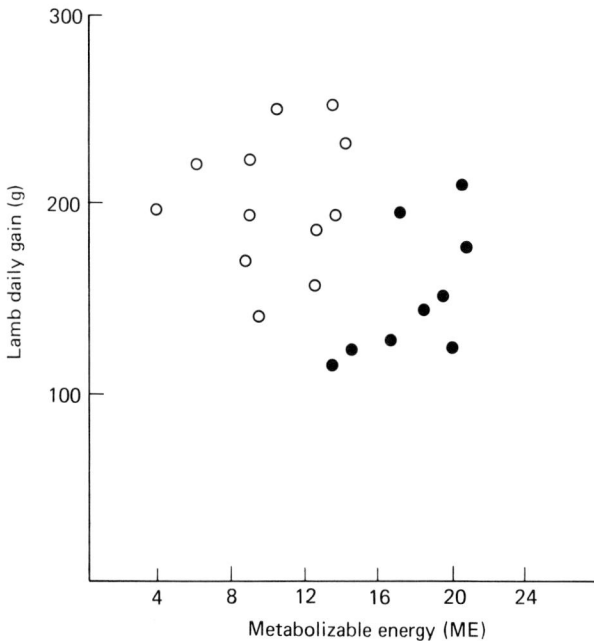

Figure 18.1 Growth rates of lambs and levels of supplementary feeding of ewes in early lactation in experimental (open circles) and farm flocks (solid circles). Source: Jaubert and Oglah (1985)

cropping and a lack of fertilizer. Application of nitrogen and phosphate increases yields substantially (*Table 18.1*) when applied to crops grown continuously (ICARDA, 1984). Yields can be further increased by introducing legumes, and using better varieties and improved tillage practices.

As well as higher grain yields, improvements in the efficiency of using barley are possible. Ewes in farm flocks receive more supplements than in experimental flocks, but the growth rates of their lambs are lower (*Figure 18.1*). Feed conversion efficiency of fattening lambs could be improved from the 12 to 1 found in commercial feedlots (ICARDA, 1982) to 4 to 1 found in experiments (ICARDA, 1986).

Cereal crop residues

Stubble and straw are the most abundant feed resources in the region. A 1 tonne grain crop produces 1.5 tonnes of straw; from an annual 75 million tonne grain harvest, 110 tonnes of straw are therefore available. This is sufficient to provide each sheep and goat in the region with 1 kg/day. That sheep and goats in Syria spend up to 4 months in summer on cereal stubbles, and that cereal straws provide 25% of the metabolizable energy in supplements offered during winter, shows how farmers value these feedstuffs.

TABLE 18.2 Yield, morphological characteristics and nutritive value of straw from three barley varieties (from Capper *et al.*, 1986)

		Variety	
	Arabi Abiad	ER/Apam	Beecher
Grain yield (kg/ha)	1830	3057	2720
Straw yield (kg/ha)	1009	1819	1731
Stem height (cm)	36	47	74
Leaf: stem ratio	0.57	0.56	0.42
Digestibility* (%)	44[a]	42[a]	30[a]
Voluntary intake† (g)	35[b]	27[a]	22[a]

*In vivo digestibility of organic matter
†Intake of organic matter/kg live weight$^{0.75}$; different letters in the same row show significant differences ($P < 0.05$)

Attempts to improve the availability of nutrients in straw include ammoniation, caustic soda treatment, and addition of urea. However, the levels and sources of variation in the composition of the straws from different varieties are usually ignored. Recent investigations have shown that, while straw from barley landraces can maintain sheep for 4 months (Thomson, unpublished data), new varieties selected for grain yield have only inferior straw quality (Capper *et al.*, 1986). The difference is explained in terms of plant height and leaf/stem ratio—tall plants have less leaf relative to stem than short plants. The higher proportion of more digestible leaf in the landraces explains the higher voluntary feed intake compared with the genotypes selected for grain yield (*Table 18.2*). Since plant height and grain yield are not related, breeders should be able to select new genotypes with both high yield and good straw quality.

Introducing new feed resources

Forage legumes

Common vetch (*Vicia sativa*), woolly pod vetch (*V. villosa* subsp. *dasycarpa*) and chickling (*Lathyrus sativus*) are ideal legumes for use on fallow lands which cover 30 million hectares in the region (Carter, 1978). These legumes are used for grazing, as hay, or as straw and grain.

Grazing vetch with lambs appears to be profitable. Live weight gains exceeding 250 kg/ha/year have been achieved using only 50 kg P_2O_5/ha (ICARDA, 1983). On farms, common vetch and chickling grown on fallow land yielded 350 kg milk/ha and 100 kg ewe live weight gain/ha in 30 days (Thomson, Jaubert and Oglah, 1985). Although these were below the performance levels achieved on medic pastures (*see below*), daily performance levels were similar.

Using forage legumes for grazing helps keep small ruminants off natural pasture on rangelands where the plants need to flower and set seed, a process essential to their productivity and persistence. However, if forages are cut the hay can be fed in autumn and winter, the period of greatest feed deficit. Forage legumes yield up to 6 tonnes of dry matter/ha at the 'hay' stage, although yields of hay are usually lower in the field (*see Table 18.1*). Harvest efficiency can be low because much of the highly nutritious leafy material is lost. Even in these semiarid countries rain can prolong the harvest operation and spoil the hay.

Another option is to allow the crop to mature and then harvest the seed and straw. Part of the seed can be used the next year and the rest fed to livestock. The straw usually contains sufficient metabolizable energy and protein to serve as a maintenance ration. This option may be the more profitable.

Pasture legumes

The major species of pasture legumes adapted to the region are *Trifolium subterraneum*, suitable for acid soils, and annual *Medicago* species (medics), suitable for the alkaline soils which predominate in the region. The advantages of medics over annual forage crops are their ability to withstand grazing and to self-regenerate after cereals in cereal/pasture rotations. They are therefore easy to manage and cheap to maintain as they do not need resowing in the pasture phase of rotations.

The medic system was developed in Australia but has not been widely adopted in the region. This is because of a lack of adapted cultivars—Australian cultivars are unable to withstand severe frosts—poor nodulation, inadequate grazing management, dependence on specialized machinery, and lack of appreciation of socioeconomic constraints. Most of these problems have been overcome, and adapted medics (Abd El-Moneim and Cocks, 1986) have now been successfully established using technologies available to farmers.

Initial results on sheep performance are now on hand. In one study, a farmer who had discovered the value of medics, grazed his lactating ewes at 20/ha for about 150 days. They produced 1400 kg milk/ha which generated 40% more sales revenue than wheat, the next most profitable crop. In other on-station trials, lambs grazing medics with their mothers showed daily live weight gains exceeding 250 g. In that experiment ewes at a low stocking rate also gained weight.

Integrated feeding systems

This section attempts to show how research on individual feed resources can be integrated into farming systems which generate more feed per unit land and thereby enable the farm to carry more livestock (Nordblom and Thomson, unpublished data).

For 6 years, scientists managed two crop rotations (barley/fallow and barley/forage crop), and three flocks of Awassi sheep (subjected to low, medium, and high nutritional regimes), each flock having access to 10 ha of unimproved or improved marginal land (marginal land is non-arable land which in north Syria is widely used for grazing). Using the biological and economic data generated, a linear programme was used to integrate the major elements of the system— cropped land, marginal grazing, and nutritional regime. The model estimated

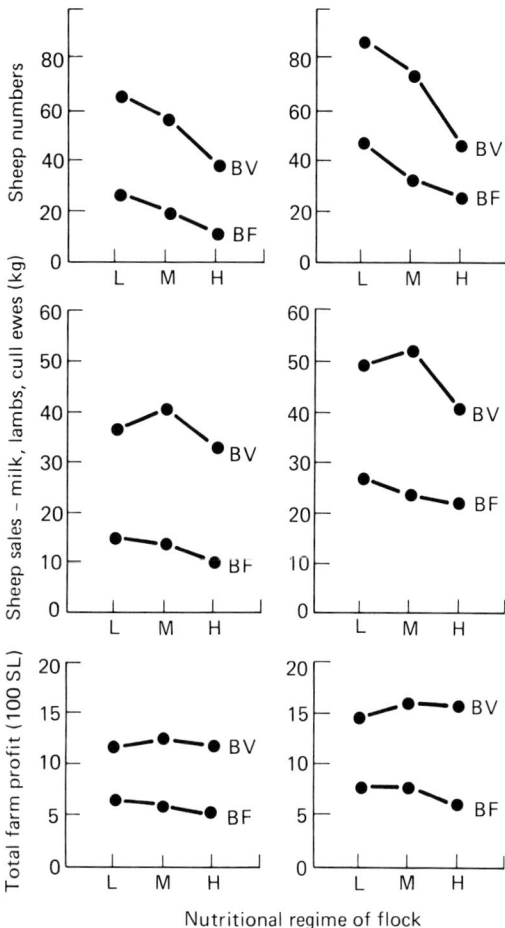

Figure 18.2 Sheep numbers, sheep sales (kg of product) and total farm profitability (100 SL/ha) of farms using barley-vetch (BV) and barley-fallow (BF) rotations and unimproved (left) and improved (right) marginal lands. L, M and H represent flocks of low, medium and high nutritional regimes respectively

maximum profits, and production of milk, lamb, and cull ewes for the various combinations.

Some results are presented in *Figure 18.2*. Noteworthy is the increase in sheep numbers from two to four/ha achieved through the replacement of fallow land with forage, although part of the response was due to the improved agronomic practices used in the barley/forage rotation. Sheep production also doubled, as did total farm profit. Applying phosphate to the marginal land increased productivity and profits still further.

At higher levels of nutrition the model chose fewer sheep in order to maximize farm profits. This indicates that the response of an unimproved sheep breed to increasing nutritional inputs is too small to be profitable. More productive selections of sheep are needed to derive the maximum benefit from improved feed resources.

Conclusions

To reduce rising imports of feedstuffs and livestock products in north Africa and west Asia efforts to improve feed resources must be given highest priority. If successful, this will allow numbers of small ruminants to increase and help the region move towards self-sufficiency. Improved feed resources will also assist genetically improved indigenous breeds to reach their full potential in terms of meat and milk production. Maximum benefit would also be derived from applying other new methods of management with potential to increase efficiency of small ruminant production.

There is sufficient evidence that production of traditional feed resources can be increased by using inputs already available in the region. Furthermore, new technologies based on forage and pasture legumes have been tested and adapted to constraints of the resource-poor farmers. Provided the inputs are made available, there is strong evidence that farmers will adopt the new technologies. This should result in higher incomes for producers and increased food security for the region.

References

Abd El-MONEIM, A.M. and COCKS, P.S. (1986). Adaptation of *Medicago rigidula* to a cereal-pasture rotation in north west Syria. *Journal of Agricultural Science, Cambridge*, **107**, 179–186

CAPPER, B.S., THOMSON, E.F., RIHAWI, S., TERMANINI, A. and MaCREA, R. (1986). The feeding value of straw from different genotypes of barley when given to Awassi sheep. *Animal Production*, **42**, 337–342

CARTER, E.D. (1978). Legumes in farming systems of the Near East and North Africa region. *Report to the International Center for Agricultural Research in the Dry Areas (ICARDA)*, pp. 118. Aleppo, Syria

FOOD AND AGRICULTURE ORGANIZATION OF THE UNITED NATIONS (FAO) (1984). *Production Yearbook 1983*. Rome, Italy

ICARDA (INTERNATIONAL CENTER FOR AGRICULTURAL RESEARCH IN THE DRY AREAS) (1982). *Range and Sheep Cooperatives and Fattening Cooperatives in Syria: Supplemental Feed Purchases and Credit Requirements*. p. 141. Aleppo, Syria: ICARDA

ICARDA (INTERNATIONAL CENTER FOR AGRICULTURAL RESEARCH IN THE DRY AREAS) (1983). *Livestock in the Farming System; Project IV Progress Report. Farming Systems Program*, pp. 65–67. Aleppo, Syria: ICARDA

ICARDA (INTERNATIONAL CENTER FOR AGRICULTURAL RESEARCH IN THE DRY AREAS) (1984). *Annual Report 1983*. Aleppo, Syria: ICARDA

ICARDA (INTERNATIONAL CENTER FOR AGRICULTURAL RESEARCH IN THE DRY AREAS) (1986). *Annual Report 1985*. Aleppo, Syria: ICARDA

JAUBERT, R. and OGLAH, M. (1985). Supplementary feeding of Awassi ewes in the barley zone of north-west Syria. *Proceedings of the First International Conference on Animal Production in Arid Zones*. Damascus, Syria: Arab Centre for Studies in Arid Zones and Dry Lands

KHALDI, N. (1984). Evolving food gaps in the Middle East/North Africa: prospects and policy implications. *Research Report No. 47*, pp. 74. Washington DC: International Food Policy Research Institute

SARMA, J.S. and YEUNG, P. (1985). Livestock products in the third world: past trends and projections to 1990 and 2000. *Research Report No. 49*, p. 87. Washington DC: International Food Policy Research Institute

THOMSON, E.F., JAUBERT, R. and OGLAH, M. (1985). On-farm comparisons of milk yield of Awassi ewes grazing introduced forages and common village lands in the barley zone of NW Syria. In *Proceedings of the First International Conference on Animal Production in Arid Zones*. Damascus, Syria: Arab Center for Studies in Arid Zones and Dry Lands

Part VII

Management and health control

Shepherding has always been the key element in any sheep system. Many new aids have been developed to help the modern shepherd in his task. Sheep 'scanning' is one aid that has taken a long time to develop but is now becoming a potent tool to enable differential management for ewes of different needs within a flock.

A breakthrough in the difficult task of easing lambing management is still under study but the day when lambing is planned to the compass of a few hours is already discernible. Few have as yet grasped the full significance of such a management procedure applied to such a key area of the shepherd's concern.

Less dramatic but nonetheless potent improvements can come from developments in promoting the health of the flock; some of these have the great merit of being easily applied without disrupting the normal shepherding procedure.

Another aspect covered is the way traditional wisdom on mating behaviour is being extended to influence the timing and effectiveness of mating.

Chapter 19

Pregnancy diagnosis and fetal number determination

I.R. White and A.J.F. Russel

Summary

Real-time ultrasonic scanning can be used to diagnose pregnancy and determine fetal numbers in sheep and has advantages over other approaches in terms of safety, accuracy and rapidity. Applied between 50 and 100 days of gestation the accuracy of diagnosis of pregnancy should be virtually 100%, and that of fetal number determination of the order of 96–98% or greater. Rates of scanning vary from 80 to 120 or more ewes per hour depending on expected litter size.

The principles of ultrasound and the use and merits of linear-array and sector scanning instruments are discussed, and factors affecting image quality and hence accuracy are considered.

The principal benefits of the use of the technique include the ability to relate feed inputs in late pregnancy to fetal numbers, thus improving the efficiency of use of feedstuffs. Feeding in relation to fetal numbers also offers a means of manipulating lamb birthweights and can lead to substantial reductions in levels of mortality of both lambs and ewes.

The uptake of the technique by the sheep industry in the United Kingdom has been rapid, and it is estimated that the number of ewes scanned increased from about 1000 in 1983 to 2 000 000 in 1986.

Accurate scanning demands a high degree of skill on the part of the operator, but the benefits of the technique to sheep production stem from the changes in management implemented on the basis of the scanning results.

Introduction

Until recently the inability to diagnose pregnancy and determine fetal numbers in sheep constituted a major limitation to progress in sheep farming. Information on the considerable differences in nutrient requirements between pregnant and non-pregnant ewes, and between ewes carrying different numbers of fetuses has been available for some time (Thomson and Aitken, 1959; Russel, Doney and Reid, 1967) but for many years the lack of a satisfactory technique for diagnosing pregnancy and determining fetal numbers prevented this knowledge from being applied.

The consequence of this inability to apply existing knowledge has been a serious economic loss to the sheep industry. The feeding of non-pregnant ewes as if they were in-lamb is clearly inefficient and wasteful. Without a knowledge of fetal numbers the feeding of pregnant ewes is a matter of guesswork or, at best, compromise, with detrimental effects to both production and economic efficiency. Lamb mortality is closely related to birthweight, with both underweight and overweight lambs being at risk. Twin or triplet lambs from ewes fed as if they were carrying only singles will be underweight and likely to die because of weakness and excessive heat loss, while single lambs from ewes fed as if they were carrying multiples will be overweight and at risk as a result of a prolonged and difficult birth. In flocks where all ewes are treated alike those carrying more than the average number of lambs are also more likely to develop metabolic disorders, such as pregnancy toxaemia, in late pregnancy.

The task of the flockmaster or shepherd at lambing is also made difficult if ewes cannot be managed according to the number of fetuses carried. In extensive systems of management in particular, much time and effort can be spent in dealing with the problems of weak and underweight lambs, orphan lambs, ketotic ewes and ewes with little or no milk, all resulting from inadequate nutrition in late pregnancy.

Non-ultrasonic techniques

Blood metabolites and hormones

Various attempts have been made over the years to diagnose pregnancy and determine fetal numbers in sheep by estimating concentrations of circulating metabolites and hormones. In theory the concentration in the blood of metabolites such as non-esterified fatty acids and ketone bodies (for example, 3-hydroxybutyrate), which serve as indices of the severity of undernourishment (Russel, 1978), should also serve as a means of classifying ewes according to fetal number. In practice, however, the variation in stage of pregnancy found within most flocks and differences in food intake between ewes militate against the use of such techniques.

Similarly, the determination of blood progesterone concentration, which can be used to diagnose pregnancy in many species, should theoretically be capable of use as an index of fetal numbers in sheep, but for a variety of reasons the technique has been shown to be less than wholly satisfactory in practice (Russel and Foot, 1971).

A major disadvantage of any technique based on the determination of blood metabolites or hormones is its failure to provide an immediate diagnosis. The need for laboratory analysis requires all ewes to be individually identified and handled twice—once at sampling and again later when the results become available. Thus, although metabolites or hormones, other than those mentioned above, may be found to be better indices of fetal number, it is unlikely that they will find application in practice. It is also unlikely that the recently developed rapid assay methods for certain hormones, and notably progesterone, which are finding favour in the dairy cow industry will be used to any extent in sheep flocks.

Radiological techniques

X-ray techniques, such as those used by Wenham and Robinson (1972) and reviewed by Wilson (1981), and the more sophisticated radiological technology described by Beach (1982), are capable of providing high levels of accuracy of both pregnancy diagnosis and determination of litter size. The capital cost of such equipment is high, it is generally mobile but not portable, and there is always a potential health hazard to operators working with perhaps tens of thousands of sheep each year. For these reasons and because such techniques can be used successfully only after about 80 days of gestation, it is unlikely that they will find widespread application in commercial sheep farming.

Other non-ultrasonic techniques

Non-ultrasonic techniques such as measurement of the viscosity of cervical mucus, arborization patterns of cervical-vaginal mucus, abdominal palpation, and others considered by Richardson (1972) as unsatisfactory would still be judged such at the present time. Similarly, the rectal-abdominal palpation technique described by Hulet (1973) is inaccurate (Trapp and Slyter, 1983) and entails the severe risk of rectal perforation (Turner and Hindson, 1975).

Principles of ultrasound

Ultrasound is simply sound of a very high frequency, generally of about 1.5–5 MHz, far beyond that which can be detected by the human ear. In instruments used for diagnosing pregnancy and determining litter size in sheep, ultrasound is produced in a device known as a transducer containing a piezoelectric crystal or crystals which convert applied electrical energy to ultrasonic energy, or alternatively convert mechanical energy (ultrasonic vibrations) into electrical energy. Thus, the transducer can be used both to transmit ultrasound and to receive returning echoes.

When a single crystal transducer is placed in close contact with skin, a beam of ultrasound is transmitted at right angles to the surface of the crystal into the underlying tissues. When the beam meets a boundary or interface between two different types of tissue, or between tissue and a fluid, part of the energy is reflected back as an echo. The level of energy returning to the transducer is determined by the difference in mechanical properties of the tissues forming the reflecting interface (that is, by the acoustic impedance), and by the nature of the tissues and distance through which the energy travels (that is, by the degree of attenuation).

The three main ways in which these basic principles are used in instruments designed for use with pregnant ewes are described below.

The Doppler shift principle

When waves of ultrasound meet a moving target the frequency of the returning echoes is different from that of the transmitted signal. If the target (that is, a heart or blood in a major vein or artery) is moving towards the signal, the frequency of the echoes is increased; if the target is moving away from the signal the frequency is reduced. This phenomenon, known as Doppler shift, also occurs in the audible range.

This principle can be used to diagnose pregnancy in sheep using either a transducer mounted in a slim probe inserted into the rectum (Deas, 1977) where it can detect the flow of blood in the uterine artery, or a transducer applied externally to the abdomen to detect fetal heartbeats (Fukui, Kimura and Ono, 1984). In theory the detection of fetal heartbeat using the Doppler principle should provide a means of determining fetal numbers, but in practice the results of numerous trials have been disappointing. The use of intrarectal Doppler probes also gives cause for concern, as Tyrrell and Plant (1979) reported that about 50% of ewes so examined suffered some form of rectal damage.

The difficulty with this approach is that the heartbeats or pulses in the major blood vessels of one fetus can be picked up from different positions and angles of the transducer on the ewe's abdomen. To be certain that signals obtained from two different positions are coming from different fetuses, and not from two parts of the one fetus, it is necessary to measure the rate of the pulses and to determine that they are different from each other and from any maternal pulses which may also be detected. Such instruments have no range or depth discrimination, and consequently there is no way of knowing that an area of interest has not been fully examined as a result of gas or bone shadows or simply due to lack of sensitivity.

A- and B-mode ultrasound

With a single-crystal transducer echoes coming from the interfaces or boundaries of the tissues through which the pulses of ultrasound are passing can be displayed on a screen in such a way that the depth of the interfaces beneath the skin surface are represented as distances in one dimension on the screen. This principle is used in A-mode scanning in which the transducer is generally placed on the bare area of the abdomen lateral to the udder and a single-point reflection of ultrasound from a particular depth, usually 9 cm or greater, is regarded as a positive diagnosis of pregnancy. The information obtained from such single-crystal transducers is obviously very limited, and many of the shortcomings of Doppler shift instruments apply equally to A-mode scanners. Results from the use of this approach have been variable and generally leave much to be desired (Madel, 1983; Trapp and Slyter, 1983; Langford et al., 1984).

An extension of this approach is the B-mode scanner in which a single crystal transducer is driven on a fixed rail and a series of one-dimensional signals from different positions are recorded photographically to give a two-dimensional representation of the tissues scanned. This type of instrument is capable of diagnosing pregnancy with a high degree of accuracy but results on the determination of fetal numbers have been disappointing (Lindahl, 1976).

Real-time ultrasonic scanning

Two types of real-time ultrasonic scanners can be used to determine fetal numbers in sheep. The 'linear array' instrument has an array or row of crystals, perhaps 60–80 in number, arranged in a line some 10 cm long. The elements or groups of elements are fired sequentially and the signals or echoes are displayed as a series of close parallel lines on the screen. The 'sector' scanner has a small number of crystals which rotate or oscillate beneath a protective shield. The area of skin contact is small and the crystals are energized over a wide arc (up to 170°). The echoes are displayed as a series of diverging lines on the screen. The effect of either principle of use of ultrasound is to give a two-dimensional 'picture' or image of the underlying tissues from which it is possible to determine, for example, the shapes of various organs or parts such as the liver, heart or limbs. As the interfaces between tissues through which the ultrasound is travelling move, so movement is seen on the screen and this 'real-time' or live attribute of the displayed image is an important aid to image interpretation.

The above description is very simplistic and the instruments now available are in fact very sophisticated, incorporating microcomputers and other electronic devices to give images in varying shades of grey as well as black and white, and to improve generally the quality of the image displayed on the screen.

Acoustic coupling
Ultrasound will not pass through air, and to ensure good transmission into the tissues, some form of coupling agent must be used between the transducer and the skin surface. Many different substances can be used, but proprietary gels are the most convenient and the most widely employed. With linear array instruments the gel is generally applied with a brush or sponge immediately before scanning, while with the only sector instrument commonly used to scan sheep, the gel is delivered from a pressurized container to the probe at the beginning of the scanning operation.

Linear array scanning
The use of a linear array instrument for diagnosing pregnancy and determining fetal numbers in sheep has been reported by White, Russel and Fowler (1984).

The linear array instruments have a 'field of view' of approximately 8–10 cm wide by 20 cm deep—that is of 160–200 cm^2. To ensure that the whole of the area of potential interest is examined thoroughly it is necessary to turn the ewe on its back, or at least into a reclined sitting attitude, and to move the transducer over the full width of the abdomen for a distance of some 20 cm in front of the udder. To achieve this the wool should first be clipped from that area. Some scanning operators now use this form of scanning without removing the wool, scanning only on the naturally bare areas of skin in that region. This practice may, however, lead to failure to detect some fetuses, particularly if multiple-bearing ewes are scanned at a more advanced stage of pregnancy.

Whether or not wool is clipped to extend the potential area of examination, it is essential that the transducer, normally held at right angles to the longitudinal axis of the ewe, is moved in a systematic pattern of searching from one side of the abdomen to the other, while maintaining good contact with the skin surface throughout. During the course of movement across the abdomen the angle at which the transducer is held against the skin is altered, making the ultrasound beam sweep or 'scan' areas of particular interest. In this manner the operator builds up a mental three-dimensional concept from the series of two-dimensional images viewed on the screen.

Sector scanning
Although any sector scanner can in theory be used to determine fetal numbers in pregnant ewes, there is only one instrument (Oviscan 3, BCF Technology Limited, Livingston, Scotland) known to the authors which has been specifically designed for this purpose and which is sufficiently robust to withstand the rigours of repeated use under the conditions usually encountered in practice. This instrument can be operated on an 85° or 170° sector and has a choice of depths of field ranging from 5 to 25 cm. Thus the area visualized at any one time can be up to 927 cm^2, in other words, approximately five times that of most linear array instruments. This means

that in many, but by no means all cases, the whole of the area of potential interest can be viewed from a single point of contact. This allows the ewe to be scanned in a standing position and thus affords a considerable advantage in terms of labour and handling equipment over the normal linear array systems. Where pregnancy is more advanced, it may be necessary to move the transducer to the opposite side of the ewe, but in all cases it should be possible to image the entire uterus and its contents from contact with the skin on the naturally bare areas in front of the udder, thus obviating the need to clip wool from the abdomen.

As in linear array scanning, a mental three-dimensional picture is built up by altering the angle at which the probe is held against the skin, thus sweeping or scanning across the entire width of the uterus from the posterior aspect through both horns to the anterior limits.

Handling systems

With linear array instruments the ewe is either held in a sitting position or placed in a cradle like a low deckchair, as shown in *Figure 19.1*, so that the operator can move the transducer freely over the shorn area immediately in front of the udder and over the uterus. Catching the ewes, turning them up, shearing wool from the abdomen and restraining them for scanning is hard physical work, particularly with large ewes, and a team of three to four people in addition to the scanning operator, is required to ensure an efficient operation.

Mobile handling systems have been developed in which ewes in an elevated race are turned onto a cradle which rotates horizontally, first to a position for shearing and secondly to a further position for scanning. The ewes are presented for shearing and scanning at waist height which reduces fatigue and makes the operation more efficient and somewhat less labour demanding.

In sector scanning the ewes are generally restrained in a crate as illustrated in *Figure 19.2*. This has the considerable advantage that ewes enter and leave the crate themselves without having to be physically handled, thereby reducing the labour requirement and increasing the efficiency of the operation.

Stage of pregnancy

The size of the fetus at different stages of gestation can be judged from the illustration shown in *Figure 19.3* in which the diameter of the coin is 2 cm. The skilled operator will be able to diagnose pregnancy by about 30 days post-conception from the identification of the fluid-filled uterus. At this stage of pregnancy, however, individual fetuses are too small to allow them to be identified with confidence. Fetuses can generally be counted accurately between 50 and 100 days of gestation. At later than 100 days the size of individual fetuses, the position of the uterus within the abdomen, and the shadows cast by the calcified fetal skulls and scapulae, combine to make the accurate determination of numbers difficult. With experience the recommended range of 50–100 days of gestation may be extended by 5 days at either end without affecting accuracy adversely.

Except in the few flocks where oestrus synchronization is practised, there will be a range in stage of gestation of ewes in any flock. The majority of ewes will, however, be mated within two oestrus cycles, that is, within about 35 days. Thus scanning of a ewe flock should be carried out after the latest mated ewe is 45 days pregnant and before the first mated ewe has reached 105 days. This gives an

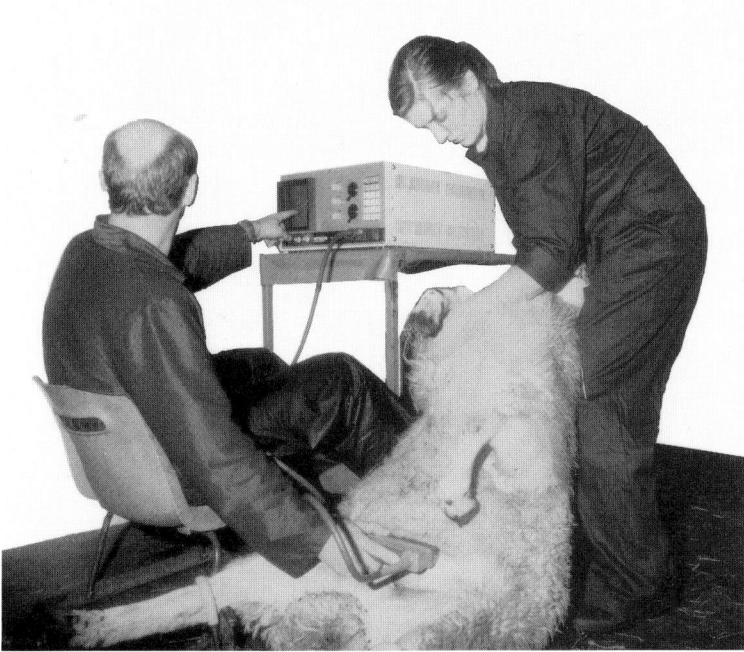

Figure 19.1 Ewe scanning with a typical linear array instrument

Figure 19.2 Ewe scanning with a sector instrument

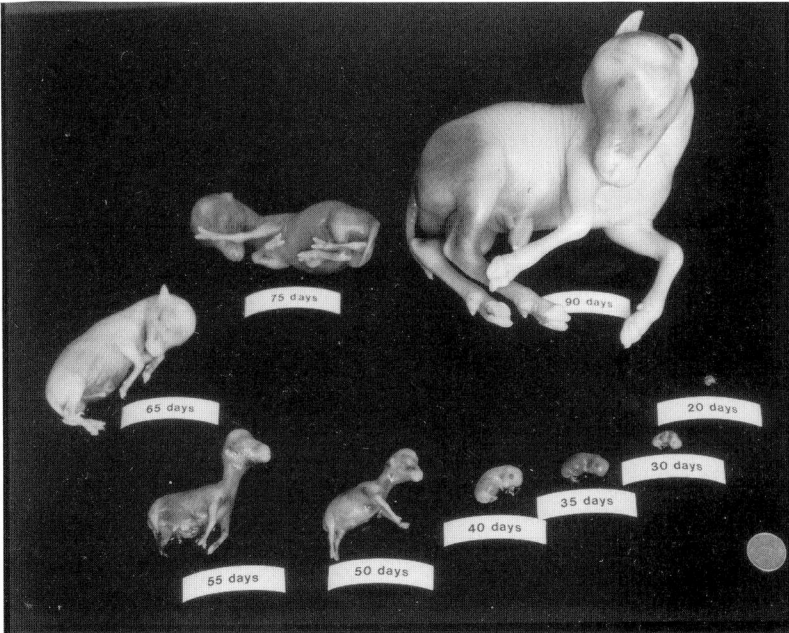

Figure 19.3 Ovine fetuses of different ages; sizes may be judged in relation to the coin which is 2 cm in diameter

interval of between 3 and 4 weeks during which any one flock is scanned, that is, from 80 to 105 days after the beginning of mating.

Image interpretation

The image viewed on the scanner screen is that of a section of the ewe lying immediately below (in linear array scanning) or above (in sector scanning) the transducer. As an aid to interpretation it can be useful to imagine that the ewe has been cut across in the plane of the ultrasound beam and that the operator is looking into the cut surface. The image on the screen can be visualized as part of this cut surface, either rectangular or semicircular in shape, depending on the type of instrument used.

The areas of black, white and grey on the screen represent different tissues and their interfaces. Fluid-filled structures appear on the screen as black areas. Occasionally the bladder is seen as a black circular area with well-defined smooth margins and lying in the midline position. The fluid-filled uterus of the pregnant ewe also appears as a black area but, being convoluted in the early stages, may have an irregular outline and appear as several apparently discrete sacs, which, by moving the transducer, can be shown to be parts of one structure.

Cotyledons can be distinguished from about 40 days as white circular structures with hollow centres which appear and disappear as the angle of the transducer is altered. Individual fetuses can be identified by 45–50 days as white structures, frequently moving independently of the surrounding tissues. At that stage the head is about half the size of the body, the limbs are present as small buds, and the fetal length is about 4 cm.

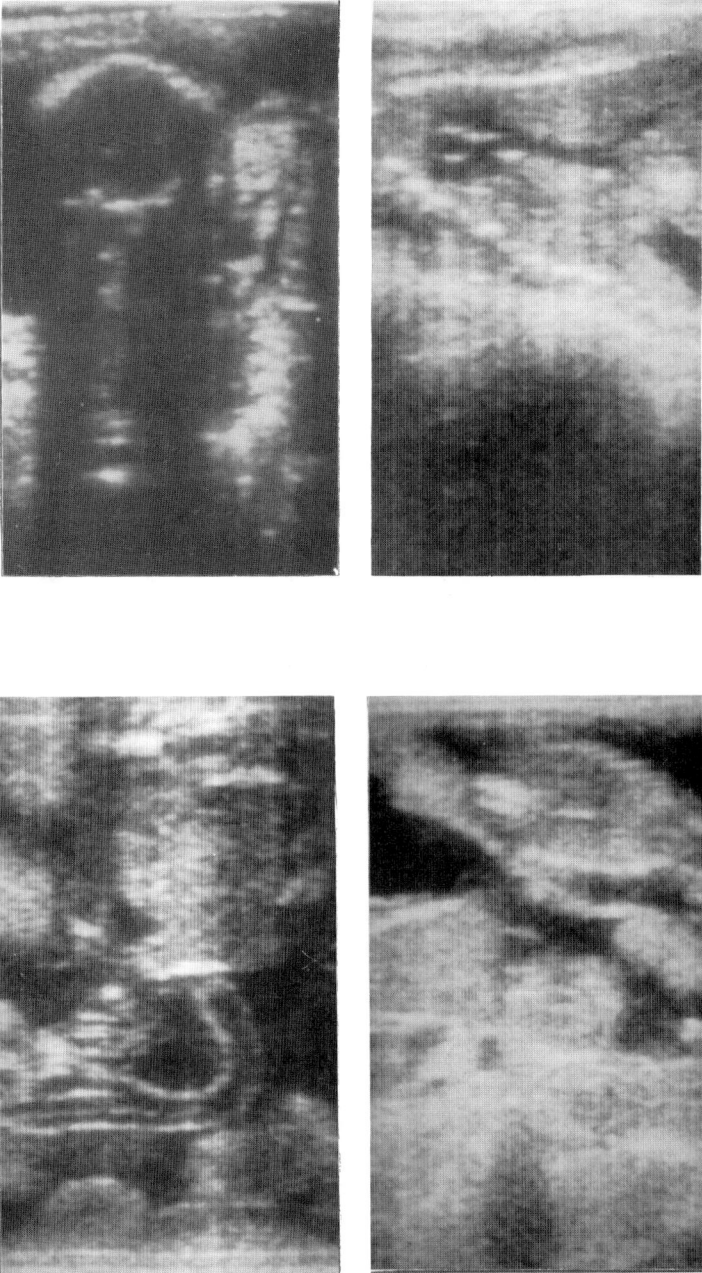

Figure 19.4 Images taken from a linear array instrument screen showing: (*a*) fetal head at about 100 days showing heavy shadowing cast by skull; (*b*) fetal thorax at about 100 days with neck to left side and showing rib cage, heart and aorta; (*c*) transverse section of fetal trunk at about 90 days (*lower centre*) showing section of umbilical cord (*centre*) and cotyledons (*top*); (*d*) longitudinal sections through rib cages of 65 day twin fetuses (*upper right* and *lower left*), showing membranes between fetuses

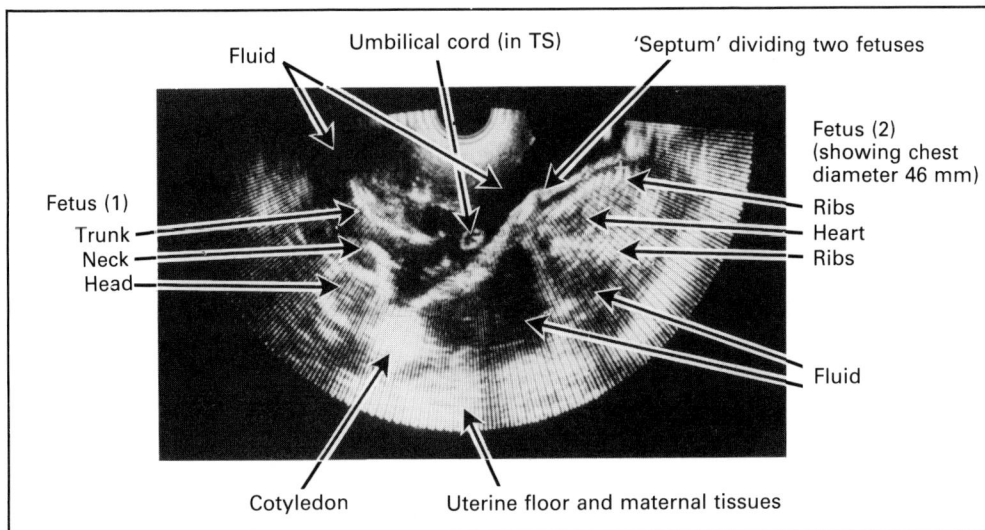

Figure 19.5 Image taken from a sector instrument screen

Fetal bones appear as bright white images from about 75 days. Ultrasound cannot pass through the calcified skeleton and thus black shadows appear underneath well-formed bony structures such as the skull and scapulae. Sections through the rib-cage show characteristic striated shadowing with black lines below each rib interspersed with white lines where ultrasound passes through the intercostal spaces.

By 100 days images of individual fetuses more than fill the entire screen and only parts of each fetus can be viewed from any one position of the transducer. Shadows from the relatively large and dense bones make it difficult and often impossible to 'see through' one fetus to others which may be lying beyond.

Photographs of images from linear array and sector instruments are shown in *Figures 19.4* and *19.5* respectively. The movement seen in 'live' real-time images is an invaluable aid to interpretation and is, of course, lost in still photographs.

Image quality

In the same way that air trapped beneath the skin surface and the transducer will prevent the transmission of ultrasound into the ewe's tissues, so gas in the rumen and intestines in the path of the ultrasound beam will hinder the passage of ultrasound and cause poor image quality. Such conditions are most commonly found in ewes which have had recent access to roughages such as hay, silage or straw. Recent trials in Hill Farming Research Organization have shown that the adverse effects of intestinal gas on image quality can be minimized by withholding roughages from ewes for 8–10 h before scanning. Starvation for more than 24 h is not recommended as, in addition to possible harmful effects to the ewe, image quality appears to deteriorate again if food is withheld for longer periods.

These same trials showed that in ewes given roughages, image quality was markedly better in ewes scanned in the standing position with the sector instrument than in ewes scanned lying on their backs with a linear array instrument. This is considered to be more an effect of the disposition of the uterus in relation to the intestines than to the type of instrument *per se*, but is still a very considerable advantage in favour of the sector scanner.

Accuracy and throughput

Accuracy is clearly a function of the scanning operator rather than the instrument. Given proper instruction most operators are capable of achieving very high levels of accuracy, particularly once they have gained some experience and, in certain cases, confidence. Most operators at the end of a 2 day course can scan accurately at a rate of 25–30 ewes/h. With experience the rate of scanning generally increases to perhaps 80–100 crossbred ewes/h where all fetuses are to be counted. Where the flockmaster merely wishes to know whether ewes are carrying zero, one or more than one lamb, as is often the case with hill flocks in the United Kingdom, rates of 100–120 ewes or more per hour are commonly achieved.

The accuracy of diagnosis of pregnancy, if it can be guaranteed that no ram has been running with the ewes for the previous 45–50 days, should be virtually 100%. Certainly no ewe should be wrongly diagnosed as non-pregnant in these circumstances. Inevitably there will be instances of fetal death and resorption after scanning, and thus there will be a few cases where ewes correctly diagnosed as pregnant may fail to lamb. Where ewes are to be classified as carrying one or more than one fetus, a level of accuracy of 98% would generally be regarded as acceptable, although a number of operators can achieve consistently better results. Where the requirement is to count the exact number of fetuses in all ewes, the level of accuracy may fall to 96–97% where most ewes have twins or triplets and to lower levels where there is a significant frequency of even greater fetal numbers.

Benefits of scanning

Ewe feeding costs now constitute a major part of the inputs in most sheep enterprises. Typical costs vary from about £3 per ewe for supplementary feeding in hill flocks to £7 or £8 per ewe in intensively managed lowground flocks. An increasing number of ewes are being housed during winter, and in such situations feed costs are likely to be of the order of £14 per ewe or more. Non-pregnant grazing ewes do not require supplementary feeding and in housed flocks it is unlikely that the flockmaster would wish to keep any non-pregnant ewes indoors on expensive feeding. Thus, there are economies in feeding to be achieved by identifying non-pregnant ewes at an early stage and removing these animals from the main flock. If the proportion of barren ewes is not high (say 3–4%), the financial benefits to be gained by identifying these ewes will not be great, but can still make an appreciable contribution towards the cost of scanning.

At the time when scanning is likely to be carried out, the price of ewes (for mutton) is generally at its highest, and if the flockmaster is prepared to sell the barren ewes, the returns, together with feed savings, can more than pay the cost of scanning the whole flock.

Eight weeks before lambing the fetal lamb weighs approximately 15% of its weight at birth, and the nutrition of the ewe from that time until lambing has a major effect on the growth rate and consequently the birthweight of the fetus. If during the last 6–8 weeks of pregnancy a ewe carrying twin or triplet fetuses is fed at a level appropriate for a ewe with only one fetus, these twin or triplet lambs will have low birthweights and are more likely to die than are heavier lambs from better-fed ewes. Ewes carrying single fetuses and fed at a level appropriate for twins are likely to produce lambs which are overweight at birth and at risk of dying as a result of a prolonged and difficult lambing.

Ewes inadequately fed in late pregnancy are also at risk from pregnancy toxaemia. This metabolic disorder is often fatal and occurs most often in fat ewes carrying two or more fetuses. It is caused by a combination of a high energy requirement and a low energy intake.

The use of real-time ultrasonic scanning offers a means of separating the single-bearing and multiple-bearing ewes and, if need be, of determining actual fetal numbers. This enables the flockmaster to regulate feeding during late pregnancy according to the ewe's requirements and consequently reduces the incidence of overweight single lambs and underweight twins or triplets, as well as lessening the risk of pregnancy toxaemia within the ewes.

It is difficult to put monetary values on the above-mentioned benefits resulting from scanning, and indeed it is unlikely that the probable financial savings would be the same in any two flocks. It has, however, been estimated (Russel, Maxwell and Sibbald, personal communication) that under United Kingdom conditions net benefits might range from about £2.50 per ewe in hill flocks to around £4.75 per ewe in intensive low ground flocks. Reports in the farming press indicate that these estimates have been achieved in practice.

A further benefit from scanning is the aid to management which a knowledge of fetal numbers affords. For example, in a traditionally managed hill sheep flock where twin lambs are often underweight at birth and have to be brought off the hill with their mothers, the separation of twin-bearing ewes from the main flock for better prelambing feeding also allows attention to be devoted to these animals at lambing, thereby further reducing lamb losses. At the other extreme, the separate penning in the inwintering house of ewes carrying one, two or more than two fetuses also makes for easier and more efficient management at lambing. The financial benefits of the 'aid-to-management' advantages of scanning are difficult to quantify in monetary terms.

Development of scanning in the United Kingdom

The real-time ultrasonic scanning of sheep was pioneered in Australia (Fowler and Wilkins, 1980; 1982) but has not been taken up as extensively there as in the United Kingdom, probably because the opportunities for savings in feed costs and the value of lambs and ewes are less.

The first trials on the use of the technique in sheep in the United Kingdom were conducted in early 1983 (White, Russel and Fowler, 1984). Because it believed that the technique had something useful to offer to the sheep industry and that it was important that those operators pioneering the use of the technique should be as well equipped as possible, the Hill Farming Research Organization offered a series of courses of instruction on the use of real-time ultrasonic scanning to determine fetal numbers. In the first 'scanning season' in the 1983–84 winter, six commercial operators scanned a total of about 100 000 ewes (Russel, 1984). In the following 1984–85 winter some 60 commercial operators scanned about one million ewes

(Russel, 1985), and it is estimated that in the recent 1985–86 winter some two million ewes, representing about 13% of the national flock, were scanned.

Most, but not all operators, have achieved a high degree of proficiency and many are now scanning between 30 000 and 40 000 ewes in a 10-week season, achieving, where flock sizes and physical conditions permit, throughputs of more than 1000 ewes per day on occasion. The proficient operators now have as much work as they can deal with in the course of the scanning season, and those few who, for various reasons, have not achieved consistently good results, find it difficult to attract business. The signs are that scanning is now accepted as a significant aid to sheep management and production in the United Kingdom, and that the number of ewes scanned will continue to increase each year for some time to come.

Conclusions

The technique of real-time ultrasonic scanning has much to offer to the flockmaster, but it must always be borne in mind that it is only an aid to management and not a substitute for good husbandry. Scanning demands a high degree of skill on the part of the operator, but it is the action taken by the flockmaster on the basis of the scanning results which brings the benefits of increased output and improved profitability.

References

BEACH, A.D. (1982). Multiple pregnancy diagnosis in sheep using pulsed video-fluoroscopy and a semiconductor video store. *New Zealand Journal of Experimental Agriculture*, **10**, 291–295

DEAS, D.W. (1977). Pregnancy diagnosis in the ewe by an ultrasonic rectal probe. *Veterinary Record*, **101**, 113–115

FOWLER, D.G. and WILKINS, J.F. (1980). The identification of single and multiple bearing ewes by ultrasonic imaging. *Proceedings of the Australian Society of Animal Production*, **13**, 492

FOWLER, D.G. and WILKINS, J.F. (1982). The accuracy of ultrasonic imaging with real-time scanners in determining litter number in pregnant ewes. *Proceedings of the Australian Society of Animal Production*, **14**, 636

FUKUI, Y., KIMURA, T. and ONO, H. (1984). Multiple pregnancy diagnosis in sheep using an ultrasonic method. *Veterinary Record*, **114**, 145

HULET, C.V. (1973). Determining foetal numbers in pregnant ewes. *Journal of Animal Science*, **36**, 325–330

LANGFORD, G.A., SHRESTHA, J.N.B., FISER, P.S., AINSWORTH, L., HEANEY, D.P. and MARCUS, G.J. (1984). Improved diagnostic accuracy by repetitive ultrasonic pregnancy testing in sheep. *Theriogenology*, **21**, 691–698

LINDAHL, I.L. (1976). Pregnancy diagnosis in ewes by ultrasonic scanning. *Journal of Animal Science*, **43**, 1135

MADEL, A.J. (1983). Detection of pregnancy in ewe lambs by A-mode ultrasound. *Veterinary Record*, **112**, 11–12

RICHARDSON, C. (1972). Pregnancy diagnosis in the ewe: a review. *Veterinary Record*, **90**, 264–275

RUSSEL, A.J.F. (1978). The use of measurements of energy status in pregnant ewes. *British Society of Animal Production Occasional Publication*, **1**, 31–39

RUSSEL, A.J.F. (1984). Ultrasonic scanning. *The Sheep Farmer*, **4(1)**, 30–31

RUSSEL, A.J.F. (1985). Sheep scanning. *The Sheep Farmer*, **5(2)**, 31–32

RUSSEL, A.J.F., DONEY, J.M. and REID, J.L. (1967). Energy requirements of the pregnant ewe. *Journal of Agricultural Science, Cambridge*, **68**, 359–363

RUSSEL, A.J.F. and FOOT, J.Z. (1971). The use of circulating progesterone concentrations in the identification of twin bearing ewes. *Animal Production*, **13**, 377 (abstract)

THOMSON, W.A. and AITKEN, F.C. (1959). *Diet in Relation to Reproduction and Viability of the Young. Part II: Sheep.* Commonwealth Agricultural Bureaux, England

TRAPP, M.J. and SLYTER, A.L. (1983). Pregnancy diagnosis in the ewe. *Journal of Animal Science*, **57**, 1–5

TURNER, M.J. and HINDSON, J.C. (1975). An assessment of a method of manual pregnancy diagnosis in the ewe. *Veterinary Record*, **96**, 56–58

TYRRELL, R.N. and PLANT, J.W. (1979). Rectal damage in ewes following pregnancy diagnosis by rectal-abdominal palpation. *Journal of Animal Science*, **48**, 348–350

WENHAM, G. and ROBINSON, J.J. (1972). Radiographic pregnancy diagnosis in sheep. *Journal of Agricultural Science, Cambridge*, **78**, 233–238

WHITE, I.R., RUSSEL, A.J.F. and FOWLER, D.G. (1984). Real-time ultrasonic scanning in the diagnosis of pregnancy and the determination of foetal numbers in sheep. *Veterinary Record*, **115**, 140–143

WILSON, I.A.N. (1981). Ovine pregnancy diagnosis. *Grassland Research Institute Technical Report No. 28*

Chapter 20

Synchronized breeding and lambing

W.M. Tempest and C.M. Minter

Summary

An outline is given of the endocrinological basis for the artificial manipulation of the oestrous cycle and parturition. The techniques of controlling these activities are reviewed and it is determined that the current optimum practices are the use of a progestagen-impregnated sponge implanted in the vagina of the ewes for 12–14 days, with or without pregnant mare serum gonadotrophin (PMSG) according to circumstances, for synchronizing breeding, and a single intramuscular injection of 16 mg betamethasone suspension 48 h before required peak lambing time to synchronize lambing into a 24 h period. The benefits of these techniques are essentially managerial and these are detailed, together with a consideration of the potential disadvantages. It is concluded that these techniques are not a substitute for better management, but that it is absurd when other technology has been adopted to still leave the breeding and lambing activities of sheep to external influences and chance when both events can be controlled.

Introduction

Synchronization of oestrus in sheep using progestagen sponges and PMSG, no matter what the physiological state of the ewe nor the time of the year, is now a well-established technique (Cognié and Mauleon, 1983). Natural variation between ewes in gestation length, however, means that groups of oestrus-synchronized ewes may not show the same degree of synchrony at lambing. Because progestagen/ PMSG-treated ewes are likely to be more prolific, and because mortality at and around birth is a major cause for lamb losses, adequate supervision at lambing time is a necessary management activity to ensure that the benefits of synchronized breeding are not lost. This activity can be more easily and efficiently carried out if the oestrus-synchronized ewes are additionally induced to lamb in synchrony. In this chapter the techniques of synchronizing both oestrus and lambing are reviewed and their application in sheep management discussed.

Synchronization of oestrus

The oestrous cycle

Sheep are seasonally polyoestrous with oestrus cycles commencing in late summer in regular 16–17 day cycles (Short, 1982). Oestrus normally lasts for 24–36 h, with ovulation occurring 24–26 h after the onset of oestrus (Hunter, 1982).

The oestrus cycle is controlled by a sequence of hormone changes which are governed by the hypothalamus and pituitary, and which interplay with the ovaries and uterus. The experimental investigations which have contributed knowledge and understanding of this control have been comprehensively reviewed by Haresign, McLeod and Webster (1983). It is not the purpose of this chapter to review the original experimental work, but to state simply what the current understanding is, and this has been done largely by reference to Hunter (1982).

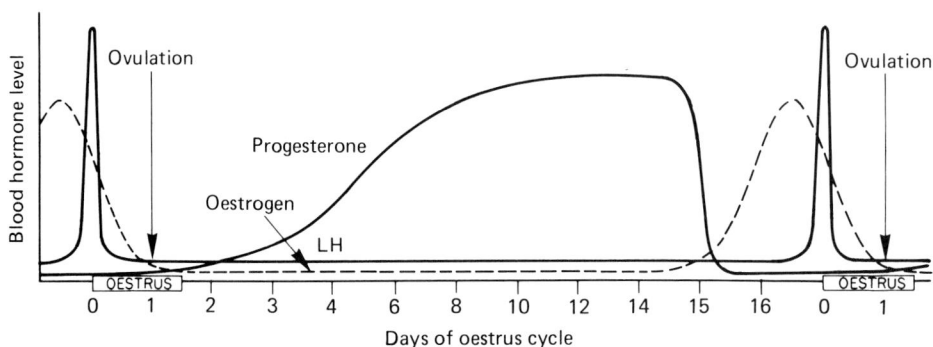

Figure 20.1 Diagrammatic representation of the pattern of changing hormone concentrations during the oestrus cycle of the ewe (adapted from Hunter, 1982 and Short, 1982)

A simple representation of the changing hormonal concentrations is shown in *Figure 20.1*. Starting at day 0, the convention for describing the day of onset of oestrus (heat), the ewe comes into heat mainly under the influence of oestrogen, in the form of oestradiol, a steroid hormone produced in the ovary. Some 24–36 h later the egg is released from a Graafian follicle in which it has been developing. This is referred to as ovulation, an event which requires a peak secretion of luteinizing hormone (LH), a gonadotrophic hormone. In the collapsed follicle, a corpus luteum begins to form, and this secretes another steroid hormone, progesterone, in increasing amounts. The purpose of this is to prepare the uterus to receive a fertilized egg in the event of a successful mating. Progesterone is secreted for approximately two-thirds of the oestrous cycle, known as the 'luteal phase'. If the ewe has not been successfully mated and a fertilized egg is not present in the uterus, then the uterus secretes a hormone known as prostaglandin—specifically prostaglandin $F_{2\alpha}$ ($PGF_{2\alpha}$). This causes the regression of the corpus luteum and the cessation of progesterone secretion. The follicles then begin to develop under the influence of follicle stimulating hormone (FSH) a gonadotrophin secreted by the pituitary. This 'follicular phase' is in time about one-third of the complete oestrous cycle. The developing follicles secrete oestrogens in increasing concentrations and the ewe returns back to day 0—the onset of oestrus or receptive heat.

The overall monitoring of this cycle is controlled in the brain by the hypothalamus which secretes gonadotrophin releasing hormones (GnRH) which in turn mediate the production of LH and FSH by the pituitary. Regulation of GnRH secretion is by the feedback actions of the steroid hormones—progesterone has a negative influence on the secretion of GnRH and therefore on pituitary gonadotrophin release, in order to prevent follicular growth during the luteal phase; oestrogen has a positive feedback effect to stimulate more pituitary gonadotrophin release for follicle growth and ovulation. These feedback actions are sometimes known as long loop feedbacks, for additionally there are also short loop feedbacks of the gonadotrophins to positively stimulate greater production of themselves.

The interacting pathways are shown in *Figure 20.2*.

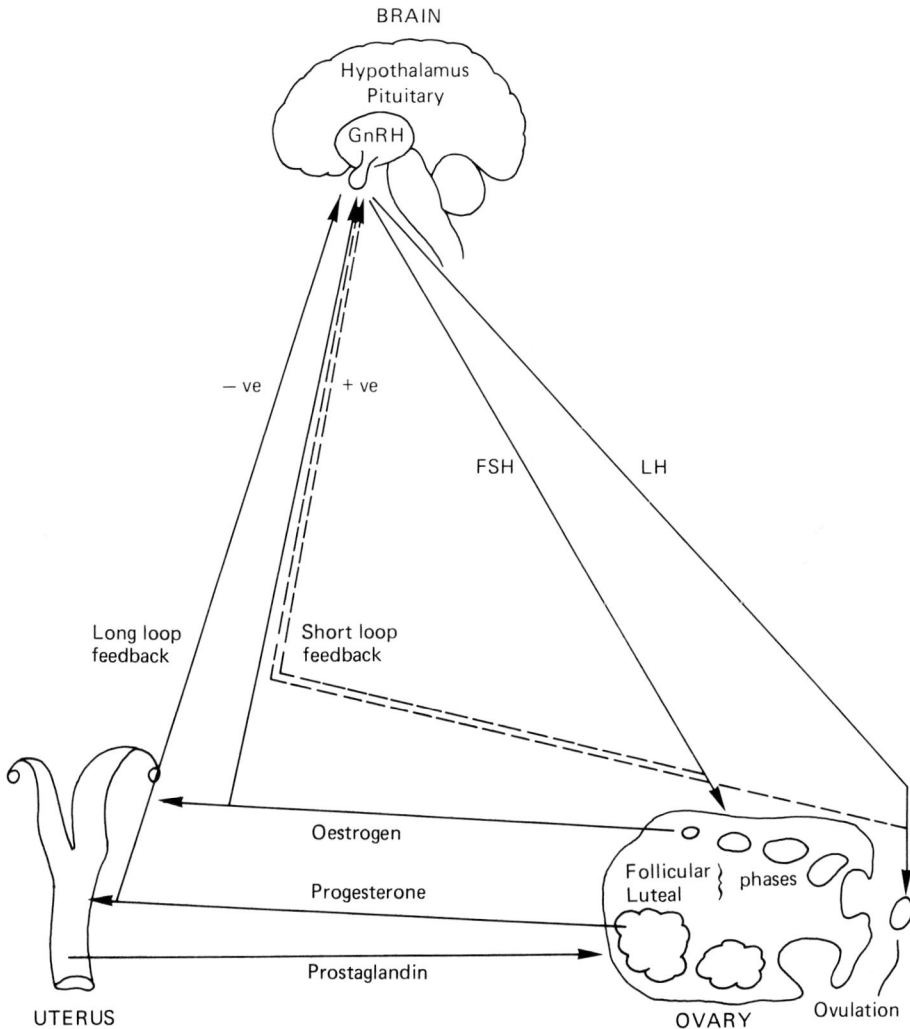

Figure 20.2 Interacting pathways of hormone regulation of the oestrus cycle in ewes (after Hunter, 1982)

Manipulation of oestrus

Manipulating the oestrus cycle of ewes involves techniques which artificially control the hormone status either by the addition of extraneous hormones or by methods which will lead the ewe to alter her own hormone status, with the purpose of bringing together into heat groups of animals in response to the treatment. Whatever method is adopted Chesworth (1974) has specified the following requirements of any treatment. It must be a simple procedure, cheap, use minimal labour, have the same effect on all animals and not interfere with sperm transport, reduce fertility or have any side-effects.

In artificially controlling the hormone status, because progesterone can be considered as the 'organizer' of the oestrous cycle, manipulation of the progesterone status should provide a convenient means of controlling oestrus (Haresign, McLeod and Webster, 1983). The period of progesterone secretion can be ended synchronously in a group of ewes by either shortening or lengthening the period artificially.

Prostaglandin treatment
Shortening the period of progesterone secretion is achieved by administration of extraneous prostaglandins to augment those produced naturally by the uterus. The corpus luteum will regress, and all the ewes in a treated group begin the follicular phase and return to oestrus at the same time. In order for this treatment to be effective, the ewes must possess an active corpus luteum (Haresign, 1978; Thimonier, 1981), and because in any group of ewes in which the oestrus cycles are occurring at random some ewes will not possess an active corpus luteum at the time of initial treatment, a double treatment is necessary to ensure that all animals are reprogrammed in synchrony (Hunter, 1982). In ewes the corpus luteum is sensitive to prostaglandin from day 4 to day 14 of the cycle, and two injections 8–9 days apart will cause most ewes to be on heat 2 days after the second injection. Lightfoot, Croker and Marshall (1979) showed a highly significant degree of synchronization using the $PGF_{2\alpha}$ analogue ICI80996, with 89% of treated ewes served compared to 44% of controls. There was no apparent effect of dose rate at levels between 40 and 120 μg. As anoestrous ewes do not possess an active corpus luteum, and in many instances oestrus control is needed in anoestrous ewes, prostaglandins are of limited use (Cognié and Mauleon, 1983). Even in seasonally active ewes, prostaglandin treatment is a less effective means of synchronizing oestrus than progestagen sponges (Henderson *et al.*, 1984).

Progestagen sponges
Lengthening the period of progesterone dominance is achieved by extraneous administration of progesterone, which extends the luteal phase thus suppressing follicular development, so that on synchronous removal of this negative feedback 'blockade' the ewes rebound into a compact follicular phase followed by synchronized oestrus (Hunter, 1982).

Up to 1964, progesterone was administered by daily injections or by feeding (Gordon, 1975). Daily injections have a high labour demand and are therefore impractical (Cumming, 1979), and feeding gives variable dose rates, and therefore a wider spread of oestrus occurrence, unless it is done individually and in higher quantities and therefore at greater cost, to ensure that all ewes receive an adequate dose (Cognié and Mauleon, 1983).

The practical breakthrough in progesterone administration came with the development of polyurethane sponges which could be impregnated with a synthetic progesterone, implanted intravaginally to release the progesterone through the vaginal wall into the blood-stream, and with the availability at the same time of synthetic progesterones—progestagens—which are more active than progesterone itself (Robinson, 1965).

Since that time much work has been done investigating the type of progestagen, its dose rate, the length of administration and its effect within the normal breeding season and on anoestrous and lactating ewes. This work has been comprehensively reviewed by Cognié and Mauleon (1983), and led to the development of a standard controlled breeding technique as described by Gordon (1975). The technique involves the intravaginal administration by sponges of 30 mg fluorogestone acetate (FGA, Cronolone, Chronogest sponges, Intervet Ltd) or 60 mg medroxyprogesterone acetate (MAP, Veramix sponges, Upjohn Ltd) for between 12 and 14 days, to promote synchronized oestrus in groups of treated ewes 2 days after sponge withdrawal. Additionally FSH in the form of PMSG administered either 2 days before or at sponge removal may be required in certain situations—particularly in anoestrus ewes (Haresign, 1978). Cyclic ewes normally show oestrus after sponge withdrawal without additional PMSG (Gordon, 1975). Henderson (1985) has proposed a guide for the requirement of PMSG—in any unsynchronized flock the date by which 50% of the flock has lambed should be calculated; if mating is to take place in the future more than 150 days prior to this date, then PMSG should be used. However, even with cyclic ewes it does appear that the use of PMSG at low levels—best defined as non-superovulatory levels—does give a more precise and reliable synchronization of oestrus (Cola et al., 1973), being slightly earlier and more compact (Hunter, 1982; Cognié and Mauleon, 1983). This extra precision may be needed to achieve accurate timing for the joining of rams or for artificial insemination in relation to ovulation in the ewe.

Generally the interval from the end of progestagen treatment (sponge removal) to the onset of oestrus is 36 h (Cognié and Mauleon, 1983) and from onset of oestrus to ovulation 21–26 h (Cumming et al., 1971, 1973; Hunter, 1982), giving a total time of between 57 and 62 h from sponge removal to ovulation. The general recommendation is to join rams 48 h, or for set-time artificial insemination at 56 h, after sponge removal (Joyce, 1972; Bryant and Tomkins, 1976; Boland and Gordon, 1979), but if greater variability in onset of oestrus results when PMSG is not used, it may be prudent to cover that greater spread by earlier ram joining at 24 h after sponge removal, despite the slightly later mean time of ovulation.

Mating management
Synchronizing oestrus in ewes requires a different approach to ram management. The first priority, whatever mating method is used is that the rams are capable both in libido and fertility, that they should be sexually experienced and have a good achievement in producing pregnancies (Gordon, 1977), and that they should be properly prepared by adequate nutrition (Cognié and Mauleon, 1983) and health routines prior to mating.

Commercial practice requires a simple but effective mating system, which is realistic under most conditions (Tempest, 1983). Paddock mating systems can give acceptable conception rates if the number of ewes per ram is restricted. Although Colas, Brice and Guerin (1974) suggested five ewes per ram, Cognié and Mauleon (1983) five to seven ewes per ram, and Bryant and Tomkins (1975) achieved conception rates to the first synchronized oestrus of 87% with six ewes per ram, very acceptable conception rates of 73% to the first oestrus were achieved at 12 ewes/ram (Bryant and Tomkins, 1976) and 89% to the first and second synchronized oestruses combined at ten ewes per ram by Tempest (1983). Although these higher ewe:ram ratios are low compared to normal commercial practice of 37:1 (Meat and Livestock Commission (MLC), 1984), it still does not mean that a formidable array of rams is needed to ensure adequate mating power (Gordon, 1975) if the flock can be split into staggered mating subflocks. Tempest (1983) used one ram to 30 ewes in the flock, and Galindez, Prud'hon and Reboul (1977) found that it was possible for one ram to serve 48–64 ewes without any decrease in fertility by spreading the timing of sponge withdrawal. This may appear to be against the overall objective of synchronization, but it enables groups of ewes to lamb at specified preset times and intervals, rather than have one long protracted lambing period.

As an alternative to paddock mating, the technique of hand mating (Jennings and Crowley, 1970, 1972) can be used, but Gordon (1975) is unclear what this means in terms of ram numbers and labour involved in its application. McClelland and Quirke (1971) and Robinson (1974) used hand mating at 48 and 60 h after sponge withdrawal to achieve satisfactory results, and Joyce (1972) in using hand mating at 48 h followed by exposure to rams in the field did not improve substantially on the conception rates achieved by paddock mating.

Problems associated with progestagen sponge treatment and alternative forms of administration

In the initial development of progestagen-impregnated sponges, fertility at the first synchronized oestrus was quite variable (Cognié and Mauleon, 1983). Quinlivan and Robinson (1969) found impairment of sperm transport and survival due to residues left in the vagina from the synthetic progestagens, and Hawk and Cowley (1971) showed that sperm removal from the female tract was more rapid following vaginal compared with oral administration of progestagens.

Several components of the treatments were modified. Most notably mating was delayed until the second half of heat or artificial insemination limited to over 55 h after the end of progestagen treatment (Colas, 1975), or mating was delayed to the second synchronized oestrus (Boaz and Tempest, 1975). Alternative methods of administering the progestagen by non-vaginal routes were investigated. Oral administration (Lindsay *et al.*, 1967; Cognié and Mauleon, 1983) produced fertility as good as that with vaginal sponges but required twice as much progestagen. Subcutaneous implants convenient when placed in the ear (Boland, Kelleher and Gordon, 1979) allow very low dose rates to be used—3 mg Norgestomet (SC21009, Searle)—and also result in very rapid onset of oestrus after the end of treatment, requiring earlier mating. Fertility was similar to sponge treatments, but prolificacy was lower (Cognié, Mariana and Thimonier, 1970; Cognié, Folch and Alonso de Miguel, 1976).

Many of the early problems were due to lack of standardization in vaginal sponge production in terms of type of progestagen, activity of the progestagen and dispersal of the required dose of the progestagen evenly throughout the sponge. These problems have now been largely overcome, and the sponge treatment is now widely accepted as the standard treatment (Cognié and Mauleon, 1983) (*Table 20.1*).

TABLE 20.1 Synchronization of breeding. (After Tempest, 1983)

Day	Operation
−14	Progestagen sponges inserted
−2	Progestagen sponges withdrawn
	PMSG injected if required
0	Rams joined

Synchronization of lambing

Because perinatal lamb losses constitute a major source of loss in sheep production, control of the time of lambing enabling maximum supervision would generally be beneficial. Developments in the knowledge of the endocrinology of spontaneous parturition (Challis et al., 1979) have indicated ways in which parturition may be induced artificially.

Spontaneous parturition

Hunter (1982) has described simply the current state of knowledge of the endocrinological mechanisms involved in spontaneous parturition. Once again, progesterone is the major hormone involved, having been secreted to maintain pregnancy by suppressing contractile activity of the uterus. Termination of pregnancy requires some means of curtailing the secretion of progesterone and overcoming the 'progesterone block'. The major source of progesterone is the corpus luteum, but additionally in the ewe this is substantially supplemented by secretion from the placenta.

It is thought that a change of hormonal secretions, referred to as the hormonal switch, is initiated in the hypothalamus of the fetus in response to conditions of fetal stress. This triggers a series or 'cascade' of hormonal steps, firstly stimulating the secretion of a releasing hormone which stimulates the fetal pituitary to produce increasing amounts of adrenocorticotrophic hormone (ACTH). This in turn causes enhanced secretion of corticosteroids into the fetal circulation, which mediates $PGF_{2\alpha}$ secreted in the placenta which in its turn regresses the corpus luteum and decreases progesterone secretion from that source. The progesterone produced by the placenta still has to be curbed and this is done under the influence of fetal cortisol, which enzymatically converts the progesterone to oestrogen. The oestrogen sensitizes the uterus to resume contractile activity under the influence of prostaglandins and uterine muscular activity begins. The sequence of hormonal steps is illustrated diagrammatically in Figure 20.3 and the changing levels of the major constituents shown in Figure 20.4.

Induced parturition

It has been shown experimentally that infusion of ACTH or cortisol into the fetal circulation initiates parturition (Hunter, 1982), but this is impractical in the field. However, it would appear that the administration of $PGF_{2\alpha}$ or corticosteroids to the dam should induce parturition. In those species where placental progesterone secretion is not a complication—such as pigs—a single injection of a prostaglandin analogue (for example, cloprostenol; Planate, ICI) will induce farrowing by prompt regression of the corpora lutea (Hunter, 1982). In the ewe the situation is

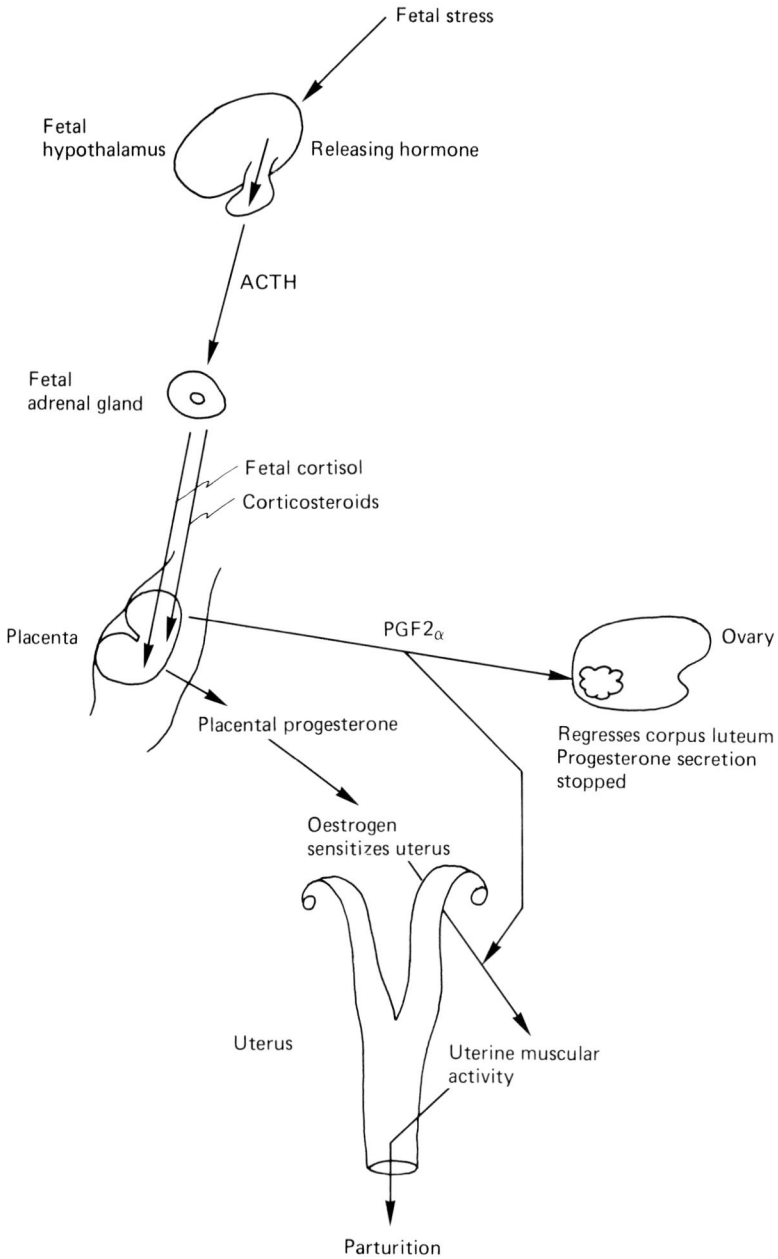

Figure 20.3 The sequence of hormonal steps involved in spontaneous parturition (after Hunter, 1982)

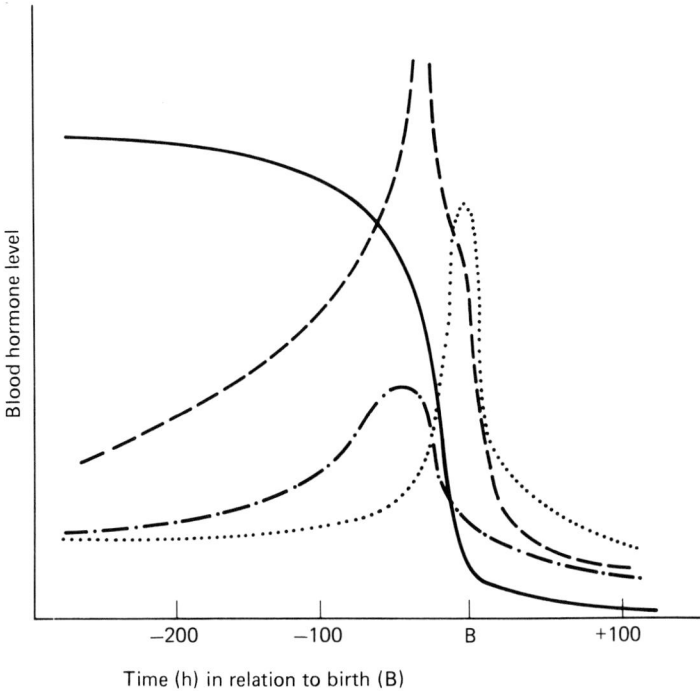

Figure 20.4 The pattern of changing hormone concentrations during spontaneous parturition; ————
progesterone; ---- oestrogen; –·–·– prostaglandin; ...... corticosteroid (after Challis *et al.*, 1979 and
Hunter, 1982)

complicated by the additional secretion of progesterone by the placenta and
prostaglandin treatment is relatively ineffective (Boland, Crosby and Gordon,
1982). Treatment of the ewe involves injection with a synthetic corticosteroid drug
at a known stage of gestation (Bosc, 1973; Harrison, 1982), the objective of which
is to simulate the fetal switch for the termination of pregnancy.

There are two main synthetic corticosteroids—betamethasone and dexamethasone—and these are usually available in long and short-acting forms, depending on
the method of preparation. Long-acting forms are suspensions or insoluble esters of
the synthetic corticosteroid, short-acting forms are in free alcohol or soluble ester
(Hunter, 1982). The length of activity is relative within each steroid, as from
product descriptions it appears that betamethasone might be relatively quicker
acting than dexamethasone. Details of the commercial preparations available are
shown in *Table 20.2*, in which the form of preparation, the concentration of active
ingredient, the proprietary name of the product and the manufacturer are shown.

Hunter (1982) describes the effect of a single injection of 16 mg short-acting
dexamethasone on day 144 of gestation (for ewes with gestation lengths between
144 and 153 days) in inducing lambing 45–48 h after treatment, with 75% of the
lambing concentrated in a 24 h period. At a lower dose of 12 mg, the response is
delayed until 52–55 h. Clarkson and Faull (1983) used 16 mg betamethasone or
dexamethasone on 33 ewes selected to be injected at 9 pm on day 142 of pregnancy.
Two ewes lambed before treatment, three within 24 h, three between 24 and 36 h,
24 between 36 and 48 h (9 am to 9 pm of day 144) and one 55 h after treatment.

TABLE 20.2 Types of synthetic corticosteroids

	Betamethasone	*Dexamethasone*
Short-acting	Clear aqueous solution 2 mg active/ml Betsolan soluble (Glaxovet Ltd)	Clear aqueous solution 2 mg active/ml Dexadreson (Intervet Ltd)
Long-acting	White aqueous suspension 2 mg active/ml Betsolan injection (Glaxovet Ltd)	White aqueous suspension 1 mg short, 2 mg long, active/ml Dexafort (Intervet Ltd)

Webster and Haresign (1981) found no significant difference between 6 mg and 12 mg of dexamethasone administered to Clun Forest ewes on day 142 of gestation, but in attempting to treat a whole flock of synchronized ewes without reference to returns to service, of those ewes which were at 125 days of gestation when treated 42% aborted and their lambs died at birth and the remaining ewes subsequently lambed normally at term. Box (1982, personal communication) recommends day 139 of gestation as being the earliest time for successful induced lambing, and has successfully induced non-oestrus synchronized ewes by taking all the ewes served naturally within a 7 day period and inducing them 140 days after the last ewe was served. This gave a span of lambings over a 24 h period with gestation ages of 141–148 days.

Tempest, Minter and Dodman (1987) have established a comprehensive body of data on the effect of various dose rates and timings of betamethasone and dexamethasone in inducing lambing in oestrus synchronized ewes. In trial 1, 211 Finn–Dorset ewes (Finnish Landrace × Dorset Horn) whose mean natural gestation length had been previously determined as 144 days (slightly shorter than the norm of 147 days taken for British breeds) were treated with either betamethasone suspension (as Betsolan injection) or dexamethasone solution (as Dexadreson) by single intramuscular injection of 16, 12 or 8 mg at 18.00 h on day 141 of gestation, or by 'double' intramuscular injection of 8 mg at 18.00 h on day 141 followed by 8 mg at 06.00 h on day 142. The results are shown in *Table 20.3*.

There was no significant effect of the dose rate of dexamethasone on the interval from treatment to lambing, but the higher the dose of betamethasone the shorter was the treatment to lambing interval ($P<0.05$). At all dose rates, betamethasone

TABLE 20.3 The effect of different synthetic corticosteroids at different dose rates on induced parturition in the ewe (interval from initial treatment to lambing in hours) (from Tempest, Minter and Dodman, 1987)

Drug		*Dose rate (mg)*				*Signifi-cance*
		16	*12*	*8*	*8 + 8*	
Betamethasone	Mean	46.17	44.66	49.34	56.94	*
(suspension)	Range	26.85–63.73	15.33–60.67	21.25–86.25	43.37–67.40	
	SD	8.62	16.14	14.18	8.80	
Dexamethasone	Mean	57.85	56.13	54.37	60.91	NS
(solution)	Range	11.50–180.18	30.47–176.12	18.25–116.48	40.93–151.52	
	SD	33.85	29.29	23.46	21.90	
Significance		***	**	*	***	

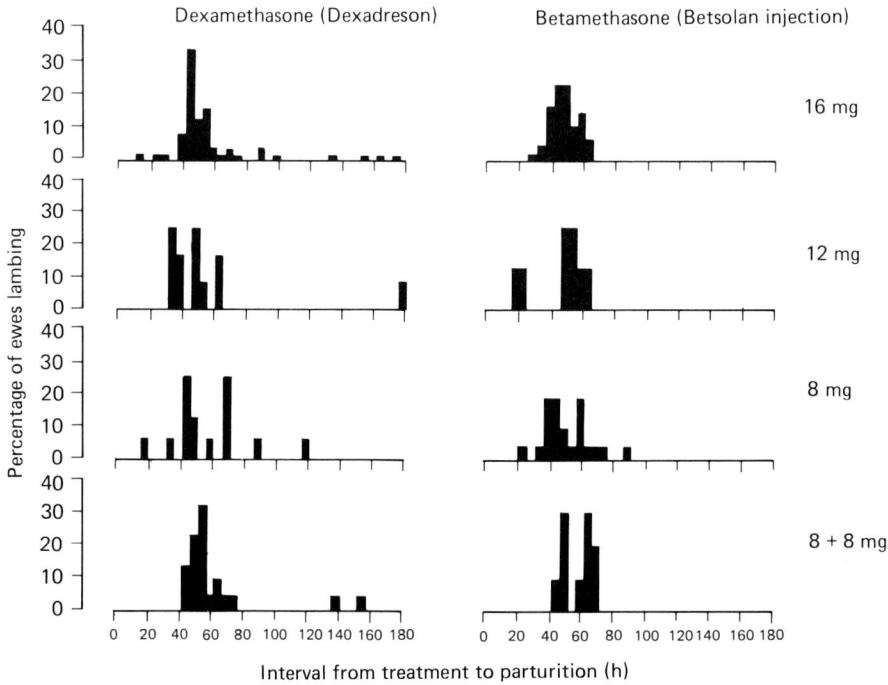

Figure 20.5 The pattern of lambing in ewes induced with various dose rates of betamethasone and dexamethasone (from Tempest, Minter and Dodman, 1987)

suspension (the longer-acting betamethasone preparation) gave significantly shorter treatment to lambing intervals than dexamethasone solution (the shorter-acting dexamethasone preparation).

Of greater practical importance than the interval between treatment and lambing is the spread and pattern of lambing. The spread of lambing is shown by the standard deviation (SD), all betamethasone treatments having a shorter spread than the dexamethasone treatments, with very compact lambings when 16 mg was used either singly or as two 8 mg injections. The single 16 mg treatment had a slightly smaller SD than the 8 + 8 mg treatment, despite a longer spread of 37 h compared to 24 h, indicating a very compact lambing for the majority of ewes but with odd ewes at a slightly wider spread. The patterns of lambing for the various treatments are shown in Figure 20.5.

TABLE 20.4 The effect of different forms of 16 mg betamethasone on treatment to lambing interval (in hours) (from Tempest, Minter and Dodman, 1987)

Form	Suspension		Solution
Number of ewes	32		31
Mean interval	49.52		40.90
Range	34.83–66.68		13.12–57.95
SD	6.86		10.68
SE	1.58		1.61
Significance		***	

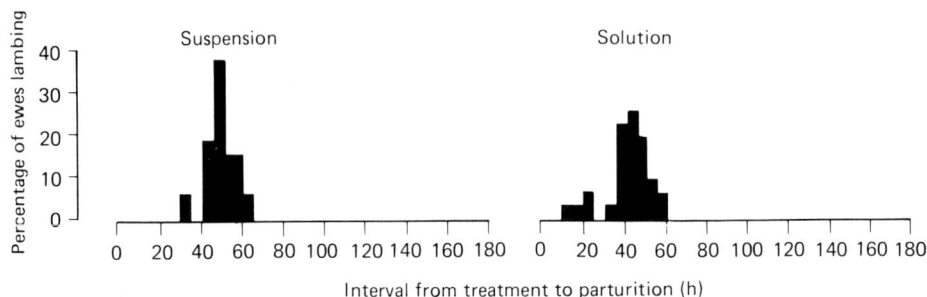

Figure 20.6 The pattern of lambing in ewes induced with 16 mg betamethasone in suspension form (Betsolan Injection) or solution form (Betsolan Soluble). (From Tempest, Minter and Dodman, 1987)

The long-acting nature of betamethasone suspension relative to betamethasone solution was tested in a second trial involving 63 Finn–Dorset ewes, at a dose rate of 16 mg given by single intramuscular injection at 18.00 h on day 141 of gestation. The results are shown in *Table 20.4* and *Figure 20.6*.

The results confirmed the relative speed of activity of the two forms of betamethasone, the solution having a significantly ($P<0.001$) shorter mean interval from treatment to lambing by nearly 9 h, but a greater spread of 45 h compared to the suspension's 32 h.

Dexafort, the mixture of one-third short-acting and two-thirds long-acting dexamethasone, was not used in any of the trials of Tempest, Minter and Dodman (1987), because of the relative lack of success of the short-acting dexamethasone on its own and because the product information indicates that although 'the short-acting fraction produces a rapid response which is maintained for approximately 48 h, the long-acting fraction does not produce its maximum effect for 48 h but extends the total period of significant activity to at least eight days'. However Dýrmundsson (1986, personal communication) has used 5 ml Dexafort, supplying 5 mg short-acting and 10 mg long-acting dexamethasone, on either day 140 or 141 of gestation in ewes mated at a synchronized oestrus in December in two trials in Iceland. Mean treatment to lambing interval was 40 h in one trial and 60 h in the other, with a spread of approximately 48 h in each case. Long-acting corticosteroids have been found to have benefits in cattle where they lead to a lower incidence of retained placenta (although stillbirths were more common) (Hunter, 1982). None of the aforementioned trials mention any such problems with the exception of Tempest, Minter and Dodman (1987) where retained cleansing was found in 8% of the ewes treated with 16 mg of short-acting dexamethasone. Although the potential exists when inducing lambing to produce premature lambs of lower birthweight and delayed onset or depressed lactation, both Webster and Haresign (1981), and Tempest, Minter and Dodman (1987) reported no significant effect on these parameters.

A third synthetic corticosteroid, flumethasone, has been used in trials in South Africa, and although successful in inducing parturition did not give good synchrony, with a mean interval of 82 h from treatment to lambing, and a range of from 18 to 276 h (Skinner, Jochle and Nel, 1970).

These various results confirm the earlier view of Tempest (1983) that the optimum treatment for an induced 24 h synchronized lambing is 16 mg

TABLE 20.5 Synchronization of lambing (after Tempest, 1983)

Day	Hour	Operation
+140	—	Ewes penned
+141	18.00	Ewes induced by single intramuscular injection of 16 mg betamethasone suspension
+143	06.00	Commencement of lambing
+144	06.00	Lambing finished

betamethasone suspension. The programme for Finn–Dorset ewes (mean natural gestation length 144 days) is shown in *Table 20.5*.

Advantages and disadvantages of synchronized breeding and lambing

A scheme in which groups of ewes come on heat and are mated together, go through pregnancy together and lamb together has advantages which are essentially managerial. Likewise, disadvantages will also be of a managerial nature. The advantages are probably best stated as a list, attempting to give some order or priority.

Advantages

(1) Increased supervision of ewes in labour and then of newborn lambs should reduce perinatal mortality (Hunter, 1982); 41% of lamb deaths are due to mismothering and starvation, and 19% of deaths due to disease are due to insufficient colostrum intake—causes which are avoidable by supervision (Meat and Livestock Commission (MLC), 1979). Lambing supervision also gives more opportunity for easier cross-fostering.

(2) Feeding groups of ewes in which the individual ewes are all at the same stage of production should enable more accurate feeding of the type and level of diet appropriate to the stage of production at all times; for example, pretupping, during pregnancy and in lactation. An indication of the potential saving of feed can be gained from a comparison of the ewe concentrate use in Tempest's (1983) frequent lambing flock of 50 kg per ewe for 1.5 lamb crops per year, with the use in annual lambing flocks recorded by the MLC of the same quantity but for one lamb crop per ewe per year (Meat and Livestock Commission (MLC), 1984), the frequent lambing flock practising synchronized breeding and lambing and performing at nearly twice the number of lambs per ewe per year.

(3) Lambs can be weaned in even batches, replacements reared to adequate body development for mating and slaughter lambs finished to specific requirements and timed to meet specific markets (Tempest, 1983).

(4) Ovulation can be precisely predicted, enabling the development of artificial insemination and thereby the use of better rams, and the development of embryo transfer for genetic improvement (Cognié and Mauleon, 1983).

(5) The use of resources such as buildings, and especially labour, can be planned, creating optimum use and avoiding excessive seasonal demands. Part-time labour use can also be planned effectively; 40% of the fixed costs of sheep production is labour, and 46% of the labour cost is directly associated with management at lambing (Meat and Livestock Commission (MLC), 1980). In a

100 ewe flock, it is estimated that it requires 1.82 h of labour costing £4.69 at lambing time per ewe. Even in a 1000 ewe flock the cost of £2.27 for 0.88 h per ewe is still substantial (Nix, 1986).

(6) Flock health programmes can be planned to ensure optimum efficacy of disease prevention measures, such as vaccinations, worming and/or optimum timing to avoid any potential depressing effects on performance from activities such as dipping.

(7) Early mating of ewe lambs can be induced (Quirke, Adams and Hanrahan, 1983).

(8) Frequent lambing systems can be developed (Thimonier and Cognié, 1971; Robinson, Fraser and McHattie, 1977; Tempest, 1983).

(9) The forward planning which is possible by controlling the date of lambing some 5½ months in advance must be attractive in large-scale systems of production (Hunter, 1982; *Table 20.6*).

TABLE 20.6 Forward planning using synchronized breeding and lambing, to avoid weekend work, and give a full working week to deal with the main group of newly lambed ewes and newborn lambs (After Tempest, 1983)

Operation	Day	Weekday
Sponges in	−14	Friday
Sponges out	− 2	Wednesday
Rams in	0	Friday
Rams out	+3	Monday
Rams in for repeats	+14	Friday
Repeats tupped	+16	Sunday
Rams out	+18	Tuesday
Pen first group of ewes	+140	Friday
Induce	+141	Saturday
Lamb	+143	Monday
Pen repeat ewes	+157	Monday
Induce repeats	+157	Monday
Lamb repeats	+159	Wednesday

Disadvantages

The major disadvantage of synchronized breeding is probably the need for more rams, until such a time as an effective sheep artificial insemination service can be commercially developed. Gordon (1975) has outlined the potential for the development of 'controlled breeding centres'.

According to Hunter (1982) the potential area for disastrous consequences in synchronized lambing is premature induction of a fetus which by its rapid growth in the later stages of gestation is building up its energy reserves in preparation for birth. The risk is of weak and dead lambs. Accurate mating records are required and a higher degree of management skill. Greater shepherding skill is also required to cope with the potential significant problem of mismothering in a condensed lambing period. Thus indoor lambing, more lambing pens and more facilities (such as lamb warming boxes) are essential.

Generally more effort is required in planning and precisely executing a plan. There is the additional cost of the drugs needed. A synchronized breeding programme with PMSG, and induced lambing may currently cost £4 per ewe. If lambs are valued at £40, then an extra 0.1 lambs per ewe need to be reared and sold to cover the additional costs; ten lambs extra from 100 ewes is more than possible by reduced mortality at supervised lambing, and there are additional savings of feed and labour.

Conclusions

There is no better summing-up of the techniques of synchronized breeding and induced lambing which go hand-in-hand than that which the authors have adapted from Hunter (1982): that it is 'manifestly absurd in the 1980s' to leave the breeding and lambing of sheep to external influences and chance when both events can be controlled without decreasing, and in most cases enhancing, performance. The potential benefits of all the other new techniques detailed in this volume will not be achieved unless there is the management care offered by synchronized lambing in conjunction with synchronized breeding.

References

BOAZ, T.G. and TEMPEST, W.M. (1975). Some consequences of high flock prolificacy in an intensive grassland sheep production system. *Animal Production*, **20**, 219–232

BOLAND, M.P. and GORDON, I. (1979). Effect of timing of ram introduction on fertility in progestagen-PMSG treated anoestrous ewes. *Journal of Agricultural Science, Cambridge*, **92**, 247–249

BOLAND, M.P., CROSBY, T.F. and GORDON, I. (1982). Induction of lambing: comparison of the effects of prostaglandin, oestradiol benzoate and dexamethasone. *Journal of Agricultural Science, Cambridge*, **98**, 391–394

BOLAND, M.P., KELLEHER, D. and GORDON, I. (1979). Comparison of control of oestrus and ovulation in sheep by an ear implant (SC21009) or by intravaginal sponge. *Animal Reproduction Science*, **1**, 275–283

BOSC, M.J. (1973). A review of methods of inducing parturition in the ewe and the cow. *Revue Médicale et Veterinaire*, **149**, 1463–1480

BRYANT, M.J. and TOMKINS, T. (1975). The flock mating of progestagen-synchronised ewes. 1. The influence of ram-to-ewe ratio upon mating behaviour and lambing performance. *Animal Production*, **20**, 381–390

BRYANT, M.J. and TOMKINS, T. (1976). The flock mating of progestagen-synchronised ewes. 2. The influence of time of ram introduction upon mating behaviour and lambing performance. *Animal Production*, **22**, 379–384

CHALLIS, J.R.G., CARSON, G.D., GLICKMAN, J.A., MACLENNAN, E.A., MANCHESTER, L., PATRICK, J.E. and WORKEWYCH, J. (1979). The peripartal endocrine system in sheep. *Animal Reproduction Science*, **2**, 29–42

CHESWORTH, J.M. (1974). Recent advances in methods of manipulating oestrus activity. *Proceedings of the Symposium on Detection and Control of Breeding Activity in Farm Animals*, edited by J.B. Owen, pp. 26–36. University of Aberdeen

CLARKSON, M.J. and FAULL, W.B. (1983). Control of breeding in sheep. In *Notes for the Sheep Clinician*, pp. 6–8. Liverpool: Liverpool University Press

COGNIÉ, Y. and MAULEON, P. (1983). Control of reproduction in the ewe. In *Sheep Production*, edited by W. Haresign, pp. 381–392. London: Butterworths

COGNIÉ, Y., FOLCH, J. and ALONSO DE MIGUEL, M. (1976). Utilisation des implants sous-cutanés de SC21009 pour la synchronisation des chaleurs chez la brebis. *Deuxièmes Journées de la Recherche Ovine et Caprine*, pp. 288–294

COGNIÉ, Y., MARIANA, J.C. and THIMONIER, J. (1970). Étude du moment d'ovulation chez la brebis normale ou traité par un progestagène associé ou non á une injection de PMSG. *Annales de Biologie Animale Biochimie Biophysique*, **10**, 15–24

COLAS, G. (1975). The use of progestagen SC9880 as an aid for artificial insemination in ewes. *Annales de Biologie Animale Biochimie Biophysique*, **15**, 317–327

COLAS, G., BRICE, G. and GUERIN, Y. (1974). Recent progress in sheep artificial insemination. *Bulletin Technique d'Information de Ministère d'Agriculture*, **294**, 795–800

COLAS, G., THIMONIER, J., COUROT, M. and ORTAVANT, R. (1973). Fertility, prolificacy and fecundity during the breeding season of ewes artificially inseminated after treatment with fluorogestone acetate. *Annales de Zootechnie*, **22**, 441–451

CUMMING, I.A. (1979). Synchronisation of ovulation. In *Sheep Breeding*, 2nd edition, edited by G.L. Tomes, D.E. Robertson and R.J. Lightfoot, revised by W. Haresign. pp. 403–422. London: Butterworths

CUMMING, I.A., BROWN, J.M., DE-BLOCKEY, M.A., WINFIELD, C.G., BAXTER, R.W. and GODING, J.R. (1971). Constancy of interval between luteinizing hormone release and ovulation in the ewe. *Journal of Reproduction and Fertility*, **24**, 134–135

CUMMING, I.A., BUCKMASTER, J.M., DE-BLOCKEY, M.A., GODING, J.R., WINFIELD, C.G. and BAXTER, R.W. (1973). Constancy of interval between luteinising hormone release and ovulation in the ewe. *Biology of Reproduction*, **9**, 24–29

GALINDEZ, F.J., PRUD'HON, M. and REBOUL, G. (1977). Reproductive performance of group synchronised Merino d'Arles and Romanov crossbred ewes. 1. A note on the effect of lactation on fecundity. *Animal Production*, **24**, 113–116

GORDON, I. (1975). Hormonal control of reproduction in sheep. *Proceedings of the British Society of Animal Production*, **4**, 79–93

GORDON, I. (1977). Application of synchronisation of oestrus and ovulation in sheep. *Proceedings of the Symposium on Management of Reproduction in Sheep and Goats*, pp. 15–30. Madison, Wisconsin: University of Wisconsin

HARESIGN, W. (1978). Ovulation control in the sheep. In *Control of Ovulation*, edited by D.B. Crighton, N.B. Haynes, G.R. Foxcroft and G.E. Lamming, pp. 435–451. London: Butterworths

HARESIGN, W., McLEOD, B.J. and WEBSTER, G.M. (1983). Endocrine control of reproduction in the ewe. In *Sheep Production*, edited by W. Haresign, pp. 353–379. London: Butterworths

HARRISON, F.A. (1982). Dexamethasone induced parturition in sheep. *British Veterinary Journal*, **138**, 402–409

HAWK, H.W. and COWLEY, H.H. (1971). Loss of spermatozoa from the reproductive tract of the ewe and intensification of sperm 'breakage' by progestagen. *Journal of Reproduction and Fertility*, **27**, 339–347

HENDERSON, D. (1985). Control of the breeding season in sheep and goats. *In Practice*, **7**, 118–123

HENDERSON, D.C., DOWNING, J.M., BECK, N.F.C. and LEES, J.L. (1984). Oestrus synchronisation in ewes: a comparison of prostaglandin F$_2$alpha tham salt with a progestagen pessary. *Animal Production*, **39**, 229–233

HUNTER, R.H.F. (1982). *Reproduction of Farm Animals*. London: Longman

JENNINGS, J.J. and CROWLEY, J.P. (1970). The mating of hormone treated sheep. *Animal Production*, **12**, 357 (abstract)

JENNINGS, J.J. and CROWLEY, J.P. (1972). The influence of mating management on fertility in ewes following progesterone-PMS treatment. *Veterinary Record*, **90**, 495–498

JOYCE, M.J.B. (1972). A comparison of three different mating systems. *Proceedings of the Seventh International Congress of Animal Reproduction and Artificial Insemination*, Munich, **2**, 935–938

LIGHTFOOT, R.J., CROKER, K.P. and MARSHALL, R. (1979). Use of a prostaglandin analogue (ICI80996) for the synchronisation of oestrus and lambing in Merino ewes. In *Sheep Breeding*, 2nd edition, edited by G.L. Tomes, D.E. Robertson and R.J. Lightfoot, revised by W. Haresign. pp. 451–456. London: Butterworths

LINDSAY, D.R., MOORE, N.W., ROBINSON, T.J., SALOMON, S. and SHELTON, J.N. (1967). The evaluation of an oral progestagen (Provera, MAP) for the synchronisation of oestrus in the entire cyclic Merino ewe. In *The Control of the Ovarian Cycle in the Sheep*, edited by T.J. Robinson, pp. 3–13. Sydney: Sydney University Press

McCLELLAND, T.H. and QUIRKE, J.F. (1971). Artificial insemination and natural service at a predetermined time in cyclic sheep treated with SC9880/progesterone sponges. *Animal Production*, **13**, 323–327

MEAT AND LIVESTOCK COMMISSION (1978). *Sheep Facts—A Manual of Economic Standards*. Bletchley: Meat and Livestock Commission

MEAT AND LIVESTOCK COMMISSION (1979). *Data Summaries on Upland and Lowland Sheep Production*. Bletchley: Meat and Livestock Commission

MEAT AND LIVESTOCK COMMISSION (1980). *Commercial Sheep Production Yearbook 1979–80*. Bletchley: Meat and Livestock Commission

MEAT AND LIVESTOCK COMMISSION (MLC) (1984). *Sheep Yearbook October 1984*. Bletchley: Meat and Livestock Commission

NIX, J. (1986). *Farm Management Pocketbook*, 16th edition. Ashford: Farm Business Unit, School of Rural Economics, Wye College

QUINLIVAN, T.D. and ROBINSON, T.J. (1969). Numbers of spermatozoa in the genital tract after artificial insemination of progestagen treated ewes. *Journal of Reproduction and Fertility*, **19**, 73–86

QUIRKE, J.F., ADAMS, T.E. and HANRAHAN, J.P. (1983). Artificial induction of puberty in ewe lambs. In *Sheep Production*, edited by W. Haresign, pp. 409–429. London: Butterworths

ROBINSON, J.J. (1974). Intensifying ewe productivity. *Proceedings of the British Society of Animal Production*, **3**, 31–40

ROBINSON, J.J., FRASER, C. and McHATTIE, I. (1977). Development of systems for lambing sheep more frequently than once per year. In *Sheep Nutrition and Management*, pp. 5–33. London: US Feed Grains Council

ROBINSON, T.J. (1965). Use of progestagen-impregnated sponges inserted intravaginally or subcutaneously for the control of the oestrus cycle in sheep. *Nature (London)*, **206**, 39–41

SHORT, R.V. (1982). Role of hormones in sex cycles. In *Hormones in Reproduction*, edited by C.R. Austin and R.V. Short, pp. 42–72. Cambridge: Cambridge University Press

SKINNER, J.D., JOCHLE, W. and NEL, J.W. (1970). Induction of parturition in Karakul and cross-bred ewes with flumethasone. *Agroanimalia*, **2**, 99–100

TEMPEST, W.M. (1983). Management of the frequent lambing flock. In *Sheep Production*, edited by W. Haresign, pp. 467–481. London: Butterworths

TEMPEST, W.M., MINTER, C.M. and DODMAN, N.W. (1987). Induction of parturition in synchronised mated ewes using synthetic corticosteroids. *Veterinary Record* (in preparation)

THIMONIER, J. (1981). Practical uses of prostaglandins in sheep and goats. *Acta Veterinaria Scandinavica*, **Supplement 77**, 193–208

THIMONIER, J. and COGNIÉ, Y. (1971). Accélération des mise-bas et conduite d'élevage chez les ovins. *Bulletin Technique d'Information du Ministère de l'Agriculture*, **257**, 187–196

WEBSTER, G.M. and HARESIGN, W. (1981). A note on the use of dexamethasone to induce parturition in the ewe. *Animal Production*, **32**, 341–344

Chapter 21

Increasing productivity by manipulating mating behaviour

I. Fayez M. Marai

Summary

The sudden introduction of rams just before the normal breeding season stimulates the breeding activity of ewes and helps in the advancement of the breeding season. Puberty of ewe lambs may also be advanced by the sudden introduction of the rams. The breeding season may be prolonged and length of the oestrous period is shortened by the introduction of the ram. Rams may contribute to variation in litter size of their mates through differences in fertilizing capacity of their semen or through the prenatal survival of their offspring.

Fertility in the flock could be stepped up by increasing the frequency of mating of the ewe through increasing the ram/ewe ratio and by mating at the optimum time, which differs according to type of insemination and to breed.

Sexual performance tests of rams have often been reported as not being closely related to the actual fertility obtained after natural mating in the flock. Replacement of rams which show little or no sexual activity during the first 1 or 2 weeks of mating may be more practical than to test the activity of all rams before they are used.

Introduction

Successful sheep production depends essentially on using efficient management techniques. This must take particular account of patterns of reproductive behaviour.

The understanding of such patterns and their influence on reproductive performance helps in developing new management techniques which can be applied in sheep production.

In this respect, much research work has been published, and some practices known for a number of years are now fully understood. However, there are a number of areas currently under study which will add to our knowledge of the subject and enable improvements in reproductive performance to be achieved.

In the present chapter, some of the areas which have a clear influence on reproductive performance of sheep, and could have importance in the management of breeding flocks, are reviewed.

Effect of the ram on reproductive performance of the ewe

Advancement of the breeding season

The breeding activity of ewes is stimulated by the sudden introduction of rams before the normal breeding season. This helps in the advancement of the breeding season which can lead to a real financial advantage by the production of lambs earlier than the regular production systems. The advancement can exceed 6 weeks in British breeds and Merino crossbred sheep (Lindsay, 1983). In the Merino sheep, this practice stimulates breeding activity at any time of the year, while in tropical sheep which cycle the entire year the introduction of rams resulted in approximately 25% of the ewes displaying oestrus the following day (Ngere and Dzakuma, 1975).

Schinckel (1954) first reported that the presence of the male can stimulate the onset of the breeding season in ewes and also assists in the synchronization of first oestrus. In the lactating ewe, a return to oestrous activity may be hastened by removal of her lambs for short periods each day before ram introduction, a practice that is termed partial weaning (Restall, 1971). The induction of oestrus does not occur when rams are kept permanently with the females; a period of isolation is necessary (Riches and Watson, 1954; Oldham, 1980b). The effective period of isolation to sensitize ewes to the ram is between 17 and 34 days (Oldham, 1980a). For the Île-de-France, it was found that 21 days isolation is adequate (Oldham, 1980b). Rams do not need to be in physical contact with the females to stimulate oestrus (Watson and Radford, 1960). Stimulation of oestrus may be mediated by the sense of smell (Morgan, Arnold and Lindsay, 1972), or it may involve a complex of signals (Signoret, 1980).

Oestrus induced by the introduction of rams to seasonally anoestrous ewes is known as the ram effect. The efficiency of the ram effect is usually measured under field conditions by the proportion of ewes which display oestrus 16–24 days after introduction of the rams (Oldham, 1980b; Murtagh et al., 1984). Knight (1980) suggests that exposure of the ewes to the ram need be only for 48 h rather than 16–24 days as advocated by others. Thus short exposure techniques enables one ram to be moved between several groups of ewes.

Murtagh et al. (1984) reported that the influence of the ram effect was greatest for adult ewes, followed by young ewes with prior experience of testosterone-primed wethers and lastly by young ewes exposed to testosterone-primed wethers for the first time. However, Oldham, Pearce and Gray (1985) reported that maiden ewes were as responsive to the ram effect as adult ewes in terms of induced oestrus. Oldham (1980b) and Murtagh et al. (1984) insisted that the ram effect varied with breed and strain of the ewe, age of rams, time of the year and other environmental factors.

The studies of Schinckel (1954) on Merino, and of Cognié et al. (1980) on Île-de-France and Préalpe sheep showed that the ewes rapidly returned to anoestrus after stimulation by rams. This problem did not appear in Romney ewes because they could only be stimulated by rams 2–6 weeks before the onset of the normal breeding season (Edgar and Bilkey, 1963).

When studying the hormonal reaction to the ram effect, it was found that the introduction of a ram to anoestrous ewes led to an increase in LH activity and subsequent ovulation (Oldham, Martin and Knight, 1978; Oldham, 1980b; Atkinson and Williamson, 1985). Within 10 min of introducing rams, the LH increased in the ewe (Martin, Oldham and Lindsay, 1980) but after 24 h, the basal concentrations of LH returned to the control group levels (Atkinson and Williamson, 1985). Within 48 h, nearly all the females ovulated (Oldham, Martin and Knight, 1978). The induced ovulation was silent and was followed in 50% of the females by an abnormal luteal phase of only 6–7 days (Martin, 1979). The percentage of multiple ovulations was higher at the silent ovulation than at the first overt oestrus or in spontaneously ovulating flock-mates (Knight, 1983). The FSH concentrations fell within 2 h of ram introduction and remained below the previously held level (Atkinson and Williamson, 1985). After the initial ram stimulus which led to ovulation, follicles continued to develop to the large follicle stage (> 4 mm), even though FSH and LH levels were depressed. Onset of oestrus occurred after about 3 weeks (Schinckel, 1954).

Knight (1983) pointed out that injection of 20 mg progesterone to ewes at ram introduction can give a high degree of synchronization 9–21 days later and eliminated the biphasic onset of the first oestrus seen in the untreated ewes due to premature regression of corpora lutea (recorded by Cognié et al., 1982). The increased ovulation rate at the silent ovulation can be capitalized on by priming with progesterone to stimulate oestrous behaviour.

Recent studies have shown that wethers or ewes, treated with 1 mg oestradiol benzoate or 105 mg testosterone propionate three times at weekly intervals, provided a reasonable response (at least as good as vasectomized rams) in stimulating anoestrous ewes to ovulate about 6 days after introduction; such animals are able to seek out and mount ewes in oestrus and identify them for subsequent artificial insemination (Signoret, Fulkerson and Lindsay, 1982). Treated wethers or ewes were recommended as substitutes for vasectomized rams, since the vasectomy operation is relatively costly, and such animals require more care in management.

It is known that pheromones, which are present in rams' wool, wax and urine, are the major component of the ram effect. Knight and Lynch (1980a) and Knight, Tervit and Lynch (1983) reported that anoestrous ewes, subjected to the application of wool and wax, taken from rams, three times a day for 2 days (in bags over the nose), exhibited the same response as when the rams themselves were introduced to the ewes. Urine collected from rams and sprayed over ewes showed a limited effect in stimulating anovular ewes to ovulate. It was also found that fatty acid pheromones isolated from the goat buck were as effective as rams in stimulating ewes (Knight, Tervit and Lynch, 1983). The pheromones are present in aqueous and petroleum spirit extracts of the wool and wax and can be absorbed into petroleum jelly which has been smeared over a ram's back for 24 h (Knight and Lynch, 1980b). The use of pheromones might overcome any unreliability of the ram effect, due to the between-breed and between-ram differences in ability to stimulate ewes, and possible seasonal variations in pheromone production.

As shown above, it is clear that the ram has an effect on the ewe; what then is the effect of the presence of cycling females on mature rams? Illius, Haynes and Lamming (1976) found that when mature rams were kept in close contact with females, the rams had larger testes and higher levels of testosterone than did rams which were kept separate from females. When the rams were tested with ewes, the same authors found that the rams kept in close proximity to ewes had more ejaculations than isolated animals. Knight (1985) recommended the introduction of one to two oestrous ewes per ram with the rams to increase the effectiveness of the rams in stimulating the anoestrous ewes that had been primed with progestagens.

In brief, the breeding activity of both ewes and rams could be stimulated by the sudden introduction of rams to ewes before the breeding season. With such practice, the breeding season could be advanced by nearly 6 weeks in the British breeds and Merino crossbred sheep. In the Merino sheep, the breeding activity is stimulated at any time of the year, with such practice.

Advancement of puberty, prolongation of the breeding season and shortening of the oestrous period

Puberty of ewe lambs may also be advanced by the sudden introduction of rams, and results in a high degree of synchronization at first mating (Dýrmundsson, 1986). However, Moore and McMillan (1981) found no effect of ram introduction on ovulation in pubertal ewe lambs. The breeding season may also be prolonged by the introduction of the ram to ewes (Riches and Watson, 1954).

Rams also influence the length of the oestrous period. Ewes remain on heat for longer periods if rams are absent, or present only intermittently, than if the rams are present continuously (Parsons and Hunter, 1967; Fletcher and Lindsay, 1971). Lindsay et al. (1975) confirmed that during oestrus itself, the presence of the male reduced markedly the duration of sexual receptivity and hastened ovulation by advancing the timing of the pre-ovulatory LH discharge.

In conclusion, the sudden introduction of the ewe lambs to rams may cause the advancement of puberty. In young and mature ewes, the breeding season may be prolonged and the length of the oestrus period is shortened by such practice.

Embryo survival

The ram may have a direct effect on the birth rate of ewes to which it has been mated (Bradford, 1972; Moore and Whyman, 1980). Parker (1972) reported that rams of different breeds tend to have a specific combining effect with ewes dependent on the breed of female. Davis et al. (1983) confirmed that mating with rams of the highly prolific Booroola breed (which have a major gene for ovulation rate as reported by Piper and Bindon, 1982), or its crosses, achieves a high lambing rate. However, Weiner (1967), Parker (1968), and Barker and Land (1970) reported that the breed of ram did not have a significant effect on the litter size of the ewes to which it was mated.

Within breeds, Newton and Betts (1968), and Turner and Young (1969) reported the possibility of a differential ram effect on the litter size of ewes in the Suffolk and Merino breeds. In the same respect, Hanrahan and Owen (1985) found in the Cambridge breed that the number of lambs born was much greater than expected from the ovulation rate, which suggested that embryo survival in the Cambridge was superior to other breeds. However, Moore and Whyman (1980) claimed that differences in prolificacy between two genotypes did not guarantee differences in the ram's ability to produce multiple births.

The rams may contribute to variation in litter size of their mates through differences in fertilizing capacity of their semen (Bradford, 1972), differences that stem from fertilizing more than one ovum, arising from variation in the viability of their spermatozoa within the ewe's reproductive tract and the frequency or duration of mating activity (Moore and Whyman, 1980), or through the prenatal survival of their offspring (Bradford, 1972). Vakil, Botkin and Roehrkasse (1968) pointed out that this ram effect is correlated with the prolificacy of its female relatives.

In general, mating with rams from certain breeds or with certain rams from the same breed, may result in an increase in the litter size of the ewes to which they have been mated.

Increasing fertility by regulating ram–ewe contact

Increasing the frequency of mating by increasing ram/ewe ratio

It has been shown that fertility is directly related to the number of services, that is, to the number of spermatozoa deposited in the ewes (Mattner and Braden, 1967; Entwistle and Martin, 1972). The number of times that a ewe may be inseminated during oestrus has been reported as six to seven times (Hulet, 1966), 12–15 services a day (Lambourne, 1956) or about once an hour (Gibson and Jewell, 1982). Also, ewes are usually served by more than one ram during oestrus (Gibson and Jewell, 1982), particularly where the ratio of rams to ewes is more than 1 : 100. However, sexual behaviour in the flock is determined by social interactions among the males (Lindsay et al., 1976).

Synnott and Fulkerson (1984) confirmed that when more than one ram was mated with a flock of ewes, modified social interaction between rams resulted in more ewes being served than if the same rams were joined individually. Mattner, Braden and George (1971) recorded similar numbers of services by Merino rams in single-sire flocks as by two rams per flock. In contrast, Bourke (1967) and Lindsay et al. (1976) demonstrated that the sexual performance of subordinate rams was adversely affected by competition. However, Mattner, Braden and Turnbull (1967), and Mattner, Braden and George (1973) reported that such an effect may be reduced under field conditions. When studying the effect of early social environment on the sexual behaviour of lambs, no differences were found between the behaviour of rams raised in isolation and those which were raised in a group (Bryant, 1975). On the other hand, it was found that fewer two-tooth females were mated when present in a group with mature ewes. Furthermore, these differences became greater as the ram : ewe ratio decreased (Allison, 1977). This may be due to the shorter oestrus of the young ewes and their subordination to older ones (Lambourne, 1956).

In general, when mating of the flock is required during a limited period of time, it is recommended that the ram(s) be changed (Thiery and Signoret, 1978), since this practice restores the libido and interest of the ram (Beamer, Bermant and Clegg, 1969).

A problem arises when the ram or rams who occupy the top positions in the dominance rating are totally infertile. In such cases, flock fertility is depressed, especially when the ram : ewe ratio is lower than that which ensures high flock fertility, as the number of subordinates present would be too few to modify the effects of the dominant ones (Fowler and Jenkins, 1976).

Hand-mating under certain farm conditions or perhaps with some sheep and rams (that is, with rams and ewes accustomed to pastoral conditions) may constitute an additional stress factor which may reduce rather than enhance conception at the synchronized oestrus (Gordon, 1983). Belschner (1965) reported that hand-mating was generally less effective in getting ewes pregnant than paddock mating, as well as being expensive and time-consuming. Joyce (1972) added that paddock matings can give conception rates following gonadotrophin treatments equal to those achieved by a hand-mating routine, provided that rams are kept away from ewes until 48 h after sponge removal.

Resolution of the problem of keeping a minimum, safe ratio of rams to ewes, which allows ewes to be covered in a reasonable period of time, is fundamental.

In practice, the ratio of rams to ewes varies widely. It has been reported in Australia that the ram : ewe ratio is usually 1 : 40–50 ranging to 1 : 100 with mature rams (Dawe *et al.*, 1970; Allison, 1972), though a ratio of 1 : 25 has been used for 1.5 year old Merino rams in paddock mating systems (Lightfoot and Smith, 1968). In New Zealand, 2 : 100 is typical (Haughey, 1959; Allison, 1975). In progestagen-treated ewes, Bryant and Tomkins (1975) used a ratio of 1 : 6, and among synchronized sheep Galindez, Prod'hon and Reboul (1977) used a ratio of 1 : 50. In most of the controlled breeding applications in Ireland a ratio of 1 : 10 is used in groups of not more than 50 ewes (Gordon, 1983). When two-tooth ewes were included in the flock, barrenness increased when a ram to ewe ratio of only 1 : 100 was used (Allison and Davis, 1976). This could be accounted for by the shorter period of oestrus (Badawy, El-Bashary and Mohsen, 1973) and reduced sexual activity of the younger ewes.

Rams served ten ewes a day over a 4 day period (Hulet, 1966) or eight to ten ewes in 24 h under field conditions, when allowed access to ewes in active oestrus (Lindsay and Ellsmore, 1968). Some rams, whose sexual motivation was maintained by provision of fresh ewes, ejaculated from 11 to 17 times per day with adequate sperm counts still recorded after 6 days, while in other rams, the number of spermatozoa per ejaculate was reduced below the level believed to be adequate for fertilization (Synnott, Fulkerson and Lindsay, 1981; Fulkerson, Synnott and Lindsay, 1982). Some studies have shown that flock fertility was not depressed when rams were performing six to eight matings a day over 17 days (Raadsma and Edey, 1984). However, Synnott, Fulkerson and Lindsay (1981) reported that their rams never mated more than five ewes in one day.

Decreasing the number of ewes joined from 100 to 25 per ram was associated with more rams serving each oestrous ewe and a greater proportion of ewes covered during the first 2 weeks of joining (Lightfoot and Smith, 1968). When synchronization is practised in the flock, short-term endurance of the ram becomes more important and the ram : ewe ratio must be increased. Jennings (1976) reported that the majority of matings of synchronized ewes occurred within 12 h after ram introduction and that harem formation was evident, but this only lasted 1 h. Under commercial conditions whatever ratio is chosen, it is recommended that at least three to four rams should be joined with the flock to compensate for variability in libido and fertility among individual rams.

One problem that arises when decreasing the number of ewes joined per ram is that of the farmer finding enough rams to meet his needs. Ram-sharing schemes have been set up to overcome this difficulty. Rams are moved to a farm for a day or two before the ewes are due to breed and work in the flock on the day of mating. With such systems, rams can be employed once weekly for a period of several weeks (Gordon, 1983).

In conclusion, the increase in ram : ewe ratio is associated with an increase in the lambing rate of the flock.

Mating at optimum time

Two daily peaks of mating activity have been identified. These were from 04.00 to 08.00 h (Blockey and Cumming, 1970; Cahill, Blockey and Parr, 1975) and from 16.00 to 20.00 h (Cahill, Blockey and Parr, 1975) coinciding with sunrise and sunset, respectively. Some of the mentioned authors attributed the decrease in mating activity between 8.00 and 16.00 h to the high ambient temperature experienced during the day. Blockey and Cumming (1970) found no such pattern of behaviour in a field study carried out in the cooler months of late autumn.

The proportion of ewes exhibiting the onset of oestrus at various periods throughout the day showed the same trend as the mating activity, with a high proportion of ewes exhibiting onset at sunrise and sunset (Robertson and Rakha, 1965; Cahill, Blockey and Parr, 1975). The highest proportion of ewes exhibiting the onset of oestrus with the shortest mean duration of oestrus were in the period from 16.00 to 20.00 h. However, Schindler and Amir (1972), and Jewell, Hall and Rosenberg (1986) found that there was no preferred time for the onset of oestrus.

The timing of ovulation was found to be 25.5 and 23.6 h after onset of oestrus in untreated and PMSG-treated ewes, respectively, by Whyman et al. (1979), while others reported that it was 24–40 h after onset of natural oestrus or at the oestrus following progestagen withdrawal (Schindler and Amir, 1972; Boshoff, van Niekerk and Morgenthal, 1973). When allowing continuous or frequent contact of ewes with rams, the mean time of ovulation was found to be several hours after the end of oestrus (Schindler and Amir, 1972; Whyman et al., 1979). In FSH-treated ewes, most ovulations occurred between 54 and 60 h after sponge removal (Walker, Smith and Seamark, 1986).

The maximum length of time that spermatozoa remain in the oviducts and retain their capacity to penetrate and fertilize ova was found to be 24–48 h in sheep (Green, 1947; Dauzier and Wintenberger, 1952).

During oestrus, the optimum time for a ram to inseminate is at 9–12 h after onset of oestrus (Slee, 1964; Jewell, Hall and Rosenberg, 1986). When using a single artificial insemination at various times during natural oestrus, the highest conception rate was obtained between 16 and 24 h after the onset of oestrus (Schindler and Amir, 1973). However, differences amongst breeds in optimum time for mating have been reported by Kelly (1937). Following gonadotrophin treatment the timing of insemination was found to be 48 h after sponge removal in natural mating (Joyce, 1972; Boland and Gordon, 1979), and in intrauterine insemination (Evans and Armstrong, 1984; Walker, Smith and Seamark, 1986). Ryan et al. (1984) obtained good fertilization rates when they inseminated ewes treated with FSH and PMSG with fresh semen 24 h after sponge removal.

Parturition may occur during the same half of the day as that when the ewes come into oestrus (Jewell, Hall and Rosenberg, 1986). However, some studies showed that most births took place during the hours of daylight (Lindahl, 1964; George, 1969).

From the above, it is clear that fertility in the flock could be increased by inseminating the ewes at a suitable time. This time differs according to breed, type of insemination, that is, natural or artificial and to the type of breeding, that is, controlled or uncontrolled.

Prediction of fertility and fecundity

Sexual performance tests

The level of sexual performance of the ram as measured in sexual behaviour tests has often been reported as not being closely related to the actual fertility obtained after natural mating in the flock, when the male is associated with a reasonable number of females.

Of the different sexual performance tests, the semen quality (Mickelsen, Paisley and Dahmen, 1981; 1982) mating competence, that is, unsuccessful/successful mounts (Cahill, Blockey and Parr, 1975; Winfield and Cahill, 1978) and libido score (Walkley and Barber, 1976) seemed to be not highly related to fertility. The studies of Hulet, Blackwell and Ercanbrack (1964) and Mattner, Braden and George (1973) showed, respectively, that 83% and 82% of rams that failed to show initial libido became active later. On the other hand, the studies of Mattner, Braden and George (1971) and Kilgour (1980) showed such tests to be related to fertility.

Winfield and Kilgour (1977) suggested that in some circumstances pen tests may enable the least fertile rams to be detected. Fletcher (1979) suggested that it was more practical for single-sire matings to replace rams which showed little or no sexual activity during the first 1 or 2 weeks of mating, than to test the activity of all rams before they are used. For multiple-sire matings, prior testing of ram activity would be appropriate only when there was evidence or inference that sexual inhibition was a cause of low lamb production. Mattner, Braden and George (1971) suggested that a simple form of libido test, based principally on a count of services performed, may provide a useful indication of the subsequent service activity of young rams under flock mating conditions.

One of the problems which affects breeding efficiency in sheep flocks is that some adult rams, although physiologically and behaviourly normal in every other respect, exhibit very little interest in oestrous ewes. Such rams fail to recognize the oestrous ewe as a positive stimulus which elicits sexual behaviour (Zenchak, Anderson and Schein, 1981). The sex behavioural deficiency in these rams may be related to lack of expression of sex-like behaviour while in all-male groups, engaging in higher levels of sex-like behaviour with other rams during rearing than do the normal rams (Zenchak and Anderson, 1980), or to their preference for rams as sexual partners (Zenchak, Anderson and Schein, 1981). Another problem that affects breeding efficiency in extensive range-sheep operations is the shyness of ewe lambs (Gonyou, 1983) and of 18 month old maiden ewes. The ewe lambs wander away from the rams during courting and after mating. Hulet and Bond (1984) reported that only about two-thirds of ewe lambs joined with ram lambs early in the breeding season conceived. The deficiencies in the behaviour of ewe lambs emphasizes the need for active and dextrous rams in flocks comprised of young females (Gonyou, 1983).

In summary, sexual performance tests conducted for rams at the beginning of the breeding season do not provide a definite indication of fertility. The least fertile rams could be detected by simple sexual performance tests.

Ram–ewe contact during mating

Recent studies show that mating activity in sheep involving ram–ewe contact during mating could be used in prediction of fertility and fecundity.

Fowler (1975) and Fowler and Langford (1976) demonstrated a close relationship between mating activity of ewes and the number of corpora lutea, of fetuses, and of lambs born in the flock, irrespective of age of the ewes and rams, live weight of ewes, nutrition, genotype, flock size, season, year, climate and location of joining. Fowler (1976, 1977) and Fowler and Langford (1976) made an attempt to use mating activity in an equation for predicting fertility and fecundity. The measurements used for mating activity were the time from the first to last observed mount for each ewe, summed over all ewes mounted more than once (T) and the total number of mounts observed in 9 h (N). The expected numbers of corpora lutea or fetuses estimated from the equation including the $T : N$ ratio were found to be no different from those observed when the ewes were slaughtered.

The studies of Land (1970) on the mating behaviour of Finnish Landrace and Scottish Blackface rams showed that the Finnish Landrace rams consistently mounted the ewes more often than did the Blackfaces, and that the reproductive behaviour in both males and females of each of the two breeds was positively correlated. The average litter size of each of the same breeds was found to be 3.4 in the Finnish Landrace (Donald and Read, 1967) and 1.9 in the Blackface (Weiner, 1967).

In short, fertility is directly related to mating activity. The use of fertile rams of high mating activity will be accompanied by high lambing rate in the flock.

References

ALLISON, A.J. (1972). The effect of mating pressure on characteristics of the ejaculate in rams on reproductive performance in ewes. *Proceedings of the New Zealand Society of Animal Production*, **32**, 112–113

ALLISON, A.J. (1975). Flock mating in sheep. 1. Effect of number of ewes joined per ram on mating behaviour and fertility. *New Zealand Journal of Agriculture Research*, **18**, 1–8

ALLISON, A.J. (1977). Flock mating in sheep. 2. Effect of number of ewes per ram on mating behaviour and fertility of two-tooth and mixed-age Romney ewes run together. *New Zealand Journal of Agricultural Research*, **20**, 123–128

ALLISON, A.J. and DAVIS, G.H. (1976). Studies on mating behaviour and fertility of Merino ewes. 1. Effects of number of ewes joined per ram, age of ewe and paddock size. *New Zealand Journal of Experimental Agriculture*, **4**, 259–267

ATKINSON, S. and WILLIAMSON, P. (1985). Ram-induced growth of ovarian follicles and gonadotrophin inhibition in anoestrous ewes. *Journal of Reproduction and Fertility*, **73**, 185–189

BADAWY, A.M., EL BASHARY, A.S. and MOHSEN, M.K.M. (1973). A study of the sexual behaviour of the female Barky sheep. *Alexandria Journal of Agriculture Research*, **21**, 1–9

BARKER, J.D. and LAND, R.B. (1970). A note on the fertility of hill ewes mated to Finnish Landrace and Border Leicester rams. *Animal Production*, **12**, 673–675

BEAMER, W., BERMANT, G. and CLEGG, M.T. (1969). Copulatory behaviour of the ram, *Ovis aries*. 2. Factors affecting copulatory satiation. *Animal Behaviour*, **17**, 706–711

BELSCHNER, H.G. (1965). *Sheep Management and Diseases*. Sydney: Angus and Robertson

BLOCKEY, M.A. de B. and CUMMING, I.A. (1970). Mating behaviour of ewes. *Proceedings of the Australian Society of Animal Production*, **8**, 344–347

BOLAND, M.P. and GORDON, I. (1979). Effect of timing or ram introduction on fertility in progestagen-PMSG treated anoestrous ewes. *Journal of Agricultural Science, Cambridge*, **92**, 247–249

BOSHOFF, D.A., van NIEKERK, C.H. and MORGENTHAL, J.C. (1973). Time of ovulations in the Karakul ewe following synchronization of oestrus. *South African Journal of Animal Science*, **3**, 13–17

BOURKE, M.E. (1967). A study on mating behaviour of Merino rams. *Australian Journal of Experimental Agriculture and Animal Husbandry*, **7**, 203–205

BRADFORD, G.E. (1972). Genetic control of litter size in sheep. *Journal of Reproduction and Fertility*, **Supplement 15**, 23–41

BRYANT, M.J. (1975). A note on the effect of rearing experience upon the development of sexual behaviour in ram lambs. *Animal Production*, **21**, 97–99

BRYANT, M.J. and TOMKINS, T. (1975). The flock mating of progestagen-synchronized ewes. 1. The influence of ram-to-ewe ratio upon mating behaviour and lambing performance. *Animal Production*, **20**, 381–390

CAHILL, L.P., BLOCKEY, M.A. de B. and PARR, R.A. (1975). Effects of mating behaviour and ram libido on the fertility of young ewes. *Australian Journal of Experimental Agriculture and Animal Husbandry*, **15**, 337–341

COGNIÉ, Y., GAYERIE, F., OLDHAM, C.M. and POINDRON, P. (1980). Increased ovulation rate at the ram-induced ovulation and its commercial application. *Proceedings of the Australian Society of Animal Production*, **13**, 80–82

COGNIÉ, Y., GRAY, S.J., LINDSAY, D.R., OLDHAM, C.M., PEARCE, D.T. and SIGNORET, J.P. (1982). A new approach to controlled breeding in sheep using the 'ram effect'. *Proceedings of the Australian Society of Animal Production*, **14**, 519–522

DAUZIER, L. and WINTENBERGER, S. (1952). Recherches sur la fécondation chez les mammifères: dureé du pouvoir fécondant des spermatozoides de bélier dans le tractus génital de la brebis et dureé de la période de fécondité de l'oeuf après l'ovulation. *Comptes Rendus des Séances de Société Biologique*, **146**, 660–663

DAVIS, G.H., KELLY, R.W., HANRAHAN, J.P. and ROHLOFF, R.M. (1983). Distribution of litter size within flocks at different levels of fecundity. *Proceedings of the New Zealand Society of Animal Production*, **43**, 25–28

DAWE, S.T., BENNETT, N.W., DONNELLY, F.B., FERGUSON, B.D., RIVE, J.P., ROBERTS, B.C. and TRIMMER, B.I. (1970). The comparative reproductive performance of ewes joined to one or three per cent of rams. *Proceedings of the Australian Society on Animal Production*, **8**, 317–320

DONALD, H.P. and READ, J.L. (1967). The performance of Finnish Landrace sheep in Britain. *Animal Production*, **9**, 471–476

DÝRMUNDSSON, Ó.R. (1987). Advancement of puberty in male and female sheep. In *New Techniques in Sheep Production*, edited by I.F.M. Marai and J.B. Owen, pp. 65–76. London: Butterworths

EDGAR, D.G. and BILKEY, D.A. (1963). The influence of rams on the onset of the breeding season in ewes. *Proceedings of the New Zealand Society of Animal Production*, **23**, 79–87

ENTWISTLE, K.W. and MARTIN, I.C.A. (1972). Effects of the number of spermatozoa and of volume of diluted semen on fertility in the ewe. *Australian Journal of Agricultural Research*, **23**, 467–472

EVANS, G. and ARMSTRONG, D.T. (1984). Reduction of sperm transport in ewes by superovulation treatments. *Journal of Reproduction and Fertility*, **70**, 47–53

FLETCHER, I.C. (1979). Sexual activity in merino rams. In *Sheep Breeding*, 2nd edition, edited by G.J. Tomes, D.F. Robertson and R.J. Lightfoot, revised by W. Haresign, pp. 487–493. London: Butterworths

FLETCHER, I.C. and LINDSAY, D.R. (1971). Effect of rams on the duration of oestrous behaviour in ewes. *Journal of Reproduction and Fertility*, **25**, 253–259

FOWLER, D.G. (1975). Mating activity and its relationship to reproductive performance in Merino sheep. *Applied Animal Ethology*, **1**, 357–368

FOWLER, D.G. (1976). Predicting the number of corpora lutea in Merino ewes. *Journal of Reproduction and Fertility*, **46**, 525

FOWLER, D.G. (1977). Behaviour patterns of young rams from flocks selected for differences in reproductive performance. *Theriogenology*, **8**, 148

FOWLER, D.G. and JENKINS, L.D. (1976). The effects of dominance and infertility of rams on reproductive performance. *Applied Animal Ethology*, **2**, 327–337

FOWLER, D.G. and LANGFORD, C.M. (1976). The prediction of fertility and fecundity from the mating activity of ewes. *Applied Animal Ethology*, **2**, 277–281

FULKERSON, W.J., SYNNOTT, A.L. and LINDSAY, D.R. (1982). Numbers of spermatozoa required to effect a normal rate of conception in naturally mated Merino ewes. *Journal of Reproduction and Fertility*, **66**, 129–132

GALINDEZ, F.J., PRUD'HON, M. and REBOUL, G. (1977). Reproductive performance of group-synchronized Merino d'Arles and Romanov crossbred ewes. *Animal Production*, **24**, 113–116

GEORGE, J.M. (1969). Variation in the time of parturition of Merino and Dorset Horn ewes. *Journal of Agricultural Science, Cambridge*, **73**, 295–299

GIBSON, R.M. and JEWELL, P.A. (1982). Semen quality, female choice and multiple mating in domestic sheep: a test of Trivers' sexual competence hypothesis. *Behaviour*, **80**, 9–31

GONYOU, H.W. (1983). The role of behaviour in sheep production: a review of research. *Applied Animal Ethology*, **11**, 341–358

GORDON, I. (1983). *Controlled Breeding in Farm Animals*. Oxford: Pergamon Press

GREEN, W.W. (1947). Duration of sperm fertility in the ewe. *American Journal of Veterinary Research*, **8**, 299–300

HANRAHAN, J.P. and OWEN, J.B. (1985). Variation and repeatability of ovulation rate in Cambridge ewes. *Animal Production*, **40**, 529

HAUGHEY, K.G. (1959). Preliminary report on a tupping survey: Ashburton County. *New Zealand Sheepfarming Annual*, 17–26

HULET, V.C. (1966). Behavioural, social and physiological factors affecting mating time and breeding efficiency in sheep. *Journal of Animal Science*, **Supplement 25**, 5–20

HULET, C.V., BLACKWELL, R.L. and ERCANBRACK, S.K. (1964). Observations on sexually inhibited rams. *Journal of Animal Science*, **23**, 1095–1097

HULET, C.V. and BOND, J. (1984). Some observations on the role of behaviour in sheep production and future research needs. *Applied Animal Ethology*, **11**, 407–411

ILLIUS, A.W., HAYNES, N.B. and LAMMING, G.E. (1976). Effects of ewe proximity on peripheral plasma testosterone levels and behaviour in the ram. *Journal of Reproduction and Fertility*, **48**, 25–32

JENNINGS, J.J. (1976). Mating behaviour of rams in late oestrus. *Irish Journal of Agricultural Research*, **15**, 301–307

JEWELL, P.A., HALL, S.J.G. and ROSENBERG, M.M. (1986). Multiple mating and siring success during natural oestrus in the ewe. *Journal of Reproduction and Fertility*, **77**, 81–89

JOYCE, M.J.B. (1972). A comparison of three different mating systems. *Proceedings of Seventh International Congress on Animal Reproduction and AI*, Munich, **2**, 935–938. Bonn: Deutschenen Gesellschaft für Züchtungskunde

KELLY, R.B. (1937). Studies in fertility of sheep. *Bulletin of the Council of Scientific and Industrial Research, Australia*, no. 112, 67 pp.

KILGOUR, R.J. (1980). Serving capacity of rams and flock fertility. *Proceedings of the Australian Society of Animal Production*, **13**, 46–48

KNIGHT, T.W. (1980). Onset of mating activity in Romney ewes after short periods of teasing. *New Zealand Journal of Agriculture Research*, **23**, 277–280

KNIGHT, T.W. (1983). Ram induced stimulation of ovarian and oestrous activity in anoestrous ewes. A review. *Proceedings of the New Zealand Society of Animal Production*, **43**, 7–11

KNIGHT, T.W. (1985). Are rams necessary for the stimulation of anoestrous ewes with oestrous ewes? *Proceedings of the New Zealand Society of Animal Production*, **45**, 49–50

KNIGHT, T.W. and LYNCH, P.R. (1980a). Source of ram pheromones that stimulate ovulation in the ewe. *Animal Reproduction Science*, **3**, 133–136

KNIGHT, T.W. and LYNCH, P.R. (1980b). The pheromones from rams stimulate ovulation in the ewe. *Proceedings of the Australian Society of Animal Production*, **13**, 74–76

KNIGHT, T.W., TERVIT, H.R. and LYNCH, P.R. (1983). The effect of boar pheromones, ram's wool and presence of bucks on ovarian activity in anovular ewes early in the breeding season. *Animal Reproduction Science*, **6**, 129–134

LAMBOURNE, L.J. (1956). Mating behaviour. *Proceedings of the Ruakura Farmers' Conference Week*, pp. 16–20

LAND, R.B. (1970). The mating behaviour and semen characteristics of Finnish Landrace and Scottish Blackface rams. *Animal Production*, **12**, 551–560

LIGHTFOOT, R.J. and SMITH, J.A.C. (1968). Studies on the number of ewes joined per ram for flock matings under paddock conditions. 1. Mating behaviour and fertility. *Australian Journal of Agricultural Research*, **19**, 1029–1042

LINDAHL, I.L. (1964). Time of parturition in ewes. *Animal Behaviour*, **12**, 231–234

LINDSAY, D.R. (1983). Mating behaviour in sheep. In *Sheep Production*, edited by W. Haresign, pp. 473–479. London: Butterworths

LINDSAY, D.R. and ELLSMORE, J. (1968). The effect of breed, season and competition on mating behaviour of rams. *Australian Journal of Experimental Agriculture and Animal Husbandry*, **8**, 649–652

LINDSAY, D.R., COGNIÉ, Y., PELLETIER, J. and SIGNORET, J.P. (1975). Influence of the presence of rams on the timing of ovulation and discharge of LH in ewes. *Physiology and Behaviour*, **15**, 423–426

LINDSAY, D.R., DUNSMORE, D.G., WILLIAMS, J.D. and SYME, G.J. (1976). Audience effects on the mating behaviour of rams. *Animal Behaviour*, **24**, 818–821

MARTIN, G.B. (1979). Ram-induced ovulation in seasonally anovular Merino ewes: effect of estradiol on the frequency of ovulation, oestrus and short cycles. *Theriogenology*, **12**, 283–289

MARTIN, G.B., OLDHAM, C.M. and LINDSAY, D.R. (1980). Increased plasma LH ewes in seasonally anovular Merino ewes following the introduction of rams. *Animal Reproduction Science*, **3**, 125–132

MATTNER, P.E. and BRADEN, A.W.H. (1967). Studies in flock mating of sheep. 2. Fertilization and prenatal mortality. *Australian Journal of Experimental Agriculture and Animal Husbandry*, **7**, 110–116

MATTNER, P.E., BRADEN, A.W.H. and GEORGE, J.M. (1971). Studies in flock mating in sheep. 4. The relation of libido tests to subsequent service activity of young rams. *Australian Journal of Experimental Agriculture and Animal Husbandry*, 11, 473–477

MATTNER, P.E., BRADEN, A.W.H. and GEORGE, J.M. (1973). Studies on flock mating of sheep. 5. Incidence, duration and effect on flock fertility of initial sexual inactivity in young rams. *Australian Journal of Experimental Agriculture and Animal Husbandry*, 13, 35–41

MATTNER, P.E., BRADEN, A.W.H. and TURNBULL, K.E. (1967). Studies in flock mating of sheep. 1. Mating behaviour. *Australian Journal of Experimental Agriculture and Animal Husbandry*, 7, 103–109

MICKELSEN, W.D., PAISLEY, L.G. and DAHMEN, J.J. (1981). The effect of scrotal circumference, sperm motility and morphology in the ram on conception rates and lambing percentage in the ewe. *Theriogenology*, 16, 53–59

MICKELSEN, W.D., PAISLEY, L.G. and DAHMEN, J.J. (1982). The relationship of libido and serving capacity test scores in rams on conception rates and lambing percentage in the ewe. *Theriogenology*, 18, 79–86

MOORE, R.W. and McMILLAN, W.H. (1981). The effect of nutrition and the time of ram introduction on the onset of puberty in Romney ewe lambs. *Annual Report 1980/81*, Whatawhata Hill Country Research Station, Private Bag, Hamilton, New Zealand

MOORE, R.W. and WHYMAN, D. (1980). Fertilizing ability of semen from rams of high and low prolificacy flocks. *Journal of Reproduction and Fertility*, 59, 311–316

MORGAN, P.D., ARNOLD, G.W. and LINDSAY, D.R. (1972). A note on the mating behaviour of ewes with various senses impaired. *Journal of Reproduction and Fertility*, 30, 151–152

MURTAGH, J., GRAY, S., LINDSAY, D.R. and OLDHAM, C.M. (1984). The 'ram effect' in Merino weaner and maiden ewes. *Proceedings of the Australian Society of Animal Production*, 15, 490–493

NEWTON, J.E. and BETTS, J.E. (1968). Factors affecting litter size in the Scottish Halfbred ewe. Pt. II. Superovulation and the synchronization of oestrus. *Journal of Reproduction and Fertility*, 17, 485–493

NGERE, L.O. and DZAKUMA, J.M. (1975). The effect of sudden introduction of rams on oestrous pattern of tropical ewes. *Journal of Agricultural Science, Cambridge*, 84, 263–264

OLDHAM, C.M. (1980a). A study of sexual and ovarian activity in Merino sheep. *PhD thesis*, University of Western Australia, Perth

OLDHAM, C.M. (1980b). Stimulation of ovulation in seasonally or lactationally anovular ewes by rams. *Proceedings of the Australian Society of Animal Production*, 13, 73–83

OLDHAM, C.M., MARTIN, G.B. and KNIGHT, T.W. (1978). Stimulation of seasonally anovular Merino ewes by rams. 1. Time from introduction of the rams to the pre-ovulatory LH surge and ovulation. *Animal Reproduction Science*, 1, 283–290

OLDHAM, C.M., PEARCE, D.T. and GRAY, S.J. (1985). Progesterone priming and age of ewe affect the life span of corpora lutea induced in the seasonally anovulatory Merino ewe by the 'ram effect'. *Journal of Reproduction and Fertility*, 75, 29–33

PARKER, C.F. (1968). Ram influence on lambing rate, lamb crop percentage and quantity of lamb weaned of Columbia and Targhee ewes. *Proceedings of Second World Conference on Animal Production*, College Park, Maryland, 1968, 437–438

PARKER, C.F. (1972). Performance of ewe and ram breeds in a multiple lambing system. *Journal of Animal Science*, 35, 181

PARSONS, S.D. and HUNTER, G.L. (1967). Effect of the ram on duration of oestrus in the ewe. *Journal of Reproduction and Fertility*, 14, 61–70

PIPER, L.R. and BINDON, P.M. (1982). Genetic segregation for fecundity in Booroola Merino sheep. *Proceedings of the World Congress on Sheep and Beef Cattle Breeding*, 1, 394–400. Palmerston North, New Zealand: The Dunsmore Press Ltd

RAADSMA, H.W. and EDEY, T.N. (1984). Dynamics of paddock mating of rams in conventional and intensified mating system. In *Reproduction in Sheep*, edited by D.R. Lindsay and D.T. Pearce, pp. 50–52. Cambridge: Cambridge University Press

RESTALL, B.J. (1971). The effect of lamb removal on reproductive activity in Dorset Horn Merino ewes after lambing. *Journal of Reproduction and Fertility*, 24, 145–146

RICHES, J.H. and WATSON, R.H. (1954). The influence of the introduction of rams on the incidence of oestrus in Merino ewes. *Australian Journal of Agricultural Research*, 5, 141–147

ROBERTSON, H.A. and RAKHA, A.M. (1965). Time of onset of oestrus in the ewe. *Journal of Reproduction and Fertility*, 10, 271–272

RYAN, J.P., BILTON, R.J., HUNTON, J.R. and MAXWELL, W.M.C. (1984). Superovulation of ewes with a combination of PMSG and FSH-P. In *Reproduction in Sheep*, edited by D.R. Lindsay and D.T. Pearce, pp. 338–341. Cambridge: Cambridge University Press

SCHINCKEL, P.G. (1954). The effect of the ram on the incidence and occurrence of oestrus in ewes. *Australian Veterinary Journal*, **30**, 189–195

SCHINDLER, H. and AMIR, D. (1972). Length of oestrus, duration of phenomena related to oestrus, and ovulation time in the local fat-tailed Awassi ewe. *Journal of Agricultural Science, Cambridge*, **78**, 151-156

SCHINDLER, H. and AMIR, D. (1973). The conception rate of ewes in relation to sperm dose and time of insemination. *Journal of Reproduction and Fertility*, **34**, 191–196

SIGNORET, J.P. (1980). Effect de la présence du mâle sur les mécanismes de reproduction chez la femelle des mammifères. *Reproduction Nutrition Developments*, **20**, 457–468

SIGNORET, J.P., FULKERSON, W.J. and LINDSAY, D.R. (1982). Effective use of testosterone-treated wethers and ewes as teasers. *Applied Animal Ethology*, **9**, 37–45

SLEE, J. (1964). Some aspects of multiple-mating and superovulation in sheep. *Journal of Agricultural Science, Cambridge*, **63**, 403–408

SYNNOTT, A.L. and FULKERSON, W.J. (1984). The influence of social interaction between rams on their serving capacity. *Applied Animal Ethology*, **11**, 283–289

SYNNOTT, A.L., FULKERSON, W.J. and LINDSAY, D.R. (1981). Sperm output by rams and distribution amongst ewes under conditions of continual mating. *Journal of Reproduction and Fertility*, **61**, 335–361

THIERY, J.C. and SIGNORET, J.P. (1978). Effect of changing the teaser ewe on the sexual activity of the ram. *Applied Animal Ethology*, **4**, 87–90

TURNER, H.N. and YOUNG, S.S.Y. (1969). *Quantitative Genetics in Sheep Breeding*. South Melbourne, Victoria: Macmillan Company of Australia Pty Ltd; London: Macmillan and Company Ltd., xviii + 332pp [B]

VAKIL, D.V., BOTKIN, M.P. and ROEHRKASSE, A.P. (1968). Influence of hereditary and environmental factors on twinning on sheep. *Journal of Heredity*, **59**, 256–259

WALKLEY, J.R.W. and BARBER, A.A. (1976). The relationships between libido score and fertility in Merino rams. *Proceedings of the Australian Society of Animal Production*, **11**, 141–144

WATSON, R.H. and RADFORD, H.M. (1960). The influence of rams on onset of oestrus in Merino ewes in the spring. *Australian Journal of Agricultural Research*, **11**, 65–71

WALKER, S.K., SMITH, D.H. and SEAMARK, R.F. (1986). Timing of multiple ovulation in the ewe after treatment with FSH or PMSG with and without GnRH. *Journal of Reproduction and Fertility*, **77**, 135–142

WEINER, G. (1967). A comparison of the body size, fleece weight and maternal performance of five breeds of sheep kept in one environment. *Animal Production*, **9**, 177–195

WHYMAN, D., JOHNSON, D.L., KNIGHT, T.W. and MOORE, R.W. (1979). Intervals between multiple ovulations in PMSG treated and untreated ewes and the relationship between ovulation and oestrus. *Journal of Reproduction and Fertility*, **55**, 481–488

WINFIELD, C.G. and CAHILL, L.P. (1978). Mating competency of rams and flock fertility. *Applied Animal Ethology*, **4**, 193–195

WINFIELD, C.G. and KILGOUR, R. (1977). The mating behaviour of rams in a pedigree pen-mating system in relation to breed and fertility. *Animal Production*, **24**, 197–201

ZENCHAK, J.J. and ANDERSON, G.C. (1980). Sexual performance levels of rams (*Ovis aries*) as affected by social experiences during rearing. *Journal of Animal Science*, **50**, 167–174

ZENCHAK, J.J., ANDERSON, G.C. and SCHEIN, M.W. (1981). Sexual partner preference of adult rams (*Ovis aries*) as affected by social experiences during rearing. *Applied Animal Ethology*, **7**, 157–167

Chapter 22

New developments in health control

K.A. Linklater

Summary

Recent developments in vaccination against footrot and pasteurellosis, in the prevention of trace element deficiencies and neonatal deaths, and in the understanding of infectious abortion in ewes, are described and discussed.

Footrot can now be prevented by the use of a vaccine which contains antigens from a wide range of serotypes of the infectious organism *Bacteroides nodosus*. This has been developed largely in New Zealand and Australia but is now available in the United Kingdom. Similarly there has been a marked improvement in the vaccines which can be used against the most common remaining cause of sudden death in sheep, pasteurellosis. These incorporate a wide range of the known serotypes of *Pasteurella haemolytica* and have been experimentally shown to protect against the pneumonic form of the disease.

Much work has also been done on trace element deficiencies especially regarding diagnosis and prevention. This has culminated in the production of an intraruminal glass bolus which incorporates all three of the trace elements most commonly found to be deficient: cobalt, copper, and selenium.

Neonatal care has now become much less subjective. Clear guide lines for the treatment of hypothermia have been well worked out, and the colostral requirements of young lambs established. Abortion continues to be a source of great loss in sheep flocks. Progress has been made on the understanding of the four most common infectious causes: enzootic abortion of ewes, toxoplasmosis, salmonellosis and campylobacteriosis (vibriosis).

Introduction

During recent years there have been several significant breakthroughs in the control of disease in sheep flocks. Foremost amongst these are the developments in vaccination against footrot and pasteurellosis, a better understanding of the trace element requirements of sheep, the development of an objective approach to neonatal care, and the elucidation of the major infectious causes of ovine abortion.

253

Footrot

Footrot is a disease of major economic importance to the sheep industries throughout the world. As it is an insidious condition it has often not demanded as much attention as the more dramatic diseases which cause acute deaths. However, a great deal of research on the subject has been carried out in Australia and New Zealand over the last 10 years. This has led to a much better understanding of the condition and the development of an efficient footrot vaccine.

Bacteroides nodosus

Bacteroides nodosus is the essential agent of ovine footrot and is an obligate parasite of the feet of sheep, goats and in some instances, cattle (Egerton, Roberts and Parsonson, 1969). It cannot survive for more than a few days in the environment. To produce lesions it requires the assistance of another organism, *Fusobacterium necrophorum*, which is present universally in soil and even in the flora of sheep gut (Roberts and Egerton, 1969).

Bacteroides nodosus has little inflammatory or destructive action of its own, but is essential to the disease process of footrot by being able to initiate the invasion of the epidermal matrix of the hoof by *F. necrophorum*. This ability is associated with high protease production. *F. necrophorum* is thought to be responsible for the large proportion of the inflammation and tissue damage. The two organisms are therefore closely synergistic in the disease process of footrot.

After an attack of footrot, sheep do not tend to be immune to a subsequent attack (Beveridge, 1941). However, immunity can be produced artificially. Virulent isolates of *B. nodosus* have characteristic hair-like surface filaments which are called fimbriae or pili. These are important in protective immunity as it has been shown that vaccination with preparations of isolated fimbriae confers a degree of protection similar to that of whole cells (Stewart *et al.*, 1986). These fimbriae are also the principal antigenic structures used to classify isolates of *B. nodosus* into at least eight major serogroups and a number of serotypes (Claxton, Ribeiro and Egerton, 1983). The serogroups are important in immunity as there is little cross protection between them (Egerton, 1972) and immunity confirmed by vaccination with purified fimbriae appears to be restricted to the serogroup involved (Stewart *et al.*, 1986). Therefore for a vaccine to be effective over a wide range of situations it must contain representatives of all the serogroups.

Vaccination

It has recently been shown that all eight known serogroups are present in flocks in Britain (Hindmarsh and Fraser, 1985). At present there are ten different strains of *B. nodosus* included in the commercial vaccine. There are single representatives of groups A,C,D,E,F,G and H with three representatives of the B group: B1, B2 and B4. The vaccine has an oil adjuvant and therefore has a tendency to produce nodules at the site of injection. For this reason it is recommended that it is administered subcutaneously high up on the neck. Two doses have to be given 4–6 weeks apart as a primary course followed by boosters at intervals of 4–12 months according to the periods of peak risk for a particular locality.

Paring and bathing

The vaccine is also helpful for treatment of footrot in sheep already affected. It therefore now plays an essential part in any programme for the control of footrot in a flock. It must, however, be backed up by other methods of foot care such as paring and bathing. For the latter there is now good evidence that zinc sulphate is preferable to either formalin or copper sulphate. Skerman *et al.* (1983) showed that a 10% solution of zinc sulphate with the inclusion of a surfactant such as Teepol at a concentration of 0.2% was more effective than a 10% solution of formalin at treating advanced lesions of footrot. It also has the advantage of being less irritant to the skin of the sheeps' feet, and to the workers using it, and does not harden the hoof horn.

Pasteurellosis

Pasteurellosis is the most common cause of death in sheep in the United Kingdom. Veterinary investigation centres in England, Wales and Scotland confirm cases from over 1500 flocks each year and this is more than twice that of any other disease listed by them (Veterinary Investigation Diagnosis Analyses II, 1977–84).

Clinical features

All ages of sheep can be affected and the signs presented are variable (Gilmour and Angus, 1983). Typically the infection is associated with outbreaks of acute pneumonia which may follow periods of stress or housing in poorly ventilated buildings. The onset of the disease may be sudden and severe and death may frequently intervene before treatment can be administered. This is the form most commonly seen in adults. Young lambs on the other hand often succumb to a septicaemic form of the disease which affects them in late spring and early summer. There are often no premonitory signs with this form and affected animals may merely be found dead. At the same time of year lactating ewes may be affected with a severe mastitis which can also lead to death. Another distinct form of the disease causes sudden deaths in weaned lambs, 6–9 months of age, in autumn and early winter. This is usually called systemic pasteurellosis and outbreaks can be quite severe with up to 10% of a group being affected unless appropriate control measures are taken (Dyson, Gilmour and Angus, 1981).

Pasteurella haemolytica

Pasteurellosis is caused by a bacterium called *Pasteurella haemolytica*. There are 15 different types of the organism which form two distinct groups or biotypes known as 'A' types and 'T' types. The 'A' types are so-called because they ferment a sugar called arabinose and are associated with pneumonia, septicaemia in young lambs and mastitis. The 'T' types ferment trehalose, another sugar, and are predominantly isolated from cases of systemic disease in weaned lambs.

These biotypes are further divided by means of serological reactions into further subtypes known as serotypes (Biberstein, Gills and Knight, 1960). The full range of serotypes of *P. haemolytica* is shown in *Table 22.1*. These serotypes are important when vaccination is considered, as there is little cross-protection between them

TABLE 22.1 Serotypes of *Pasteurella haemolytica*

	A Biotypes	*T Biotypes*
Serotypes	A1, A2, A5, A6 A7, A8, A9, A11 A12, A13, A14	T3, T4 T10, T15

either within or between the two major biogroups. Therefore vaccines must contain most of the common types if they are to be effective. Those currently available in the United Kingdom contain serotypes A1, A2, A6, A7, T3, T4 and T10.

Vaccination

Experimentally it has been shown that the pneumonic form of pasteurellosis can be successfully prevented by vaccination (Gilmour *et al.*, 1979). A primary course of two doses of the vaccine is necessary to provide a satisfactory level of immunity. Thereafter single boosters may be given either once or twice annually according to the history of the flock; for example, it is a wise precaution to give a booster 2 weeks before housing. Young lambs can be protected via the colostrum when a combined pasteurella/clostridial disease booster vaccine is given to ewes before lambing. However, colostral immunity to pasteurellosis only lasts for 4 weeks, and if there is a problem in lambs in a flock between 4 weeks of age and weaning they must be actively immunized with a two-dose course of a straight pasteurella vaccine. There is as yet no firm evidence that the systemic form of the disease can be prevented by vaccination. Nevertheless it is a sensible precaution to have weaned lambs fully vaccinated before the period of peak risk in September–November.

Trace elements

The three trace elements of most significance in the maintenance of health in sheep are cobalt, copper and selenium. During the past few years significant advances have been made in the understanding of these, especially with respect to improved diagnostic techniques and prevention of the deficiencies.

Cobalt

For a long time serum and liver vitamin B_{12} levels have been used as a guide to the intake of cobalt by ruminants. More recently other diagnostic aids which depend on the failure of some aspect of metabolism dependent on adequate vitamin B_{12} status have been developed. The principal effect of vitamin B_{12} deficiency is a depression in propionate metabolism which in turn leads to a build up in methylmalonic acid (MMA). This in turn is excreted by the kidneys and builds up in the urine. Urinary determinations of MMA may therefore be used to detect cobalt-depleted animals (Millar and Lorentz, 1979). Recent work suggests that serum MMA estimations may also be used for the determination of cobalt deficiency.

A similar metabolite accumulating in the urine of cobalt deficient animals is formiminoglutamic acid (FIGLU). It is an intermediary product in the catabolism of histidine and its conversion to glutamic acid requires tetrahydrofolic acid which is not produced unless vitamin B_{12} is present. Therefore FIGLU accumulates in vitamin B_{12}-deficient animals until it is eventually excreted in the urine, where it can be detected at levels of up to 30 times greater than that in the urine of animals on an adequate cobalt intake (Russel et al., 1975).

For optimal growth sheep require 0.11 mg cobalt per kg of their dry matter intake. If this level is not present in the ration, supplementation can be carried out in several ways (MacPherson, 1983). For housed livestock and those outside which are getting supplementary feeding, a cobalt salt can be added to the diet to bring the cobalt concentration of the whole feed to greater than 0.1 mg/kg dry matter. Mineral mixes containing as little as 100–200 ppm cobalt will provide an adequate intake when mixed at the rate of 25 kg/tonne of concentrates, provided these concentrates form at least 10% of the final dry matter intake.

In extensive grazing situations the use of cobalt bullets provides a satisfactory method of supplementation. The bullets are administered by a balling gun and lodge in the reticulum where they remain for 1–3 years giving a regular slow release of cobalt. Lambs should not be 'bulleted' until weaning to prevent the formation of an insoluble coating of calcium phosphate on the bullets which prevents the cobalt from coming into solution.

For grazing animals the most satisfactory long-term solution to the problem is to apply cobalt sulphate ($CoSO_4$ $7H_2O$) to the pasture at a rate of 2 kg/ha either as a low volume spray or incorporated in granular fertilizer at manufacture. One such application to soil should raise the cobalt content of deficient pasture to adequate levels for at least 4 years. The cobalt should not be applied too soon after liming as high pH renders cobalt much less available to the plants. In some soils 2 kg/ha of cobalt sulphate may not be adequate. In these cases it should be increased to 6 kg/ha and applied to one-third of the area.

Other methods of prevention include oral dosing with cobalt sulphate and addition of cobalt to the water supply, all of which are appropriate for certain circumstances. Where rapid response is required, injections of vitamin B_{12} may be given but have the disadvantage of requiring to be repeated frequently.

Copper

In sheep the major manifestation of copper deficiency is swayback in lambs. It takes two forms, one where symptoms are apparent at birth and another where the clinical signs do not develop until 6–8 weeks of age. The only convincing reports in this country of ill-thrift, wool changes and bone weaknesses associated with copper deficiency have come from workers at the Hill Farm Research Organization in Edinburgh (Whitelaw et al., 1979). They described these effects in Blackface lambs grazing improved hill pastures in Scotland. Whether this is a widespread phenomenon throughout the country has yet to be established, but it does appear to be common in certain breeds of sheep on that type of pasture.

The copper status of a group of sheep is usually established by blood copper estimations carried out on a sample of at least ten to 15 animals. Liver copper estimations provide a better guide to the copper reserves of animals but are obviously less readily available than blood. Recently it has been suggested that the estimation of some copper-containing enzymes may provide a more accurate assessment of when a copper deficiency situation is clinically significant. An example of these enzymes is superoxide dismutase (SOD) and indications are that a low SOD activity is directly related to the appearance of clinical signs of copper deficiency, though more work is required to confirm this (Suttle and McMurray, 1983).

Methods used for preventing and treating copper deficiency in sheep follow similar lines to that for cobalt deficiency (MacPherson, 1983). The requirements in the diet vary with the class of animal, the diet being fed and whether they are kept inside or out. Oral dosing with copper sulphate is perhaps the simplest method of prevention and therapy. Despite the fact that only about 4% ingested is absorbed from the intestines, this has over the years proved a successful method for the prevention of swayback in lambs. Ewes are dosed with 1 g copper sulphate ($CuSO_4$ $5H_2O$) in solution 8 weeks and 4 weeks before lambing.

There are a variety of copper injections which are now probably the commonest form of treatment. They should only be used on animals which are known to be low in copper and at the recommended dose rate, as toxicity may readily result from overdosage. Some of the preparations cause local lesions at the site of injection (such as Cu-methionates and Cu-glycinates). These tend to be slowly translocated to the liver and are of lower toxicity but give poorer protection in situations of simple copper deficiency. The more rapidly translocated complexes such as diethylamine copper oxyquinoline sulphonate must be given in smaller doses to avoid acute toxicity. The copper calcium complex of ethylamine diamine tetra-acetic acid occupies an intermediate position with respect to rate of translocation and tolerance.

Copper oxide needles have recently been introduced for prevention. These are administered in capsules which dissolve allowing the needles to penetrate the wall of the ovine abomasum where they slowly dissolve over 2–3 months. They provide a highly concentrated slow release source of copper of relatively low toxicity. Doses of 2 g for lambs and 4 g for adults are currently recommended and they should not be used with any other form of copper therapy.

Selenium

In this country the main manifestation of selenium deficiency in sheep is white muscle disease (WMD), otherwise known as stiff lamb disease, muscular dystrophy and nutritional myopathy. Young lambs and hoggets may be affected. As the name suggests, the deficiency which is closely interrelated with levels of vitamin E leads to muscular damage. Affected animals may die suddenly when the heart muscle is affected, breathe with difficulty when the muscles of the chest are affected, or walk stiffly when the skeletal muscles are affected. Apart from the myopathies there is no good evidence for the occurrence of other selenium-responsive conditions in sheep in this country. Claims that reproductive problems are responding to selenium therapy and that weight gains may be suppressed by selenium deficiency have yet to be substantiated in Britain.

White muscle disease is diagnosed by the clinical signs backed up by enzyme estimations on blood samples. These enzymes are released into the circulation from damaged muscles and include creatine kinase and aspartate aminotransferase. Similarly, the selenium status of a group of animals can be assessed by examination of blood samples from at least ten members of the group for the selenoenzyme glutathione peroxidase (GSH Px) and the mean compared with that of animals of adequate selenium status (Anderson, Berrett and Patterson, 1979). It must be remembered that over 90% of blood GSH Px activity is in the red cells and is synthesized during erythropoiesis. Therefore GSH Px levels are indicative of the selenium status of the animals when the red cells are being formed and not necessarily at the time of sampling.

Dietary supplementation of selenium is now widely practised. It is desirable to achieve a final concentration of 0.1 mg/kg dry matter. Excessive selenium in the diet may be toxic and other methods of selenium therapy should not be concurrently used. Injections are the most widely used methods for administering selenium to animals and can be given at strategic times to prevent white muscle disease. Selenium can also be supplemented through piped water supplies or given as drenches (MacPherson, 1983). Other products available include pellets and subcutaneous implants of selenium-containing glasses. All methods used appear to be effective in raising the selenium status of animals and choice of method depends upon the circumstances.

Where more than one deficiency coexists in a group of animals the use of recently introduced glass boluses may be appropriate. These contain cobalt, selenium and copper and appear to be adequate for supplementation of the first two. In some instances especially in severe deficiency situations, they have been found inadequate for copper supplementation of ewes for the prevention of swayback (MacPherson, 1985).

Neonatal care

It is well known that the greatest mortality of lambs occurs around lambing time. The results of a survey which was carried out in the Scottish Borders in seven Halfbred and Greyface flocks are shown in *Table 22.2*.

TABLE 22.2 Summary of causes of perinatal mortality in seven Halfbred and Greyface flocks in the Scottish borders 1974–76

Diagnosis	Percentage of deaths
Stillborn	36.5
Starvation/chilling	32.7
Infectious disease	17.3
Misadventure	4.9
Congenital abnormalities	3.7
Others	4.9

Stillbirths

Stillbirths are the largest category. Most of these were associated with dystocia despite intensive shepherding at lambing time. The most common form of dystocia involved large single lambs which were presented with both forelegs back resulting in the so-called 'hung lamb'. These often result from overfeeding of ewes carrying singles as the whole flock is normally fed as if all the ewes are carrying twins.

Identification of single-bearing ewes by ultrasound methods of pregnancy diagnosis and subsequent separate treatment of them and the multiple-bearing ewes obviously help to obviate this problem. Alternatively, if the twinning rate in the flock can be increased, these deaths can also be reduced. However, more emphasis should be placed on the selection of breeding stock from strains with low incidence of dystokia. Selection for fashionable conformations in certain breeds also probably exacerbates the condition.

Neonatal deaths

Hypothermia
Of lambs born alive it can be seen that by far the greatest number die from starvation and chilling or hypothermia. Physiological effects of this have recently been quantified (Eales *et al.*, 1982). Similarly, recent work from the Moredun Research Institute shows that over 80% of these lambs could be saved by careful management at birth (Eales, Small and Gilmour, 1982). This includes providing adequate shelter with facilities for warming if necessary in a flow of warm air at 40–45 °C. Hypothermia occurs at two main periods: first, before they are adequately dried through direct loss of body heat; and after 10 h of life when body energy resources are quickly used up. In the latter instances injections of 20% glucose must be given before warming and adequate supplies of milk must be available afterwards or hypothermia will recur. Use of a stomach tube on the end of a large syringe greatly facilitates feeding weak lambs. Hence the reason for grouping starvation and chilling together.

Infectious disease
Infectious disease, while accounting for approximately half as many deaths as starvation and chilling, is still very important in a lowground situation where normally ewes are brought into closely confined quarters for lambing. The most common conditions seen are scours, navel ill and joint ill.

Scours
Scouring is due to a variety of organisms including bacteria, viruses and protozoa (Mitchell and Linklater, 1983). The bacteria most commonly involved are *Escherichia coli*, salmonellae and *Clostridium perfringens*. It should be remembered that *E. coli* is a normal inhabitant of the gut and only a few strains cause disease. Special laboratory procedures are required for the detection of these enterotoxigenic strains which normally affect very young lambs. The effects of salmonellosis and lamb dysentery (*Cl. perfringens* types B and C) are well known and do not require further discussion here. During the past 10 years much work has been done on the role of viruses in diarrhoea of most species. Rotavirus is considered to be the most important in lambs and brings about its effects by destroying the absorptive villous epithelial cells of the small intestine. A small protozoan parasite known as *Cryptosporidium* is the most recently described cause of neonatal diarrhoea in lambs. It and the viruses are particularly serious as, unlike the bacteria, there are no specific drugs available for their treatment which therefore depends on symptomatic and supportive therapy.

Navel/joint ill syndrome
With the navel ill/joint ill syndrome, infection gains entry through the umbilicus soon after birth. The bacteria then migrate to the liver or joints to produce abscesses. In some instances these may occur in the spinal cord leading to posterior paralysis which may be mistaken for swayback unless a careful examination is carried out. In other cases, the bacteria which are introduced through the navel may multiply in the blood-stream causing fever and death unless suitable treatment is instigated. Antibiotics by injection are the most effective way of treating these infections but they must be given early in the course of the disease to be effective.

Watery mouth
Watery mouth is an interesting condition which has long been recognized in South-east Scotland but has now apparently become widespread throughout the country. Characteristic signs include dullness, drooling of saliva and distension of the belly. The lamb ceases to suck and, if not treated, its condition deteriorates quickly. The condition is due to slowing down of gut movement which in turn encourages the multiplication of bacteria within the gut. It is important to remember that the abdominal distension is partly due to gas formation in the stomach and despite appearing to be full, the affected lamb may actually be starving (Collins, Eales and Small, 1985). Therefore treatment is aimed at reversing these processes as follows:

(1) Oral and intravenous treatment with suitable antibiotic.
(2) Attempting to encourage gut movement by means of an enema—20 ml of soapy water administered by means of a syringe and cutdown stomach tube placed about 5 cm into the rectum.
(3) Feeding the lamb by stomach tube. A mixture of glucose and salt in water is recommended (10% glucose and 0.9% salt—that is 100 g glucose and 9 g common salt per litre of water). Three times daily feeding of from 50 to 150 ml of the solution is desirable, with the larger volume being required if the lamb is not sucking at all.

Treatment should be continued until the lamb is sucking well from the ewe and apparently back to normal.

Prevention

As with most sheep diseases, prevention of these neonatal problems is the best course to take. As there are few specific remedies for most of the conditions discussed, general principles of good husbandry and management are essential. Perhaps the most important factor is adequate feeding of the ewes before and after lambing to produce strong lambs at birth and adequate colostrum of good quality.

Colostrum
Having got the ewes to produce ample colostrum it is essential to see that the lambs get an adequate supply as soon as they are able to suck. This will be greatly facilitated if the ewe and her lambs are moved to a small pen after lambing and kept there for about 36 h or until the lambs are strong and the ewe/lamb bond is well established. As a general rule one small pen about 1.5 m^2 is required for every eight ewes in the flock. Ideally, pens should be cleaned out between each ewe. Should any disease occur, pens must be thoroughly disinfected after cleaning and, if possible, left empty for a few days. It is important that ewes have *ad libitum* access to clean water at all times even when restricted to the small pens.

Any lamb which has not sucked colostrum 1–1½ h after birth should be fed by stomach tube. Ewes' colostrum is obviously best but as a substitute cows' colostrum may be used. If possible, this should be collected from a cow which has been vaccinated with a multiple clostridial sheep vaccine before calving—10 ml of the same preparation as used for the ewes on three occasions, that is, 3 months, 1 month and 2 weeks before calving (Clarkson, Faull and Kerry, 1985). It is best to avoid animals of the Channel Island breeds and cows on sheep farms as problems of anaemia have occasionally been reported in lambs that have received colostrum from certain cows (Stubbings, 1983). Eales, Small and Gilmour (1982) stipulated that colostrum should be fed at a dose rate of 50 ml/kg of body weight three times per day.

Treating the navels

To prevent infections occurring, navels should be treated with a strong iodine solution as soon as possible after birth. As well as acting as an antiseptic this has astringent properties to promote sealing of the umbilical vessels. The whole navel, including the junction with the body, should be immersed in the solution and allowed to soak for several seconds.

Castration and docking

It has been shown that anything which interferes with the intake of colostrum is likely to predispose to watery mouth and possibly other neonatal conditions. Therefore manipulations such as castration and docking by rubber rings should be avoided for as long as possible and certainly not carried out within the first 2 days of life (Collins, Eales and Small, 1985). They must, however, legally be done within a week of birth.

Antibiotics

Antibiotics should only be used when required. In certain circumstances they provide the only hope in controlling diseases such as watery mouth and scours. However, if used unnecessarily they allow the infectious organisms to develop resistance to them so that when required they are no longer effective. Therefore, they should not be used indiscriminately at the start of lambing but reserved until the first signs of trouble are observed. Thereafter it may be necessary to use antibiotics routinely to every lamb born to try to prevent some of the diseases. However, it must be remembered that there are no substitutes for good husbandry and hygiene.

Abortion

The infectious condition which causes greatest concern to flock owners in the United Kingdom at present is abortion. The four most common causes are enzootic abortion of ewes (EAE), toxoplasmosis, campylobacteriosis (formerly known as vibriosis) and salmonellosis. Of these, the first two are by far the most common, accounting for approximately 20% and 14% respectively of the incidents of abortion recorded by Veterinary Investigation Centres in England, Wales and Scotland (Veterinary Investigation Diagnosis Analysis, II, 1977–84). In comparison, campylobacteriosis and salmonellosis each accounts for about 4% of the outbreaks recorded each year. Control of these four infections will be dealt with in turn.

Enzootic abortion of ewes (EAE)

This is different from all other infectious causes in that it is a specific infection of sheep. It is caused by chlamydia, microorganisms which are midway between bacteria and viruses. The infection is introduced into a flock by an infected sheep and subsequent spread is solely from sheep to sheep. Normally, susceptible ewes are infected at lambing time one year by eating bedding and foodstuffs contaminated by the placentae and discharges of ewes which have aborted, and these susceptible ewes abort during the latter part of the next pregnancy. Recently, however, it has been shown that infection and abortion in the same pregnancy can occur and may be more common than was previously thought (Blewett *et al.*, 1982). Infection takes place by mouth and, as far as is known, the ram plays no part in the spread of disease. Infection is usually therefore introduced into clean flocks by the purchase of infected females which excrete the chlamydia at the time of subsequent abortion or normal lambing. This may often go unnoticed by the shepherd and it is usually the subsequent year before the full effects are experienced in the form of an abortion storm. Ewe lambs born of infected ewes, or fostered by ewes which have aborted, may harbour infection and abort during their first pregnancy. On introduction of infection into a flock all age groups will be affected, that is at the later stages of the pregnancy following the year of introduction. Subsequently, it will be mostly ewes pregnant for the second time which are likely to be affected after picking up infection at their first parturition.

Vaccination
An effective dead vaccine has played an important role in reducing the incidence of abortion due to enzootic abortion of ewes (Foggie, 1973). However, it must be used properly to have a chance of controlling infection. Susceptible animals must therefore be vaccinated before they come in contact with natural infection. Ideally this is done before first mating. Immunity is said to last for only 3 years but most ewes are only vaccinated once in their lifetime. Recently there have been some reports of breakdowns in vaccinated flocks (Linklater and Dyson, 1979). These can often be traced to faulty vaccination routine; for example, some farmers mate ewe lambs in their first year, that is at 7–9 months of age, but may not vaccinate them until they are a year older, the more common time for starting breeding. Such sheep, therefore, may pass through lambing pens and pick up enzootic abortion before being vaccinated. However, there is now good evidence that the vaccine itself is losing its efficacy in some flocks. It appears that this is due to the vaccine not giving protection against some field isolates which are more virulent than the old strain from which the commercial vaccine is produced (Aitken, Robinson and Anderson, 1981). One of these new field isolates, S26/3, has now been incorporated in the commercial vaccine in an attempt to improve the protection given.

Antibiotic treatment
Chlamydia which cause enzootic abortion of ewes are susceptible to the tetracycline antibiotics. Recent work suggests that long-acting preparations of oxytetracycline could be useful in reducing the effects of an abortion storm due to enzootic abortion (Greig and Linklater, 1985). However, this treatment does not eradicate the chlamydia from a flock, and should be considered only as an interim measure until other forms of prevention are introduced into a flock.

Toxoplasmosis

Toxoplasmosis was first recorded as a cause of ovine abortion in the United Kingdom in 1959 (Beverley and Watson, 1961). Since then it has been found worldwide. Serological surveys indicate that infection in sheep flocks is widespread with lowground flocks being more frequently infected than hill flocks. It is caused by a protozoan known as *Toxoplasma gondii* which can infect man and many other species of animals including cats and small rodents.

Following natural infection with *Toxoplasma gondii* sheep are immune to further challenge irrespective of age or time of infection (Beverley and Watson, 1971). If an outbreak of abortion due to this organism occurs one year it is unlikely to be repeated the following year provided steps are taken to ensure that additions to the flock pick up the infection before mating. As there is as yet no reliable vaccine available, natural infection is used as a method of control and this can be done by encouraging spread to non-immune sheep before they are pregnant. Experimentally it has been very difficult to demonstrate sheep to sheep transmission apart from spread from dam to fetus (Miller, Blewett and Buxton, 1982). Rams have been shown to excrete the organism in their semen but only for a short period after infection (Spence *et al.*, 1978). At any rate venereal transmission is unlikely to be involved in cases of abortion where infection of the dam takes place in mid-pregnancy. Venereal infection could lead to early embryonic death but this appears to be a rare manifestation of toxoplasmosis in sheep.

The only identifiable source of toxoplasma oocysts for sheep remains the domestic cat which in turn probably becomes infected either by eating tissue cysts in small rodents or by oocysts from other cats (Blewett and Watson, 1983). Young cats are particularly susceptible and, although they are only temporarily infected, they can excrete a large number of oocysts over a short period of time. These oocysts can in turn remain viable in the environment for at least 6 months.

Infection of susceptible sheep at strategic times therefore remains a hit or a miss. The best that can be done is to graze replacements along with the adult ewe flock before mating in the hope that they will pick up infection from the environment. Obviously the longer replacements are on the farm the better chance they will have of becoming infected, provided of course they are not kept in isolation some way from the farm buildings. Cats should not necessarily be eliminated from farm buildings as it may in fact be beneficial to have them there as a source of infection and therefore providing continuing immunity. It is, of course, prudent to keep them away from food stores and out of houses used for housing ewes in early and mid-pregnancy.

Campylobacteriosis (vibriosis)

The species of *Campylobacter* most commonly associated with abortion in sheep is *C. foetus foetus* (previously *intestinalis*). Susceptible ewes become infected by ingesting contaminated fetal material or vaginal discharges and, unlike cattle, venereal spread does not appear to take place in sheep. The sources of infection for flocks are not fully understood. Sheep which carry the organism in their gut may introduce infection into a flock. However, vibrionic abortion does occur in closed flocks and in these situations it is thought that infection may be acquired from wild birds such as ravens, crows, sparrows and magpies. Voles have also been shown to be carriers of *C. foetus*. As this infection is highly contagious it is important that ewes which abort are quickly separated from the rest of the flock to try to limit spread. Often, however, infection has already spread within the flock and little can be done to halt the number of abortions. Oral and parenteral antibiotics have been tried with equivocal results. Experimentally it has been shown that vaccination of the rest of the flock after the first abortion has occurred may limit the losses

(Gilmour, Thompson and Fraser, 1975). However, commercial vaccines are not available in the United Kingdom.

Ewes which abort are immune and should be retained for further breeding. Because of the sporadic occurrence of *Campylobacter* abortion, and because it does not tend to become endemic in flocks or localities in the United Kingdom, there is no justification for widespread vaccination. Therefore, although efficient vaccines have been developed experimentally there is little incentive to develop them commercially in this country. The only justification for their use could be in face of an outbreak. Three serotypes of *Campylobacter foetus foetus* associated with outbreaks of ovine abortion have been identified in Britain, and if a vaccine were to be produced it would have to contain all of them as there is little cross-protection offered between serotypes (Gilmour, 1983).

Salmonellosis

Several serotypes of salmonellae have been associated with abortion in sheep. The most prevalent used to be *Salmonella abortus-ovis* but in recent years the incidence of outbreaks due to this serotype has waned while others, notably *Salmonella dublin* and *Salmonella montevideo*, have become more common. From time to time *Salmonella typhimurium* has also been associated with abortion as well as general systemic and enteric infection in sheep.

Salmonella montevideo

The clinical pictures of salmonellosis due to *S. abortus-ovis*, *S. typhimurium* and *S. dublin* in sheep have been well documented (Jack, 1971); *S. montevideo* has recently been reported to be causing serious losses in sheep flocks in South-east Scotland (Linklater, 1983). With this serotype, abortion is the predominant clinical sign. Some ewes may be off-colour the day before lambing but soon recover spontaneously after parturition. Diarrhoea is not a feature of the infection even in lambs born alive at full-term in these flocks. Few deaths can be attributed to *S. montevideo* in ewes or lambs outside the periparturient period. Abortion rates of up to 30% have been recorded in affected flocks.

Linklater (1985) has shown experimentally that the mean period between infection and abortion due to *S. montevideo* is 15 days and affected ewes excrete the organism in their faeces for a mean of 43 ± 19 after infection. This contrasts with the much more severe febrile condition produced by *S. typhimurium* which runs a much shorter course (mean of 6.4 ± 1.5 days from infection to death) with survivors excreting the organism in their faeces for a mean of 16 ± 7 days (Linklater, 1985).

Biotyping of *S. montevideo* indicates that one biogroup named 10 di has become endemic in sheep (Reilly *et al.*, 1985). This contrasts with another common biogroup named 2 d which has been shown by the same workers to be responsible for most of the human, cattle and poultry infections in England and Wales. It appears therefore that most of the spread of *S. montevideo* infection within the sheep population is by sheep themselves rather than from other animals, wild birds and the environment as has been commonly believed in the past.

Salmonella montevideo is therefore an important cause of ovine abortion in certain parts of the United Kingdom. It appears to have become established in the sheep population and is now of much more importance than *S. abortus-ovis* in this country.

References

AITKEN, I.D., ROBINSON, G.W. and ANDERSON, I.E. (1981). Enzootic abortion: experimental infection. *Proceedings of the Sheep Veterinary Society*, **5**, 53–60

ANDERSON, P.H., BERRETT, S. and PATTERSON, D.S.P. (1979). The biological selenium status of livestock in Britain as indicated by sheep erythrocyte glutathione peroxidase activity. *Veterinary Record*, **104**, 235–238

BEVERIDGE, W.I.B. (1941). Foot-rot in sheep: a transmissible disease due to infection with *Fusiformis nodosus. Bulletin of the Council for Scientific and Industrial Research, Australia*, **no. 140**

BEVERLEY, J.K.A. and WATSON, W.A. (1961). Ovine abortion and toxoplasmosis in Yorkshire. *Veterinary Record*, **73**, 6–10

BEVERLEY, J.K.A. and WATSON, W.A. (1971). Prevention of experimental and of naturally occurring ovine abortion due to toxoplasmosis. *Veterinary Record*, **88**, 39–41

BIBERSTEIN, E.L., GILLS, M. and KNIGHT, H. (1960). Serological types of *Pasteurella haemolytica. Cornell Veterinarian*, **50**, 283–300

BLEWETT, D.A. and WATSON, W.A. (1983). The epidemiology of ovine toxoplasmosis. II. Possible sources of infection in outbreaks of clinical disease. *British Veterinary Journal*, **139**, 546–555

BLEWETT, D.A., GISEMBA, F., MILLER, J.K., JOHNSON, F.W.A. and CLARKSON, M.J. (1982). Ovine enzootic abortion: the acquisition of infection and consequent abortion within a single lambing season. *Veterinary Record*, **111**, 499–501

CLARKSON, M.J., FAULL, W.B. and KERRY, J.B. (1985). Vaccination of cows with clostridial antigens and passive transfer of clostridial antibodies from bovine colostrum to lambs. *Veterinary Record*, **116**, 467–469

CLAXTON, P.D., RIBEIRO, L.A. and EGERTON, J.R. (1983). Classification of *Bacteroides nodosus* by agglutination tests. *Australian Veterinary Journal*, **60**, 331–334

COLLINS, R.O., EALES, F.A. and SMALL, J. (1985). Observations on watery mouth in newborn lambs. *British Veterinary Journal*, **141**, 135–140

DYSON, D.A., GILMOUR, N.J.L. and ANGUS, K.W. (1981). Ovine systemic pasteurellosis caused by *Pasteurella haemolytica* biotype T. *Journal of Medical Microbiology*, **14**, 89–95

EALES, F.A., SMALL, J. and GILMOUR, J.S. (1982). Resuscitation of hypothermic lambs. *Veterinary Record*, **110**, 121–123

EALES, F.A., GILMOUR, J.S., BARLOW, R.M. and SMALL, J. (1982). Causes of hypothermia in 89 lambs. *Veterinary Record*, **110**, 118–120

EGERTON, J.R. (1972). Significance of *Fusiformis nodosus* serotypes in resistance of vaccinated sheep to experimental foot-rot. *Australian Veterinary Journal*, **50**, 59–62

EGERTON, J.R., ROBERTS, D.S. and PARSONSON, I.M. (1969). The aetiology and pathogenesis of ovine foot-rot. I. A histological study of the bacterial invasion. *Journal of Comparative Pathology*, **79**, 207–216

FOGGIE, A. (1973). Preparation of vaccines against enzootic abortion of ewes. A review of the research work at the Moredun Institute. *Veterinary Bulletin*, **43**, 587–590

GILMOUR, N.J.L. (1983). Vibriosis. In *Diseases of Sheep*, edited by W.B. Martin, pp. 133–135. London: Blackwell Scientific Publications

GILMOUR, N.J.L. and ANGUS, K.W. (1983). Pasteurellosis. In *Diseases of Sheep*, edited by W.B. Martin, pp. 3–8. London: Blackwell Scientific Publications

GILMOUR, N.J.L., THOMPSON, D.A. and FRASER, J. (1975). Vaccination against *Vibrio (Campylobacter) fetus* infection in sheep in late pregnancy. *Veterinary Record*, **96**, 129–131

GILMOUR, N.J.L., MARTIN, W.B., SHARP, J.M., THOMPSON, D.A. and WELLS, P.W. (1979). The development of vaccines against pneumonic pasteurellosis in sheep. *Veterinary Record*, **104**, 15

GREIG, A. and LINKLATER, K.A. (1985). Field studies on the efficacy of a long-acting preparation of oxytetracycline in controlling outbreaks of enzootic abortion of sheep. *Veterinary Record*, **117**, 627–628

HINDMARSH, F. and FRASER, J. (1985). Serogroups of *Bacteroides nodosus* isolated from ovine footrot in Britain. *Veterinary Record*, **116**, 187–188

JACK, E.J. (1971). Salmonella abortion in sheep. In *The Veterinary Annual*, 12th issue, edited by C.S. Grunsell, pp. 57–63. Bristol: John Wright and Sons Ltd

LINKLATER, K.A. (1983). Abortion in sheep associated with *Salmonella montevideo* infection. *Veterinary Record*, **112**, 372–374

LINKLATER, K.A. (1985). Studies on the pathogenesis of salmonellosis in sheep. *Fellowship thesis*, Royal College of Veterinary Surgeons

LINKLATER, K.A. and DYSON, D.A. (1979). Field studies on enzootic abortion of ewes in south-east Scotland. *Veterinary Record*, **105**, 387–389

MacPHERSON, A. (1983). Oral treatment of trace element deficiencies in ruminant livestock. In *Trace Elements in Animal Production and Veterinary Practice*, edited by N.F. Suttle, R.G. Gunn, W.M. Allen, K.A. Linklater and G. Wiener, pp. 93–103. Occasional Publication No. 7, British Society of Animal Production

MacPHERSON, A. (1985). Methods of copper supplementation. *Veterinary Record*, **116**, 330

MILLAR, K.R. and LORENTZ, P.P. (1979). Urinary methylmalonic acid as an indicator of vitamin B_{12} status of grazing sheep. *New Zealand Veterinary Journal*, **27**, 90–92

MILLER, J.K., BLEWETT, D.A. and BUXTON, D. (1982). Clinical and serological response of pregnant gimmers to experimentally induced toxoplasmosis. *Veterinary Record*, **111**, 124–126

MITCHELL, G.B.B. and LINKLATER, K.A. (1983). Differential diagnosis of scouring in lambs. *Veterinary Record*, **Supplement 5, In Pratice**, 4–12

REILLY, W.J., OLD, D.C., MUNRO, D.S. and SHARP, J.C.M. (1985). An epidemiological study of *Salmonella montevideo* by biotyping. *Journal of Hygiene, Cambridge*, **95**, 23–28

ROBERTS, D.S. and EGERTON, J.R. (1969). The aetiology and pathogenesis of ovine foot-rot. II. The pathogenic association of *Fusiformis nodosus* and *F. necrophorus*. *Journal of Comparative Pathology*, **79**, 217–227

RUSSEL, A.J.F., WHITELAW, A., MOBERLY, P. and FAWCETT, A.R. (1975). Investigation into diagnosis and treatment of cobalt deficiency in lambs. *Veterinary Record*, **96**, 194–198

SKERMAN, T.M., GREEN, R.S., HUGHES, J.M. and HERCEG, M. (1983). Comparison of footbathing treatments for ovine footrot using formalin or zinc sulphate. *New Zealand Veterinary Journal*, **31**, 91–95

SPENCE, J.B., BEATTIE, C.P., FAULKNER, J., HENRY, L. and WATSON, W.A. (1978). *Toxoplasma gondii* in the semen of rams. *Veterinary Record*, **102**, 38–39

STEWART, D.J., CLARK, B.L., EMERY, D.L., PETERSON, J.E., JARRETT, R.G. and O'DONNELL, I.J. (1986). Cross-protection from *Bacteroides nodosus* vaccines and the interaction of pili and adjuvants. *Australian Veterinary Journal*, **63**, 101–106

STUBBINGS, D.P. (1983). Feeding cows' colostrum to newborn lambs. *Veterinary Record*, **112**, 393

SUTTLE, N.F. and McMURRAY, C.H. (1983). Erythrocyte Cu : Zn superoxide dismutase activity and hair or fleece copper concentrations in the diagnosis of hypocuprosis in ruminants. In *Trace Elements in Animal Production and Veterinary Practice*, edited by N.F. Suttle, R.G. Gunn, W.M. Allen, K.A. Linklater and G. Wiener, pp. 134–135. Occasional Publication No. 7, British Society of Animal Production

VETERINARY INVESTIGATION DIAGNOSIS ANALYSIS (VIDA) II (1977–84). Ministry of Agriculture, Fisheries and Food and Department of Agriculture and Fisheries for Scotland

WHITELAW, A., ARMSTRONG, R.H., EVANS, C.C. and FAWCETT, A.R. (1979). A study of the effects of copper deficiency in Scottish Blackface lambs on improved hill pasture. *Veterinary Record*, **104**, 455–460

Part VIII

Evaluation of new techniques

Researchers are often puzzled as to why new discoveries are not quickly acted upon by the farming industry. With hindsight it is often clear that the slow uptake of certain techniques is for valid reasons, based on sound economic sense. The first chapter in this section is important in concentrating on the seminal issue of the economic optimum in relation to reproductive performance. The complex of factors involved in assessing economic response and the changing interrelationships of input costs both amongst themselves and in relation to output prices, makes this a formidable task.

Many new techniques have also failed to achieve uptake by farmers because of the inadequacy with which the final system development phase has been carried out. Just as many, if not more, problems need solving at the system stage as in the development of the component technique. It is fitting that these considerations should provide the conclusion to a discussion on the development of new techniques into the practice of sheep-keeping.

Chapter 23

Economic response to increasing genetic potential for reproductive performance

G. Nitter

Summary

Genetic improvement of litter size in sheep has been shown to be possible both as a trait coded by many genes and by increasing the frequency of a major gene. There is, however, much contradiction and uncertainty about the economic response to improvement.

A simplified way has been shown to obtain the economic response per unit change of litter size. The approach is based on the profit for different litter types and their change of distribution following the change of the mean. The profit can either be calculated by subtracting costs from monetary values or estimated by subjective assessment if figures are not available. A general assumption of a constant coefficient of variation is discussed to obtain the distribution of litter types if many genes are involved. It is shown how this is modified by the presence of a major gene.

As an example, the economic response per unit change of litter size is derived for the Merinoland breed under different production systems and for a wide range of average litter size. It is shown how the economic response degresses rapidly with increasing mean in the population.

Introduction

Non-genetic resources to increase reproductive performance in sheep are almost inexhaustible. The most striking may be application of exogenous gonadotrophic hormones and the immunization of ewes against steroid hormones. All these efforts, however, require continuous inputs which may raise some doubts as to whether they pay under the usual environmental circumstances in the sheep industry. It therefore seems advisable to remember that the effects of genetic improvement are both cumulative and permanent, and in any circumstances they are usually accompanied by economic response as well.

This chapter tries to show that the economic response to increasing the genetic potential for reproduction not only depends on the intensity of the production environment but also on the mean level of performance already achieved. Litter size is taken as an example. A simplified way of evaluating the economic response to increasing the mean level by one unit is demonstrated.

Characters and genetic resources

Reproductive performance is a complex characteristic made up of various characters and dependent on maternal as well as non-maternal effects (Nitter, 1984). Selection for reduced seasonality and post-partum anoestrus should be worth considering if production and marketing conditions require improvement of this character (Ricordeau, 1982). Other circumstances may call for genetic improvement in lamb survival where there is also evidence of genetic variation (Cundiff, Gregory and Koch, 1982; Petersson and Danell, 1985).

For the subject of this chapter, however, it seems appropriate to choose litter size at birth as the main objective for selection. It can be measured easily and has proved to be accessible to selection in a great number of experiments (Bradford, 1985). Genetic gain can be accelerated through use of laparoscopy, measuring ovulation rate as the most limiting factor of litter size (Hanrahan, 1980). Furthermore, single major genes coding for high ovulation rate are known which can be introduced to increase litter size very rapidly (Piper and Bindon, 1985; Jónmundsson and Adalsteinsson, 1985).

Genetic improvement of litter size is assumed here to be attained by means of within-breed selection. It is not proposed to consider the economic response to improvement achieved by introducing a certain percentage of a highly prolific genotype such as the Finnish Landrace or Romanov, which would certainly cause deterioration in growth and carcass traits. If the frequency of a foreign major gene is to be increased, this is achieved by appropriate repeated backcrossing and is assumed not to change other characters of the breed.

Method of evaluation

The aim is to determine the economic gain per unit increase of average litter size, for example, 0.1 extra lambs born per lambing. Costs of recording are neglected. The question of who really benefits from the genetic improvement is disregarded whether it be (a) the breeder, or the breeders' organization, selling improved breeding stock to the producer; or (b) producers who have ewes which give birth to more lambs profiting from the increased number of lambs sold to the butcher; or (c) the consumer benefiting from lower lamb prices resulting from rising productivity at the producers' level.

An increase of average litter size is inevitably accompanied by decreased birthweights and decreased lamb survival. Lower birthweights in turn lead to reduced lamb growth, particularly in the early weeks of life. Furthermore, increased costs of feeding, labour and attention must be considered, as well as the increased degree of stress on the ewe during late pregnancy. These negative effects are all very difficult to quantify, and vary with breeds and management conditions.

An appropriate method of approach to evaluate the economic response of increased prolificacy in such a situation seems to be the following:

(1) Find the relative frequencies of litter types (singles, twins, triplets, etc.) for a given average litter size in the population (a) and for an average with 0.1 lambs more $(a + 0.1)$.

(2) Estimate or calculate the profit value from each litter type separately by subtracting its costs of production from its monetary value.

(3) Find the total profit value of the population with average litter sizes a and $(a + 1)$ by means of weighting the monetary value of each litter type by its frequency and adding up to the total.
(4) Calculate the differences between the total profit values in $(a + 0.1)$ and a.

Taking symbols v_1, v_2, v_3 etc. for profit values of singles, twins, triplets etc., and x_1^a, x_2^a, x_3^a etc. for their frequencies in an average litter size a, then the total value in a is

$$T_a = v_1 x_1^a + v_2 x_2^a + v_3 x_3^a + \ldots + v_n x_n^a \tag{1}$$

$$= \Sigma v_i x_i^a$$

The expression for $T_{(a + 0.1)}$ is similar and the economic response per unit increase of the average litter size is simply

$$T_{(a + 0.1)} - T_a \tag{2}$$

The most difficult thing to realize in this procedure is getting accurate profit values for litter types which hold under practical conditions. This can be calculated by discounting the costs and returns of each litter type under the particular environment of production. Under many circumstances, however, a more realistic approach would be just to rely on the producer's intuitive assessment of the value of multiple litters compared to those of singles. This flexibility of approach is one of the most obvious advantages of this procedure.

Much less difficult than ascertaining values of litters is assessing their distribution under various average litter sizes. Both aspects will now be treated in more detail.

Profit value of litter types

There are several observations in the literature concerning mortality and individual birthweights for various litter types. These figures, in common with growth rates directly related to birthweights, might be reasonable biological indicators of the relative profit values of different litters.

It is not surprising, however, that a review of the literature does not present any clear picture. In the main, mortality in various multiple births compared to singles varies enormously between reports. Even the difference in mortality of twins compared to singles, which is based on very extensive data, varies considerably depending on breeds and environmental conditions. Some reports show no difference at all (Wiener and Hayter, 1975; Kanter, 1977; Boaz and Tempest, 1975 in Scottish Halfbreds; Cornu and Cognié, 1985); others find mortality rates about three times higher in twins than in singles (Beetson and Lewer, 1985). While higher mortality of twins is to be expected if production environments are extensive, there may be no difference between singles and twins if adequate forage, housing and attention are offered. Furthermore, in breeds where dystocia due to large birthweights is frequent, such as in the Texel, mortality rates of singles can be even higher than those of twins.

Figures concerning relative mortality in litters higher than twins vary much more in view both of variable environments and of the scant data available. On the one hand, lamb losses in triplets and quadruplets are reported to be higher than 40% and 75% respectively (Kanter, 1977; Beetson and Lewer, 1985; Boaz and Tempest, 1975; Speedy and Fitzsimons, 1977). In another extreme, under a very intensive production system with the Romanov breed there was hardly any difference in mortality between singles, twins, triplets and quadruplets, and quintuplet mortality was not more than double that of singles (Cornu and Cognié, 1985).

Such great variability suggests that empirical figures, or even subjective appraisals concerning lamb mortality in the specific environment and breed considered, are usually most realistic.

More uniform than mortality are observations concerning relative birthweights of lambs born in various litter types. Bradford's (1972, 1985) predicted relative birthweights for individual twins and triplets compared to singles can be accepted as a general rule for most genotypes. These are roughly 78% for twins and 62% for triplets in purebred matings. In the pure Romanov breed (Cornu and Cognié, 1985), and also in a synthetic line with 50% Romanov genes (Razungles et al., 1985) rather different estimates of these relationships were reported (in pure Romanov about 86% for twins and 76% for triplets). It is not known whether this is really a breed effect or merely the result of better nutrition of ewes with multiple litters. In addition, individual birthweights of pure Romanov quadruplets and quintuplets compared to singles were relatively high (73% and 63% respectively) and much higher than those which can be deduced from Bradford's figures.

It is known that low birthweight of lambs born in multiple litters is also a handicap for their growth potential, because they simply eat less than heavier lambs. In addition, the different labour requirement of multiples and their impact on the ewe's health should be taken into account if their relative profit value is to be evaluated.

The most variable observations are those required to evaluate the profit value of litters higher than twins and triplets. For these, again, it may be argued that intuitive assessments for the environment considered could be more realistic than calculations based on rather erratic figures.

Distribution of litter types

It is a well-known fact that variances of different populations are related to their means. Populations with higher means generally have higher absolute variances. But relative variances, expressed in terms of coefficient of variation, have been shown to be fairly constant at least in populations which are comparable (Wright, 1968). This has also proved to be valid for populations which have become differentiated following selection.

Before the assumption of constant coefficient of variation is accepted as a working hypothesis for litter size as well, the specific nature of this trait should be considered. Litter size is a threshold character and far from being normally distributed around its mean. In most circumstances, however, one should assume that it is a polygenic trait affected by many genes with small effects acting and interacting at many loci. Variation of litter size in this case is expected to be different from a situation where a major gene is involved. This being so, the problems posed by these two situations will be discussed separately.

Many genes with small effects

The general rule of constant coefficient of variation expected for normally distributed traits has to be confirmed for a categorical character like litter size in sheep. Reports in the relative literature indicate that an average value lies between 30 and 35%. This figure is lower if first lambing ewes are omitted and is higher if 'zero litters' are taken into account. It is also much higher if hormonal treatment is applied to induce ovulation rate. However, various breeds may differ in uniformity of litter size as shown in an extensive field study of Norwegian breeds (Steine, 1985).

For the purpose of this chapter it is important to know whether a constant coefficient of variation holds for different levels of litter size in one population selected for litter size. Assuming a continuous underlying variable, the threshold model can be used to predict the change in frequency of litter sizes following a certain change in threshold values. Petersson and Danell (1985) have applied this approach to lamb survival. If improvement through selection means the same change of threshold levels for singles, twins etc., independent of their location on the scale, it can be shown that the variance of litter size increases following an increase of the mean. However, variance increases more slowly than the mean, and the assumption of a constant coefficient of variation does not hold precisely although this complication is ignored in the following analysis.

The coefficient of variation will be taken as being constant at a level of 34%. This has been ascertained for years in the German breed considered (Merinoland) and fits well with the predicted birth rank distribution of Davis et al. (1983) for 'local breeds' without a major gene. To find frequencies x_i^a for singles ($i = 1$), twins ($i = 2$), triplets ($i = 3$) etc. for the average litter size (a), the following information is available:

$$1 = \Sigma x_i^a$$

$$a = \Sigma i x_i^a$$

$$c = \sqrt{\Sigma x_i^a (i - a)^2}/a$$

where c represents the coefficient of variation assumed.

A further assumption is that the change of frequency of each litter type with increasing mean follows a normal distribution. The distribution pattern calculated on these relationships shows that the percentage of twins is highest near an average litter size of 1.9 and that of triplets is highest near 2.7. It also shows that frequencies of triplet, quadruplet and quintuplet litters higher than around 1% start at average levels of 1.5, 2.0 and 2.5 respectively. For a range of means from 1.2 to 2.5 the distribution of twin, triplet and quadruplet litters is shown by dashed lines in *Figure 23.1*. Here the important fact is demonstrated that the frequency of twin litters, which are economically most valuable, decreases with mean litter sizes higher than 1.9.

Major genes

If selection for litter size is based on accumulating a single major gene for fertility (F) the frequency of the three genotypes changes (FF, Ff and ff; f is the 'normal'

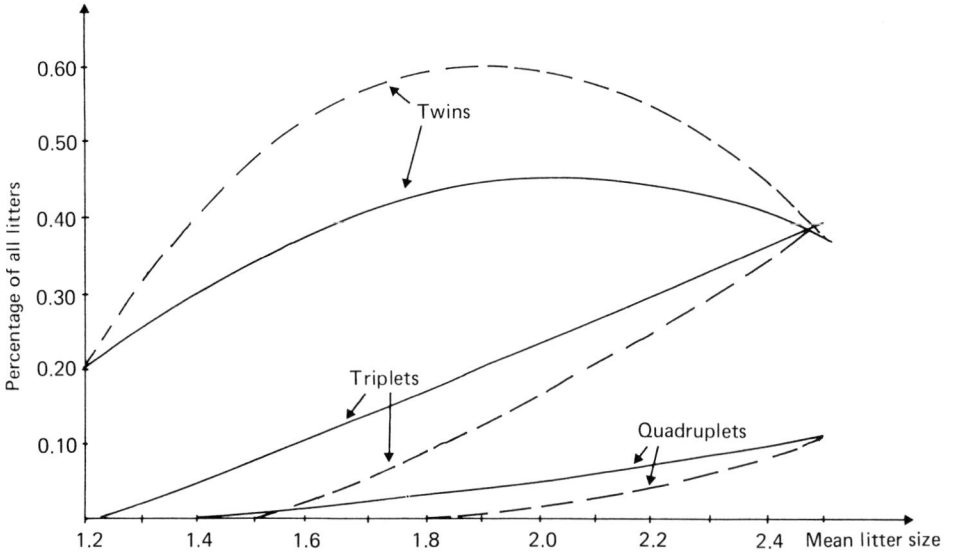

Figure 23.1 Distribution of birth ranks higher than singles with increasing average litter size of the population. ---- without a major gene, CV = 34%, constant; ————— major gene with mean litter sizes: FF 2.5, Ff 2.0, ff 1.2, CV constant within the three genotypes

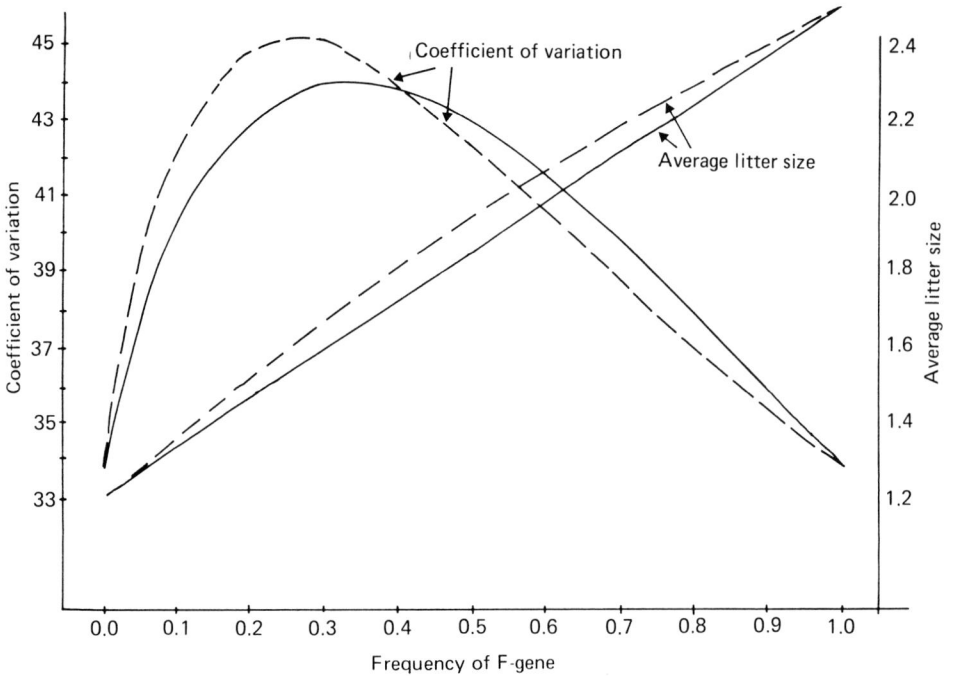

Figure 23.2 Coefficient of variation and average litter size in a population with increasing frequency of the major gene with (-----) and without (—————) dominance effect. Mean litter size of genotypes: ----- FF 2.5, Ff 2.0, ff 1.2; ————— FF 2.5, Ff 1.85, ff 1.2

gene). Hardy–Weinberg equilibrium is assumed for all generations. All animals of the same genotype are considered a subpopulation with different mean levels of litter size, but with uniform coefficient of variation within subpopulations. The coefficient of variation for the whole population, however, changes considerably with increasing frequency of the F gene. It is assumed as being 34% both at the beginning (only genotypes ff) and at the end of the scale (only genotypes FF). *Figure 23.2* shows that the coefficient of variation is found to be highest at frequencies of the major gene below 50%. This depends on the degree of dominance at the locus.

The consequence of this for our problem is indicated by solid lines in *Figure 23.1*. The frequency of the most desirable twin litter is shown to be much lower than in populations without the major gene. It is never higher than around 45% whereas it reached a maximum level of about 60% without the major gene. On the other hand, litters higher than twins occur with remarkable frequency at much lower levels of mean litter size than in breeds without the major gene. This corresponds with the observations of Davis *et al.* (1983) who found fewer twin bearing ewes and correspondingly more single, triplet and quadruplet litters in Booroola flocks than in flocks of local breeds with similar average litter size. Predicted birth rank distributions as a result of using heterozygous or homozygous Booroola rams in local breeds are straightforward and will not be discussed here.

Application to the Merinoland breed

Merinoland is the main sheep breed in southern parts of Germany. Four production systems are defined in which these sheep are mainly kept, and for which the impact of increasing the genetic potential for litter size is expected to be different.

The main criteria of these systems are listed in *Table 23.1* (Schäfer, 1981). Most of the Merinoland sheep are kept in rather big flocks continuously shepherded nearly the whole year round (systems 1 to 3). These sheep have variable lambing

TABLE 23.1 Systems of sheep production with the Merinoland breed

| | | System | | | |
		1	*2*	*3*	*4*
Continuously shepherded		yes	yes	yes	no
Spring lambing only		no	no	no	yes
Migratory		no	no	yes	no
Nursery available		yes	yes	no	yes
Labour requirement per lamb in the nursery (hours)		2	1.8	no	2.5
Per cent lamb mortality	singles	10	10	12	10
	twins	10	10	18	10
	triplets	18	18	40	18
Age of weaning (days)		95	60	95	203
Total kg concentrates per lamb		76	81	76	26
DM* per unit roughage		0.40	0.40	0.40	0.20
Slaughter weight (kg)	(singles)†	46	45	46	51
Slaughter weight per litter	single	41.4	40.5	40.5	45.9
	twin	78.6	78.6	72.2	84.6
	triplet	104.7	104.7	75.6	110.1

*Deutschmark
†Twins 2 kg less, triplets 4 kg less

TABLE 23.2 Calculated and estimated profits with four different litter types in various systems of Merinoland sheep (in Deutschmark (DM))

		System			
		1	*2*	*3*	*4*
Feed costs	singles	82.9	80.5	82.1	63.6
	twins	165.1	157.9	142.9	135.1
	triplets	227.7	220.8	157.0	198.9
Labour costs	singles	12.5	12.6	12.6	9.2
	twins	32.2	31.2	23.3	31.8
	triplets	53.7	51.2	29.8	59.0
Profit	singles	70.2	68.9	67.2	110.8
	twins	117.1	125.3	122.5	171.5
	triplets	137.4	146.8	115.6	182.7
	quadruplets (estimated)	70.0	69.9	35.0	110.0

times due to the availability of food, housing and labour. One of these systems (number 3) is the traditional extensive migratory system for which no artificial rearing, and therefore high lamb mortality and low output of lambs of slaughter weight, are assumed. In all three systems roughage for lambs is assumed to be rather expensive compared to system 4 where lambs are born in spring only and fattened to high slaughter weights on grass with a low level of concentrates. Systems 1 and 2 are very similar and differ in age at weaning only. Lamb losses for singles and twins in systems 1, 2 and 4 are taken as being equal (10%) as artificial rearing of surplus lambs is practised, and ewes of this breed tend to suffer from dystocia due to heavy single lambs. 'Surplus lambs' means all third and higher rank of lambs in a multiple litter and one-third of all twins. Labour requirement per artificially reared lamb differs between systems because of different flock size and housing facilities.

Costs and profits for single, twin and triplet litters are listed in *Table 23.2*. Profit is the difference between monetary value and costs, the value being found by weighting the litter weight at slaughter (see *Table 23.1*) with the price per kg (in this example DM 4). Costs are presented separately for both feed and labour requirement. The profit of quadruplet litters was estimated and not calculated as in the other litter types. It was assumed to be similar to the profit of a single lamb in systems with a nursery available (systems 1, 2 and 4) and about one-half of a single in system 3.

Based on these *v* values and on frequencies of litter types for a situation without a major gene—frequencies not listed, see dashed line in *Figure 23.1*—equations (1) and (2) are now applied to calculate the economic value per unit change of litter size. Results are shown in *Figure 23.3* for all four production systems and for a range between 1.1 and 2.5 in the average litter size of the population.

In all systems the economic value per 0.1 extra lambs born is high (between DM 4.50 and 6.10) as long as the average litter size is not more than about 1.5. At higher levels it drops dramatically, particularly in the extensive system 3. Even negative values are obtained if the average surpasses a level of about 2.1. The degressive response from improving litter size with increasing average level is most pronounced in system 3 and least in system 1. It is evident that there is no system where an improvement of average litter size higher than around 2.4 pays.

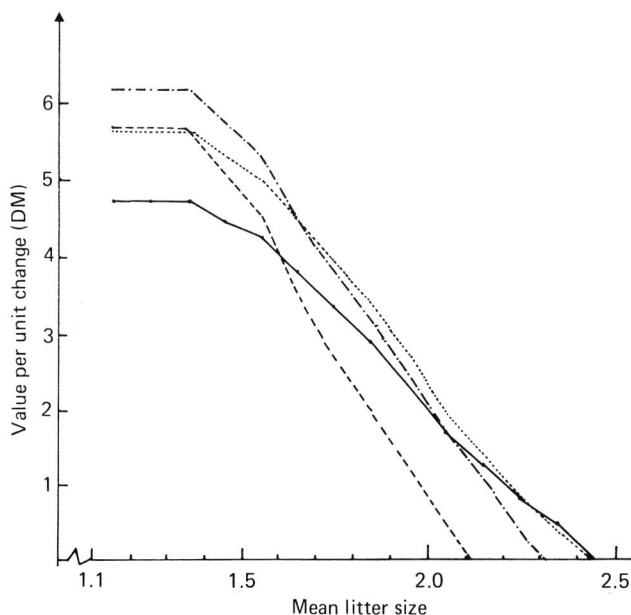

Figure 23.3 Economic value per unit change of litter size (0.1 lambs) depending on intensity of production environment and on average litter size of the population. ————— System 1; ····· system 2; ----- system 3; –·–·–·– system 4

References

BEETSON, B.R. and LEWER, R.P. (1985). Productivity of Booroola cross Merinos in Western Australia. In *Genetics of Reproduction in Sheep*, edited by R.B. Land and D.W. Robinson, pp. 391–398. London: Butterworths

BOAZ, T.G. and TEMPEST, W.M. (1975). Some consequences of high flock prolificacy in an intensive grassland sheep production system. *Animal Production*, **20**, 219–232

BRADFORD, G.E. (1972). The role of maternal effects in animal breeding: maternal effects in sheep. *Journal of Animal Science*, **35**, 1324–1334

BRADFORD, G.E. (1985). Selection for litter size. In *Genetics of Reproduction in Sheep*, edited by R.B. Land and D.W. Robinson, pp. 3–18. London: Butterworths

CORNU, C. and COGNIÉ, Y. (1985). The utilization of Romanov sheep in a system of integrated husbandry. In *Genetics of Reproduction in Sheep*, edited by R.B. Land and D.W. Robinson, pp. 383–389. London: Butterworths

CUNDIFF, L.V., GREGORY, K.E. and KOCH, R.M. (1982). Selection for increased survival from birth to weaning. *Proceedings of Second World Congress of Genetics Applied to Livestock Production*, **5**, Madrid, 4–8 October, 310–337

DAVIS, G.H., KELLY, R.W., HANRAHAN, J.P. and ROHLOFF, R.M. (1983). Distribution of litter sizes within flocks at different levels of fecundity. *Proceedings of the New Zealand Society of Animal Production*, **43**, 25–28

HANRAHAN, J.P. (1980). Ovulation rate as the selection criterion for litter size in sheep. *Proceedings of the Australian Society of Animal Production*, **13**, 405–408

JÓNMUNDSSON, J.V. and ADALSTEINSSON, S. (1985). Single genes for fecundity in Icelandic sheep. In *Genetics of Reproduction in Sheep*, edited by R.B. Land and D.W. Robinson, pp. 159–168. London: Butterworths

KANTER, R. (1977). Untersuchungen über die Lämmerverluste und ihre Ursachen bei Schafen verschiedener Rassen und Rassenkreuzungen. *Dissertation thesis*, University of Giessen

NITTER, G. (1984). Theoretical aspects of selection for reproductive performance with sheep as an example. *Zeitschrift für Tierzüchtung und Züchtungsbiologie*, **101**, 81–85

PETERSSON, C.J. and DANELL, Ö. (1985). Factors influencing lamb survival in four Swedish sheep breeds. *Acta Agriculturae Scandinavica*, **35**, 217–232

PIPER, L.R. and BINDON, B.M. (1985). The single gene inheritance of the high litter size of the Booroola Merino. In *Genetics of Reproduction in Sheep*, edited by R.B. Land and D.W. Robinson, pp. 115–125. London: Butterworths

RAZUNGLES, J., TCHAMITCHIAN, L., BIBÉ, B., LEFÈVRE, C., BRUNEL, J.C. and RICORDEAU, G. (1985). The performance of Romanov crosses and their merits as a basis for selection. In *Genetics of Reproduction in Sheep*, edited by R.B. Land and D.W. Robinson. London: Butterworths

RICORDEAU, G. (1982). Selection for reduced seasonality and post-partum anoestrus. *Proceedings of Second World Congress of Genetics Applied to Livestock Production*, **5**, Madrid, 4–8 October, 338–347

SCHÄFER, K. (1981). Ermittlung von Grenznutzen für Leistungsmerkmale beim Schaf. *Diploma thesis*, University of Hohenheim

SPEEDY, A.W. and FITZSIMONS, J. (1977). The reproductive performance of Finnish Landrace × Dorset Horn and Border Leicester × Scottish Blackface ewes mated three times in two years. *Animal Production*, **24**, 189–196

STEINE, T. (1985). Genetic studies of reproduction in Norwegian sheep. In *Genetics of Reproduction in Sheep*, edited by R.B. Land and D.W. Robinson, pp. 47–54. London: Butterworths

WIENER, G. and HAYTER, S. (1975). Maternal performance in sheep as affected by breed, crossbreeding and other factors. *Animal Production*, **20**, 19–30

WRIGHT, S. (1968). Evolution and the genetics of populations. *Genetic and Biometric Foundation*, Volume 1, Chapter 10. Chicago: University of Chicago

Chapter 24

Systems development of new techniques

J.B. Owen

Summary

Sheep systems and the effect upon them of incorporating new techniques need to be evaluated in terms of overall economic efficiency and of ease of management. The efficiency of the system and the effect of new techniques are shown to vary according to circumstances; some techniques are useful in some and not in others. Lamb production per ewe bears a curvilinear relationship to litter size at birth and to ovulation rate and the optimum value varies markedly according to husbandry conditions.

Relatively little information is available on ewe intake, an important component of system efficiency. Some data is presented which illustrates the important effect of this factor in the evaluation of new sheep genotypes.

The fact that sheep are part of mixed systems involving other species and other crop enterprises on the same land makes evaluation of new techniques more complex. The evaluation of new silvopastoral systems is an interesting development in this field.

Introduction

The object of adopting new techniques is to increase the efficiency of sheep production or to ease flock management. Efficiency depends on the ratio of output to input and both output and input usually have several components. At any time maximum overall efficiency, in terms of money, is achieved with the optimum array of inputs. Because product and input prices and, more importantly, the ratio of these to each other, vary according to circumstances, the effect of inputs and of techniques that influence them also vary.

Table 24.1 shows how economic efficiency of a sheep system depends on lambing frequency and on ewe prolificacy. In the same study (Scottish Agricultural Colleges, 1977), it was demonstrated that the evaluation of the effect of both prolificacy and lambing frequency is sensitive to changes in level of lamb mortality, lamb price and purchased feed cost.

TABLE 24.1 Net margin per ewe and per hectare (includes grass costs and interest on capital) from different sheep production systems

Lambing frequency	Litter size per ewe	Net margin* per ewe	Net margin* per hectare
Once per annum	1.5	100	100
	2.0	163	150
	2.5	193	174
Three times in 2 years	1.5	113	146
	2.0	194	239
	2.5	248	301

*Based on lambing frequency once per annum and litter size per ewe 1.5 as 100

It is therefore most important to assess carefully the consequences of the adoption of new techniques in terms of the effect on the system as a whole as shown by Nitter (Chapter 23).

Once this overall assessment has been made (and it may need repeating where circumstances are different), the other main consideration is to assess the effect of the new technique on ease of management. Occasionally a new technique or a new system which gives high overall efficiency may not be adopted because it is complex to manage and therefore not as robust in long-term practice as the established system.

Optimum prolificacy

Work carried out at Bangor extends some of the theoretical principles outlined by Nitter (Chapter 23) to actual data on the prolific Cambridge breed. *Figures 24.1*

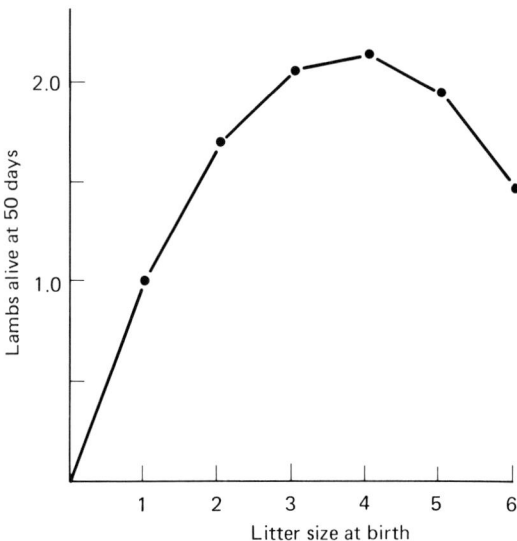

Figure 24.1 The relationship between the number of lambs alive at 50 days and litter size (2, 3 and 4 year olds)

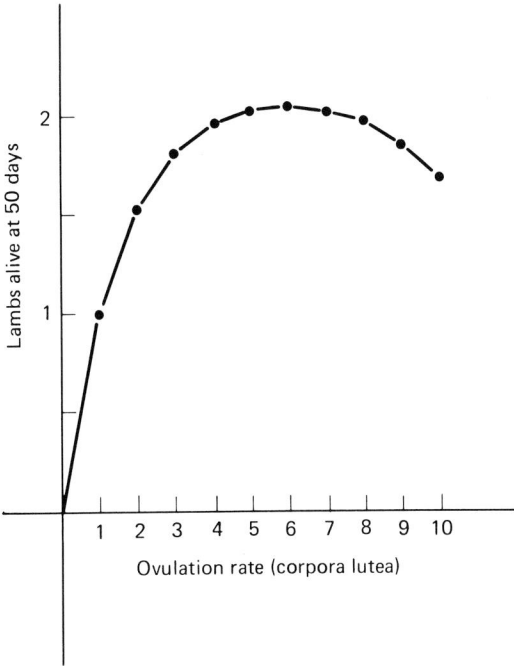

Figure 24.2 The relationship between lambs alive at 50 days and ovulation rate (based on Hanrahan and Piper, 1982)

and *24.2* show that the relationship of number of lambs weaned to litter size at birth and to ovulation rate is curvilinear and puts the upper limit of the range of optimum ovulation rate at 5 and of the litter size at 4, in this population. However, at the mean value for all breeds in widespread use, increases in ovulation rate and litter size are likely to be highly cost effective in similar lowland conditions. On the other hand, under extensive range conditions with little supplementary feeding and unsupervised lambing, the optimum rates for ovulation and litter size is much nearer to unity. One important point in relation to the assessment of the value of increased prolificacy is that it should be judged in relation to the production of a standard number of lambs. Increases in lamb production per ewe therefore results in the need for fewer ewes and consequent increases in efficiency, rather than in the common misconception that the use of prolific ewes somehow necessarily increases the market supply of lambs.

Stocking rate comparisons

One of the problems of assessing the value of new sheep genotypes, for example, is the dearth of information on the appropriate forage requirements. This contrasts sharply with the relative wealth of information normally available in relation to sheep output. Wilson (1976) studied the feed intake of six sheep genotypes varying from the Shetland (40 kg live weight) to the Greyface (Border Leicester × Scottish Blackface) (86 kg live weight), and the forage dry matter intake per ewe for these

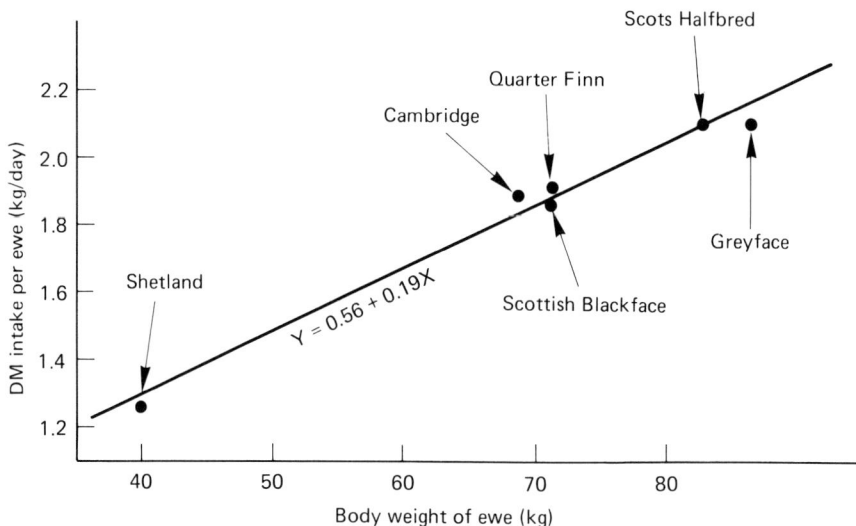

Figure 24.3 Intake and body weight in various ewe genotypes

TABLE 24.2 Expected relative appropriate stocking rates for various sheep genotypes assuming a basal stocking rate for Welsh Mountain ewes of 15 per hectare

Genotype	Ewe body weight (kg)	Relative stocking ewes per ha
Welsh Mountain	35	15
Llanwenog	50	12
Lleyn	55	11½
North Country Cheviot	74	9
Scottish Blackface	57	11
Clun Forest	54	11½
Border Leicester	82	9
Suffolk	81	9
Welsh Halfbred	57	11
Scottish Halfbred	74	9½
Mule	72	9

data was equal to 0.56 + 0.018 (body weight in kg) (*Figure 24.3*). Using this relationship Owen (1984) derived the values in *Table 24.2* for the appropriate relative stocking rate of several British sheep genotypes.

This analysis is one of the few available on measured intakes under commercial conditions and it indicates that assumptions on the relative forage requirements of different genotypes have a marked influence on the evaluation of the comparative efficiency of sheep genotypes.

In a comparison of the Border Leicester, Cambridge and Lleyn breeds as damline sires crossed with Welsh Mountain sheep (Owen and Whitaker, 1987), the incorporation of allowances for forage intake was an important element in comparing the efficiency of the three types of crossbred ewe as shown in *Table 24.3*.

TABLE 24.3 The main features of comparative data on crossbred ewes sired by three damline breeds from Welsh Mountain dams taking the Border Leicester as 100 (for traits where the overall differences were significantly different)

	Sire breed		
	Border Leicester	Cambridge	Lleyn
Lambs born/ewe lambed	100	113	89
Lambs alive at 56 days/ewe lambed	100	124	93
Mean lamb birthweight	100	79	92
Mean lamb weight at 56 days	100	82	91
Mean age at slaughter	100	115	109
Ewe weight at tupping	100	93	84
Efficiency $\dfrac{\text{relative lamb production}}{\text{relative feed intake}}$	100	128	83

Mixed systems

A major problem in assessing the economic efficiency of sheep systems, and of the effect thereon of the incorporation of new techniques, is that sheep are often part of complex mixed systems. This may involve grazing in common with other species, such as cattle and goats, or being part of a multiple land use involving livestock and crops of various kinds. In these circumstances the allocation of input costs between the various forms of land use is difficult so that assessing the effect of new techniques is also complicated.

In general the use of a mixed, rather than a specialist sheep, system is beneficial, particularly in relation to parasite control. There is also some evidence of complementarity between different uses so that the overall system efficiency is higher than for single-purpose use. A simple example is that cattle, sheep and goats tend to differ in grazing habit so that a more complete utilization of the natural range resource may be possible (Owen, 1976).

A recent development in multiple land use involving sheep is the work on silvopastoral systems, where trees are grown in wide rows to allow sheep to graze the pasture within the planted area. Evaluation of the benefits of this system is being carried out at the University of Bangor, in Chile and in New Zealand (Alcock and Hall, 1986).

Conclusions

It is important that new scientific discoveries are quickly and efficiently translated into practice. This can only be done by a partnership involving the scientist, the economist and the flockmaster working together in a sound, realistic approach. It is hoped that some of the contributions in this book will help further this aim.

References

ALCOCK, M.B. and HALL, J.B. (1986). *Agroforestry*. Occasional mimeograph School of Agriculture and Forestry, University College of North Wales, Bangor
HANRAHAN, J.P. and PIPER, L.R. (1982). Nature of the genetic control of ovulation rate and its

relationship with litter size. *Proceedings of the European Association of Animal Production,* Leningrad

OWEN, J.B. (1976). *Sheep Production.* London: Baillière Tindall

OWEN, J.B. (1984). *Defaid (Sheep).* Penygroes, Caernarfon: Cyhoeddiadau Mei

OWEN, J.B. and WHITAKER, C.J. (1986). Comparison of crossbred ewes raised from Welsh Mountain dams by three sire breeds—Cambridge, Border Leicester and Lleyn. *Journal of Agricultural Science* (in press)

SCOTTISH AGRICULTURAL COLLEGES (1977). *A Study of High Lamb Output Production Systems.* Technical Note no. 16, Scottish Agricultural Colleges

WILSON, R.M. (1976). Voluntary food intake studies with six genotypes of sheep. *PhD thesis,* University of Aberdeen

Index